Advances in Structure, Function and Molecular Targeting of DNA Topoisomerases

Advances in Structure, Function and Molecular Targeting of DNA Topoisomerases

Editors

Joseph E. Deweese
Neil Osheroff

MDPI

Basel • Beijing • Wuhan • Barcelona • Belgrade • Novi Sad • Cluj • Manchester

Editors
Joseph E. Deweese
Biological, Physical and
Human Sciences
Freed-Hardeman University
Henderson
United States

Neil Osheroff
Department of Biochemistry
Vanderbilt University
Nashville
United States

Editorial Office
MDPI
St. Alban-Anlage 66
4052 Basel, Switzerland

This is a reprint of articles from the Special Issue published online in the open access journal *International Journal of Molecular Sciences* (ISSN 1422-0067) (available at: www.mdpi.com/journal/ijms/special_issues/DNA__topoisomerases).

For citation purposes, cite each article independently as indicated on the article page online and as indicated below:

Lastname, A.A.; Lastname, B.B. Article Title. *Journal Name* **Year**, *Volume Number*, Page Range.

ISBN 978-3-0365-9701-0 (Hbk)
ISBN 978-3-0365-9700-3 (PDF)
doi.org/10.3390/books978-3-0365-9700-3

Cover image courtesy of Joseph E. Deweese

Contents

About the Editors

Joseph E. Deweese

Dr. Joseph E. Deweese is a Professor of Biochemistry and Director of Undergraduate Research at Freed-Hardeman University. Dr. Deweese teaches biochemistry and molecular biology and is involved in researching human topoisomerase II using biochemical and bioinformatic approaches.

Neil Osheroff

Dr. Neil Osheroff is a Professor of Biochemistry and Medicine (Hematology/Oncology) at Vanderbilt University School of Medicine and is the John G. Coniglio Chair in Biochemistry. Dr. Osheroff has been researching topoisomerases and teaching medical students for over 40 years.

Preface

Topoisomerases have been studied for over 50 years. Researchers continue to make discoveries about this complex family of enzymes. One driving force in the continued study of topoisomerases has been the targeting of these enzymes in anticancer chemotherapy. Numerous compounds have been found that disrupt the function of type I and type II topoisomerases, and these agents are widely used in cancer chemotherapy. This Special Issue looks at recent studies examining both fundamental biology of topoisomerase along with the therapeutic targeting and design of novel therapeutics.

Joseph E. Deweese and Neil Osheroff
Editors

International Journal of *Molecular Sciences*

MDPI

Editorial

No Time to Relax and Unwind: Exploration of Topoisomerases and a Growing Field of Study

Joseph E. Deweese [1,2,*] and Neil Osheroff [2,3]

1 Department of Biological, Physical, and Human Sciences, Freed-Hardeman University, Henderson, TN 38340, USA
2 Department of Biochemistry, Vanderbilt University School of Medicine, Nashville, TN 37232, USA; neil.osheroff@vanderbilt.edu
3 Department of Medicine (Hematology/Oncology), Vanderbilt University School of Medicine, Nashville, TN 37232, USA
* Correspondence: jdeweese@fhu.edu

With the topoisomerase field in its sixth decade [1], it is worth taking a moment to pause and consider where our knowledge has come from and where the field is headed. From the time when topoisomerases were first studied in prokaryotes to their exploration in eukaryotic systems such as yeast and *Drosophila*, the field has advanced significantly, with dozens of laboratory groups and hundreds of researchers worldwide [2–4]. The pioneering work in yeast and other model systems laid a critical foundation for the research now performed with human enzymes, cells, and disease states. Additionally, the examination of prokaryotic enzymes has continued to develop, enabling the exploration and discovery of additional enzymes and antibacterial compounds that can be used to manipulate these enzymes and cure infectious diseases [5–8]. With the approval of antibody–drug conjugates targeting topoisomerase I, there is renewed interest in identifying novel compounds and understanding the mechanisms of topoisomerase-targeted anticancer agents [9]. Work on topoisomerases has transitioned from examinations in cellular systems to studies of purified enzymes with ensemble approaches to single-molecule experiments and structural studies that allow for the detailed analysis of the mechanisms and forces involved in topoisomerase catalysis [10]. Finally, studies that use single-molecule or genomic approaches have moved back into cells in order to detail the interactions and functions of topoisomerases in living systems [10].

Several themes in the field have emerged in recent decades. First, projects are becoming increasingly interdisciplinary in nature. This truism has occurred across science, and is certainly reflected in the topoisomerase field. For instance, we can see interdisciplinary collaborations aiding in the development of novel antibacterial and anticancer agents [11,12]. From studying purified enzymes to exploration with cellular models to applications in animal models, numerous areas of expertise are needed, often requiring collaborative efforts among multiple laboratory groups.

Second, with the increased study of genome function, critical roles of topoisomerases in maintaining topology and influencing genome regulation have been revealed [13]. These studies have also required expertise in multiple fields of study to explore and integrate topoisomerase function with genomic regulation. Interdisciplinary collaborations have enabled the mapping of topoisomerase binding sites across the genome and at different stages of the cell cycle in multiple model systems [14]. This wealth of information is incredibly valuable, but our ultimate comprehensive understanding of the regulation and operation of topoisomerases will depend on our ability to collate huge volumes of information.

Third, the rise of "data science" and bioinformatics has enabled the storage and processing of vast amounts of data (such as those mentioned above). Methods such as

Citation: Deweese, J.E.; Osheroff, N. No Time to Relax and Unwind: Exploration of Topoisomerases and a Growing Field of Study. *Int. J. Mol. Sci.* **2023**, *24*, 13080. https://doi.org/10.3390/ijms241713080

Received: 11 August 2023
Revised: 18 August 2023
Accepted: 21 August 2023
Published: 23 August 2023

Hi-C that helped map interactions between and within chromosomes could be combined with topoisomerase binding maps to provide a more genome-wide vision of topoisomerase function [15]. Additional databases could also be brought into play, such as those that document post-translational modifications (e.g., Phosphosite), protein binding partners (e.g., The BIOGrid), and tissue-specific expression (e.g., Human Protein Atlas) [16–18]. We live and work in an era that provides huge quantities of data, and much of these data are freely available to the public. Accessing and interpreting data from these repositories often requires training or expertise that may require collaborative efforts. Alternatively, we may need to do a better job of training our current students and postdoctoral fellows in these areas.

Fourth, the integration of data from various sources has led to several studies that explore unknown or underappreciated aspects of topoisomerase function [19]. For instance, insights into the inter-relationships in the unstructured C-terminal domain may be explored using computational methods [20]. Granted, not all computational data are created equal. However, at the very least, there is value in the application of computational data to develop testable models for topoisomerase function and regulation.

What can we learn from the above observations? An obvious lesson is that we are stronger together rather than in siloed laboratories. The more we interact and collaborate, the more likely we are to make meaningful discoveries and advance the topoisomerase field. Because the critical roles of topoisomerases in cellular and developmental systems are coupled with their roles in disease development and therapy, new discoveries have the potential to influence our understanding of science and enhance human health. Enhanced interactions between colleagues and collaborators are supported by the increased use of video conferencing that began during the COVID-19 pandemic. Virtual discussions between researchers from around the globe now take place on a routine basis and will help us to keep in contact with one another.

Another important lesson is that we need to broaden the training of our students and postdoctoral fellows. Gone are the days when learning traditional enzymological methods and cell culturing techniques can suffice. Although these skills are still invaluable and continue to be utilized, the modern student needs to be exposed to bioinformatics, 3D genomics, and other resources and skill sets. The good news is that training in these areas is not hard to find. Whether formalized or through streaming videos, resources abound for training in these areas. Enabling students and postdocs to expand their training will foster a stronger scientific community and likely increase the collaborative nature of science in the topoisomerase and DNA topology fields. While newer methods and approaches must become a part of the training of our students, we should not lose sight of the detailed, careful enzymology experiments that have provided the basis for our current understanding of these enzymes.

Finally, we need to embrace the era of "data science" and find ways to leverage new tools to gain a deeper understanding of topoisomerase structure, function, and regulation. Utilizing data generated with these tools, we can design experiments that will help to validate and expand our knowledge of these fascinating enzymes. Once again, we will require collaboration and interdisciplinary interactions to accomplish such goals.

The topoisomerase field continues to expand and blossom in new and unexpected ways and will continue to do so if researchers incorporate new approaches, ideas, and data into their analyses. Here's to the next 60 years of topoisomerase research!

Author Contributions: Writing—original draft preparation, J.E.D.; writing—review and editing, J.E.D. and N.O. All authors have read and agreed to the published version of the manuscript.

Funding: N.O. is funded in part by National Institutes of Health grants R01 GM126363 and R01 AI170546.

Conflicts of Interest: The authors declare no conflict of interest.

References

1. Wang, J.C. Interaction between DNA and an Escherichia coli protein omega. *J. Mol. Biol.* **1971**, *55*, 523–533. [CrossRef] [PubMed]
2. Hsieh, T.-S.; Brutlag, D.L. ATP-dependent DNA topoisomerase from *D. melanogaster* reversibly catenates duplex DNA rings. *Cell* **1980**, *21*, 115–121. [CrossRef] [PubMed]
3. Goto, T.; Wang, J.C. Yeast DNA topoisomerase II. An ATP-dependent type II topoisomerase that catalyzes the catenation, decatenation, unknotting, and relaxation of double-stranded DNA rings. *J. Biol. Chem.* **1982**, *257*, 5866–5872. [CrossRef] [PubMed]
4. Osheroff, N.; Shelton, E.R.; Brutlag, D.L. DNA topoisomerase II from *Drosophila melanogaster*. Relaxation of supercoiled DNA. *J. Biol. Chem.* **1983**, *258*, 9536–9543. [CrossRef] [PubMed]
5. Desai, J.; Sachchidanand, S.; Kumar, S.; Sharma, R. Novel bacterial topoisomerase inhibitors (NBTIs)—A comprehensive review. *Eur. J. Med. Chem. Rep.* **2021**, *3*, 100017. [CrossRef]
6. Jia, Y.; Zhao, L. The antibacterial activity of fluoroquinolone derivatives: An update (2018–2021). *Eur. J. Med. Chem.* **2021**, *224*, 113741. [CrossRef] [PubMed]
7. Martín-Encinas, E.; Selas, A.; Palacios, F.; Alonso, C. The design and discovery of topoisomerase I inhibitors as anticancer therapies. *Expert. Opin. Drug Discov.* **2022**, *17*, 581–601. [CrossRef] [PubMed]
8. Okoro, C.O.; Fatoki, T.H. A mini review of novel topoisomerase II inhibitors as future anticancer agents. *Int. J. Mol. Sci.* **2023**, *24*, 2532. [CrossRef] [PubMed]
9. Conilh, L.; Sadilkova, L.; Viricel, W.; Dumontet, C. Payload diversification: A key step in the development of antibody-drug conjugates. *J. Hematol. Oncol.* **2023**, *16*, 3. [CrossRef] [PubMed]
10. McKie, S.J.; Neuman, K.C.; Maxwell, A. DNA topoisomerases: Advances in understanding of cellular roles and multi-protein complexes via structure-function analysis. *Bioessays* **2021**, *43*, e2000286. [CrossRef] [PubMed]
11. Bruno, P.M.; Lu, M.; Dennis, K.A.; Inam, H.; Moore, C.J.; Sheehe, J.; Elledge, S.J.; Hemann, M.T.; Pritchard, J.R. The primary mechanism of cytotoxicity of the chemotherapeutic agent CX-5461 is topoisomerase II poisoning. *Proc. Natl. Acad. Sci. USA* **2020**, *117*, 4053–4060. [CrossRef] [PubMed]
12. Gibson, E.G.; Bax, B.; Chan, P.F.; Osheroff, N. Mechanistic and structural basis for the actions of the antibacterial gepotidacin against Staphylococcus aureus gyrase. *ACS Infect. Dis.* **2019**, *5*, 570–581. [CrossRef] [PubMed]
13. Martínez-García, P.M.; García-Torres, M.; Divina, F.; Terrón-Bautista, J.; Delgado-Sainz, I.; Gómez-Vela, F.; Cortés-Ledesma, F. Genome-wide prediction of topoisomerase IIβ binding by architectural factors and chromatin accessibility. *PLoS Comput. Biol.* **2021**, *17*, e1007814. [CrossRef] [PubMed]
14. McKie, S.J.; Maxwell, A.; Neuman, K.C. Mapping DNA topoisomerase binding and cleavage genome wide using next-generation sequencing techniques. *Genes* **2020**, *11*, 92. [CrossRef] [PubMed]
15. Uuskula-Reimand, L.; Hou, H.; Samavarchi-Tehrani, P.; Rudan, M.V.; Liang, M.; Medina-Rivera, A.; Mohammed, H.; Schmidt, D.; Schwalie, P.; Young, E.J.; et al. Topoisomerase II beta interacts with cohesin and CTCF at topological domain borders. *Genome Biol.* **2016**, *17*, 182. [CrossRef] [PubMed]
16. Hornbeck, P.V.; Zhang, B.; Murray, B.; Kornhauser, J.M.; Latham, V.; Skrzypek, E. PhosphoSitePlus, 2014: Mutations, PTMs and recalibrations. *Nucleic Acids Res.* **2015**, *43*, D512–D520. [CrossRef] [PubMed]
17. Oughtred, R.; Stark, C.; Breitkreutz, B.J.; Rust, J.; Boucher, L.; Chang, C.; Kolas, N.; O'Donnell, L.; Leung, G.; McAdam, R.; et al. The BioGRID interaction database: 2019 update. *Nucleic Acids Res.* **2019**, *47*, D529–D541. [CrossRef] [PubMed]
18. Thul, P.J.; Akesson, L.; Wiking, M.; Mahdessian, D.; Geladaki, A.; Ait Blal, H.; Alm, T.; Asplund, A.; Bjork, L.; Breckels, L.M.; et al. A subcellular map of the human proteome. *Science* **2017**, *356*, eaal3321. [CrossRef] [PubMed]
19. Bartas, M.; Slychko, K.; Cerven, J.; Pecinka, P.; Arndt-Jovin, D.J.; Jovin, T.M. Extensive bioinformatics analyses reveal a phylogenetically conserved winged helix (WH) domain (Ztau) of topoisomerase IIa, elucidating its very high affinity for left-handed Z-DNA and suggesting novel putative functions. *Int. J. Mol. Sci.* **2023**, *24*, 10740. [CrossRef] [PubMed]
20. Townsley, T.D.; Wilson, J.T.; Akers, H.; Bryant, T.; Cordova, S.; Wallace, T.L.; Durston, K.K.; Deweese, J.E. PSICalc: A novel approach to identifying and ranking critical non-proximal interdependencies within the overall protein structure. *Bioinform. Adv.* **2022**, *2*, vbac058. [CrossRef]

International Journal of
Molecular Sciences

MDPI

Review

To Break or Not to Break: The Role of TOP2B in Transcription

Ian G. Cowell [1,*], John W. Casement [2] and Caroline A. Austin [1,*]

[1] Biosciences Institute, The Faculty of Medical Sciences, Newcastle University, Newcastle upon Tyne NE2 4HH, UK
[2] Bioinformatics Support Unit, The Faculty of Medical Sciences, Newcastle University, Newcastle upon Tyne NE2 4HH, UK
* Correspondence: ian.cowell@ncl.ac.uk (I.G.C.); caroline.austin@ncl.ac.uk (C.A.A.)

Abstract: Transcription and its regulation pose challenges related to DNA torsion and supercoiling of the DNA template. RNA polymerase tracking the helical groove of the DNA introduces positive helical torsion and supercoiling upstream and negative torsion and supercoiling behind its direction of travel. This can inhibit transcriptional elongation and other processes essential to transcription. In addition, chromatin remodeling associated with gene activation can generate or be hindered by excess DNA torsional stress in gene regulatory regions. These topological challenges are solved by DNA topoisomerases via a strand-passage reaction which involves transiently breaking and re-joining of one (type I topoisomerases) or both (type II topoisomerases) strands of the phosphodiester backbone. This review will focus on one of the two mammalian type II DNA topoisomerase enzymes, DNA topoisomerase II beta (TOP2B), that have been implicated in correct execution of developmental transcriptional programs and in signal-induced transcription, including transcriptional activation by nuclear hormone ligands. Surprisingly, several lines of evidence indicate that TOP2B-mediated protein-free DNA double-strand breaks are involved in signal-induced transcription. We discuss the possible significance and origins of these DSBs along with a network of protein interaction data supporting a variety of roles for TOP2B in transcriptional regulation.

Keywords: topoisomerase; DNA supercoiling; DNA topology; anticancer; DNA damage; transcription; chromatin; TOP2; TOP2B

Citation: Cowell, I.G.; Casement, J.W.; Austin, C.A. To Break or Not to Break: The Role of TOP2B in Transcription. *Int. J. Mol. Sci.* **2023**, *24*, 14806. https://doi.org/10.3390/ijms241914806

Academic Editors: Joseph E. Deweese and Neil Osheroff

Received: 8 September 2023
Revised: 28 September 2023
Accepted: 29 September 2023
Published: 30 September 2023

1. Introduction

DNA topoisomerases are present in all organisms and catalyze strand passage reactions required to regulate DNA topology. This enables changes in DNA supercoiling state, and in the case of type 2 DNA topoisomerases such as vertebrate TOP2A and TOP2B, permits DNA (de)catenation and (un)knotting. These reactions are carried out via a transient enzyme-linked break in one DNA strand (Type I topoisomerases) or both strands (Type 2 topoisomerases). Typically, these DNA breaks remain buried within the enzyme, covalently attached to the catalytic tyrosine through a phosphotyrosine linkage, and are efficiently re-ligated by the enzyme after passage of a second strand or duplex through the enzyme-coupled break (Figure 1A,B). As a result, topoisomerase strand passage activity does not typically activate DNA damage signaling pathways and no DNA repair is required [1–3]. However, topoisomerases can generate more persistent DNA breaks if the religation step of their reaction cycle is inhibited. This occurs if the enzyme encounters specific types of DNA repair intermediate or drugs which permit the cleavage step but block religation (termed topoisomerase poisons, e.g., camptothecin (TOP1) or etoposide (TOP2)) [4]. For TOP2, the resulting stalled topoisomerase-DNA covalent complex (CC) is processed to a protein-free DSB (pf-DSB) which activates DNA damage signaling resulting in the phosphorylation of histone H2AX ser-139 (γH2AX,

Topoisomerases are involved in many types of DNA transaction, from replication and recombination to DNA repair and gene expression [6]. In transcription, topoisomerases regulate negative superhelical torsion behind and positive torsion ahead of an elongating polymerase as described in the twin domain model [2,7,8]. Accumulation of positive supercoils ahead of an elongating polymerase ultimately impedes elongation, and polymerase stalling could also occur due to the torque associated with accumulation of negative supercoiling upstream of the polymerase. Regulation of template supercoiling appears to be particularly important for the correct expression of long and highly expressed genes [2,8–11]. In addition to this general role in transcription, TOP2B has a specific role in the correct execution of transcriptional networks associated with neural and immune cell development [2,9,12–14]. Furthermore, TOP2B is implicated in efficient induction of gene expression by nuclear hormones and other stimuli [15–21], where gene activation is accompanied by the formation of promoter-associated DNA DSBs, seemingly induced directly by TOP2B. Notably, TOP2B is enriched in the promoter regions of most actively transcribed genes and is particularly associated with CTCF (CCCTC-Binding Factor)/cohesin genomic binding sites. Indeed, 50% of conserved CTCF genomic binding sites have been reported to be co-habited by TOP2B. Since CTCF is intimately involved with cohesin loops, the presence of TOP2B at these sites suggests a role in facilitating these loops or modulating transcription-induced supercoiling at loop boundaries [22,23]. Perhaps reflecting the diversity of these gene-expression roles, a wide range of TOP2 protein–protein interaction partners have been identified, including multiple factors involved in chromatin remodeling, transcriptional regulation, DNA repair and chromosome organization and structure [2]. In this article, we will elaborate on the specific roles of TOP2B in transcription focusing on the nature of the DSBs observed at activated promoters and on the reported protein interactions of TOP2B with established protein complexes and individual proteins associated with transcription-related processes. Figure 1C) [5].

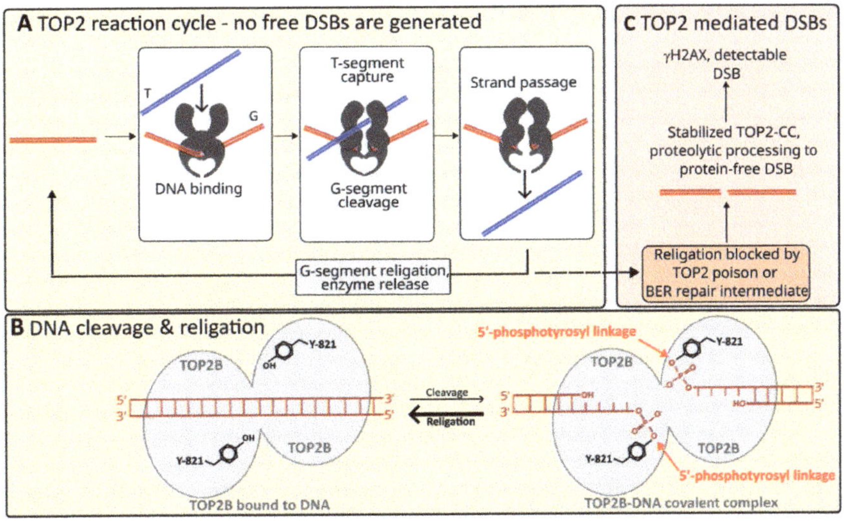

Figure 1. TOP2 strand passage and strand breaks. (**A**) Simplified representation of the TOP2 reaction cycle. The TOP2 dimer (grey) binds one double-stranded DNA segment (Gate or G segment—red), a second DNA duplex (Transported or T segment—blue) is captured and passed through a transient covalent enzyme-coupled DSB in the G segment before G segment religation and release. The transient enzyme-coupled strand breaks remain within the enzyme and the 5′-ends remain covalently coupled to the active site tyrosine of each protomer (Y821 in TOP2B) via a phosphotyrosine linkage as illustrated in (**B**); no free DSBs are generated in this process. (**B**) Representation of TOP2 DNA cleavage and religation products, TOP2 dimer represented as grey ovals. (**C**) After DNA cleavage, if religation is not completed, the TOP2 covalent complex (CC) is proteolytically processed, revealing a protein-free DSB. The TOP2 religation step can be blocked by the presence of TOP2 poisons such as etoposide or base-excision repair (BER) intermediates for example.

2. Specific Roles for TOP2B in Transcriptional Programs

Unlike its paralogue TOP2A, TOP2B is dispensable for cell proliferation and early embryonic development [24,25]. However, while mice constitutively null for TOP2B can develop in utero, they were observed to die perinatally. This was due to failure to properly innervate the diaphragm [14]. Following this original study, others demonstrated essential roles for TOP2B in the later stages of neuronal development and in the immune system, specifically in B cell development [2,9,12,13,26–30]. While the role of TOP2B in these cases appears to be in establishing the correct expression of developmentally regulated genes, the mechanism/s involved are not clear. It was recently shown in a human neuroblastoma cell line model that genes downregulated in TOP2B null cells were enriched for very long genes (>200 kb) and for genes that are highly expressed in wild type cells [9]. The link with long genes and long neural genes in particular had been made previously [2,11,26,31], and suggests that TOP2B may be required for efficient transcription of long transcripts especially in post-mitotic neurons lacking TOP2A. However, since long genes are often associated with neuronal development, and some of the longest mammalian genes are neuronal specific [32], this association may reflect a particular importance of TOP2B in developmental gene expression rather than gene length. Furthermore, *PAX5*, a key B cell transcription factor gene that is under expressed in murine TOP2B hypomorphs [12], is also very long (over 185 kb). The implied requirement for TOP2B to overcome topological constraints of generating very long transcripts does not rule out involvement of TOP2B in other aspects of transcriptional regulation. For example, ChIP and ChIP combined with microarray analyses have demonstrated the presence of TOP2B at the promoters of its neural target genes in mouse tissue [33–35]. In fact, ChIP-seq data reveal that TOPB is generally enriched in and around promoter regions of genes [22,23], including *PAX5* [12] and other B-cell regulators, including *FOXO1* (Figure 2A). As alluded to above, a large proportion of CTCF ChIP-seq peaks correspond to peaks of TOP2B binding [22]. A pattern of TOP2B distribution observed for *PAX5*, *FOXO1* and other genes consists of a sharp TOP2B peak corresponding to a site of CTCF binding, usually upstream of the TSS, embedded in a much broader region of TOP2B enrichment extending well into the body of the gene [12] (Figure 2A).

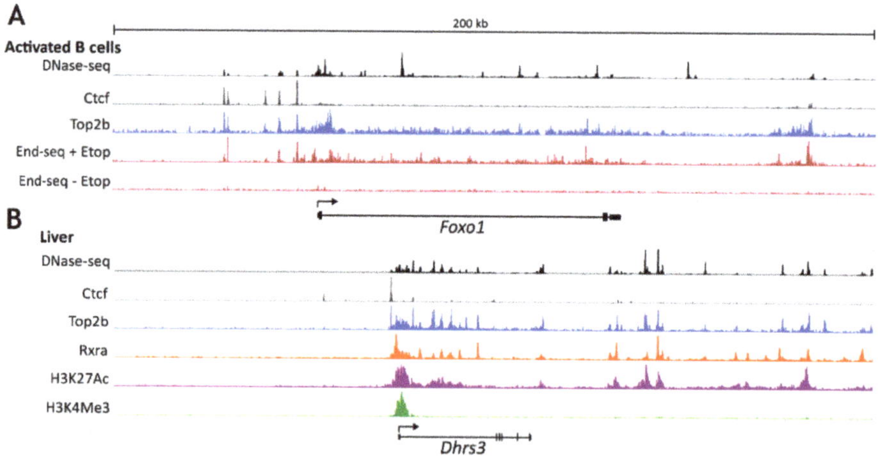

Figure 2. TOP2B chromatin distribution on example genes. (**A**) Distribution of DNAse-seq signal, Ctcf, Top2b and End-seq signal in the presence or absence of etoposide over the mouse *Foxo1* gene. Data are from [36] GSE172807 and [23] GSE99197. (**B**) Distribution of DNAse-seq signal, Ctcf, Rxra, H3K27Ac and H3H4Me3 in the region of the mouse *Dhrs3* gene. Data are from [36] GSE172807,GSE91731, [22] mouse liver Top2b ChIP-seq data, realigned to the mm10 genome assembly, [37] GSE197813, [36] GSE82825.

3. TOP2B and Stimulus-Induced Gene Expression

We describe above how TOP2B is necessary for the correct expression of genes in certain developmental settings, but TOP2B is also required in the context of stimulus-induced gene expression. For example, TOP2B is essential for efficient nuclear hormone induced transcription, activation of *Fasn* (fatty acid synthase) upon feeding in mouse and for activation of the early-response genes including *Fos* and *Npas4* following neural activity in mouse primary neurons (see [2,9,35] and references therein). Similarly, the normally large retinoic acid-stimulated induction of genes such as *CYP26A1*, *CYP26B1*, *DHRS3* and *CRABP2* is dramatically curtailed in TOP2B null neuroblastoma cells [9]. A common feature in each of these examples is a rapid move from low to high level expression, leading to the idea that TOP2B is required in some way for the speedy changes in chromatin organization that accompany this switch in transcription state. Consistent with this, TOP2B is enriched in the 5′-end of RA inducible genes such as *Dhrs3* (Figure 2B). Intriguingly, gene activation in these cases of signal-induced transcription appears to be associated with localized TOP2 dependent DNA breaks. For example, estradiol induced induction of the *TFF1* gene, retinoic acid (RA) induced expression of *RARB*, thyroid hormone (T3) induced DIO1 and TPA induced MMP12 [15] as well as feeding induced induction of *Fasn* transcriptional activation [16] was accompanied by the appearance of a DNA break (free 3′-OH) in the promoter region of the respective genes that could be labelled with biotin-dUTP through the action of terminal deoxynucleotide transferase (TdT) [15,16]. Although this assay does not specifically detect DNA double-strand breaks (DSBs) as TdT labelling with biotin dUTP efficiently detects both nicks and DSBs, ChIP-PCR assays revealed that gene activation was accompanied by recruitment of not only TOP2B, but also the DNA DSB repair factor DNA-PK and PARP1 to the promoter region of the relevant genes [15,16]. Furthermore, in a different study [21] dexamethasone treatment resulted in promoter DSBs detected with iterative primer extension in the glucocorticoid-responsive MMTV promoter. Dexamethasone also led to the recruitment of TOP2B, and the DSB repair factors KU70 (XRCC6) and KU80 (XRCC5) to the MMTV promoter, and notably, the TOP2 catalytic inhibitor merbarone inhibited both dexamethasone-induced transcription and strand breaks [21]. In other studies, it was observed that gene activation was accompanied by localized histone H2AX phosphorylation [35,38,39], an established marker of DNA DSBs [40], supporting the notion that at least some of the signal-induced strand breaks represent DSBs. As for the MMTV example above, estradiol-induced DNA breaks in the *TFF1* promoter were inhibited by merbarone, suggesting a direct role of TOP2 in the induction of these breaks. In addition, NMDA-induced induction of *Fos* and *Npas4* and localized histone H2AX phosphorylation in cultured mouse neurons were both suppressed following shRNA-based depletion of TOP2B [35]. Furthermore, the presence of NMDA-induced DSBs in the *Fos* gene was demonstrated using the independent method of COMET-FISH [41]. The appearance of TOP2-induced breaks, including DNA DSBs that initiate local H2AX phosphorylation is on the face of it surprising, since TOP2 does not produce such protein-free breaks as part of its standard reaction mechanism. Canonically, the transient enzyme-linked strand breaks formed during the reaction cycle remain buried within the enzyme as TOP2-covalent DNA complexes (TOP2-CCs) and are rapidly resolved after strand-passage (Figure 1A,B). However, if religation is impeded, TOP2-CCs can be processed to protein free DSBs (pf-DSBs) [5] which can initiate a DNA damage response including phosphorylation of histone H2AX (Figure 1C). This leads to the question of whether these promoter-associated, signal-induced, TOP2B-dependent DSBs are the incidental consequence of occasional failure of TOP2 to quickly complete its reaction cycle or are part of a mechanism that specifically requires the formation of distinct pf-DNA DSBs for efficient transcriptional induction (see Section 6).

Other studies have linked TOP2-induced strand breaks with RNA polymerase promoter proximal pause release [38,39]. Many genes, including serum-induced immediate early genes such as *FOS*, *JUN* and *EGR1* and the heat-shock protein *HSPA1B* are regulated at this stage [42], where RNA polymerase pauses approximately 30–100 bp downstream of the transcriptional start site (TSS). Stimulation relieves the pause, allowing productive elongation to commence. In one study, phospho-H2AX and PRKDC (DNA-PK_{cs}) accumulation was observed in the TSS and gene bodies of *FOS*, *JUN* and *EGR1* following release from serum starvation [43], consistent with the presence of DSBs. Notably, TOP2B also accumulated at the TSSs and the TOP2 catalytic inhibitor ICRF-193 increased pausing and led to decreased accumulation of phospho-H2AX [39]. Additional support for the association of TOP2B and DSBs in polymerase pausing, is provided by recent whole-genome DNA break analysis using the sensitive DSBCapture whole genome break mapping technique; this revealed that in unperturbed cells genomic DSBs are preferentially located around the TSSs of highly transcribed and paused genes and that pause sites are enriched in DSBs. Furthermore, treatment with the TOP2 poison etoposide (which prevents religation of the normally transient DSB, resulting in the accumulation of pf-DSBs) increased DSBs at pause sites implicating TOP2 in these DNA breaks [23,44].

4. TOP2B Protein Interactors

4.1. TOP2B/DNA-PK/PARP Connection

Several separate studies have found TOP2B to colocalize with the DNA repair proteins DNA-PK_{cs} (PRKDC), KU70 (XRCC6), KU80 (XRCC5) and PARP1 in the promoter regions of TOP2B-dependent inducible genes [2,15,16,19,20,39,45,46]. Independently, protein–protein interactions have been identified between TOP2B and PRKDC [16], XRCC6 [22,43], XRCC5 [16] and PARP1 [15,22,47,48]. Although the precise role of these repair proteins in this context remains unclear, they may relate to the induction of transcription-coupled DSBs discussed above, as it is possible to imagine that rapid repair of promoter DSBs would be required to prevent transcriptional silencing and possible genomic instability. However, it has also been suggested that the protein kinase activity of DNA-PK, activated by promoter DNA breaks and phosphorylating downstream targets, may form part of the mechanism that turns on transcription in inducible genes [16,39].

4.2. Broad Range of Protein Interaction Partners

A number of TOP2B protein–protein interactions have been identified through traditional biochemical isolation of protein complexes and subsequent identification of individual components. For example, affinity purification of the TLE complex from rat neural stem cells using TAP-tagging revealed TOP2B to be part of a complex of 13 polypeptides, including non-muscle myosin II heavy chain (MYH9), β-actin (ACTB), RAD50, PARP1, nucleolin (NCL), HSP70, p54nrb (NONO), and nucleophosmin (NPMP) [45]. In a second example, affinity purification of USF1 interacting proteins from mouse liver revealed a complex containing TOP2B, PRKDC (DNA-PK), XRCC5 (KU80), XRCC6 (KU70), PARP1, PP1, KAT2B (P/CAF) [16]. Using a similar approach both TOP2A and TOP2B were found in immunoprecipitates of HA-tagged MYCN [49]. A subsequent study demonstrated that TOP2A and TOP2B appear in functional complexes with TOP1 and either MYC or MYCN respectively. The TOP2-MYC/MYCN-TOP1 complex, termed the topoisome, is proposed to resolve DNA overtwisting and supercoiling associated with high output transcription [50]. Many additional TOP2B protein interactions have been identified via high-throughput screens including proximity-labelling (e.g., BioID) [22,43,51] or IP/Affinity purification followed by mass-spectroscopy [47]. Although the significance and biological importance of such high throughput screen hits is not always certain, evidence for direct or indirect interaction with multiple components of the same complex or pathway, or statistical filtering, can lead to higher confidence. For example, a BioID screen for TOP2B-interacting proteins revealed 25 high-confidence interactors, including CTCF and several other cohesin complex components, the chromodomain protein CBX8 and topoisomerase I. Querying protein interaction databases such as BioGRID [52] and IntACT (http://www.ebi.ac.uk/intact) revealed a total of several hundred distinct TOP2B protein interactions partially derived from high-throughput screens. Biological processes relating to chromosome and chromatin processes, DNA repair, chromatin remodeling, and transcriptional regulation scored with the greatest significance in enrichment analysis of these interactions (see Figure 3), consistent with an important role in transcriptional processes. Manual inspection and comparison with the CORUM comprehensive resource of mammalian protein complexes database [53] revealed an association of TOP2B with the TLE complex as described above, but also with multiple components of the WINAC, BAF and NuRD ATP-dependent chromatin remodeling complexes and the FACT histone chaperone complex (Table 1, Figure 4). This points to a connection between TOP2B transcriptional functions and chromatin remodeling. Also of note, interactions were detected between TOP2B and three constituents of the Mediator Complex (MED15, MED24 and MED27, Figure 4) [47] which is involved in many steps of transcriptional regulation, including facilitating promoter–enhancer interactions and nuclear receptor function [54]. TOP2B interactors from the above analysis also include factors involved in transcriptional repression and heterochromatin, including members of the PRC2 polycomb complex (EZH2, SUZ12), polycomb- and HP1- class chromodomain proteins (CBX8, CBX3, CBX5). DNA repair proteins also feature in the list of TOP2B interactors. In addition to the NHEJ-associated factors PRKDC, XRCC6 and XRCC5 (DNA-PKcs, KU70 and KU80, see above), components of the base excision and nucleotide excision repair pathways were also present amongst the set of TOPB interactors (see Figures 3 and 4). It is tempting to suppose that the presence of DNA repair proteins amongst the set of TOPB interactors reflects the capacity of TOP2 to generate DNA strand breaks either as an incidental aspect of its normal activity, or mechanistically where the formation of distinct DNA DSBs are required for efficient transcriptional induction.

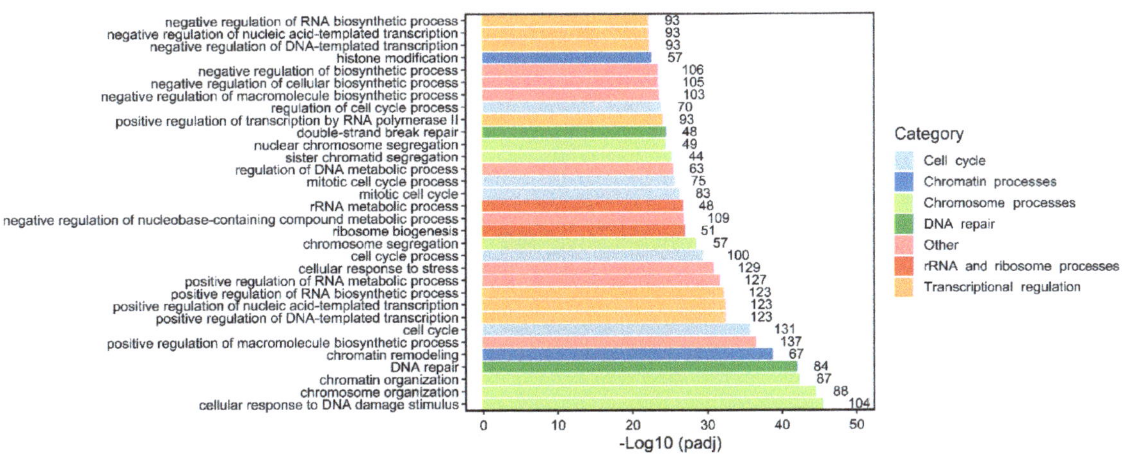

Figure 3. Gene set enrichment analysis (GSEA): Biological Processes most enriched amongst TOP2B interacting proteins. A non-redundant list of TOP2B interactors was compiled from previous publications and from querying BioGRID [52] and IntAct databases. This list was the input for GSEA analysis using g:GOSt functional profiling tool (https://biit.cs.ut.ee/gprofiler/gost (accessed on 10 January 2023)). The top 32 overrepresented Biological Processes (BPs) were plotted according to their significance (-Log10 multiple testing corrected p-value (padj)). The intersect size is shown to the right of each bar. BPs were manually placed into seven color-coded functional categories shown on the right.

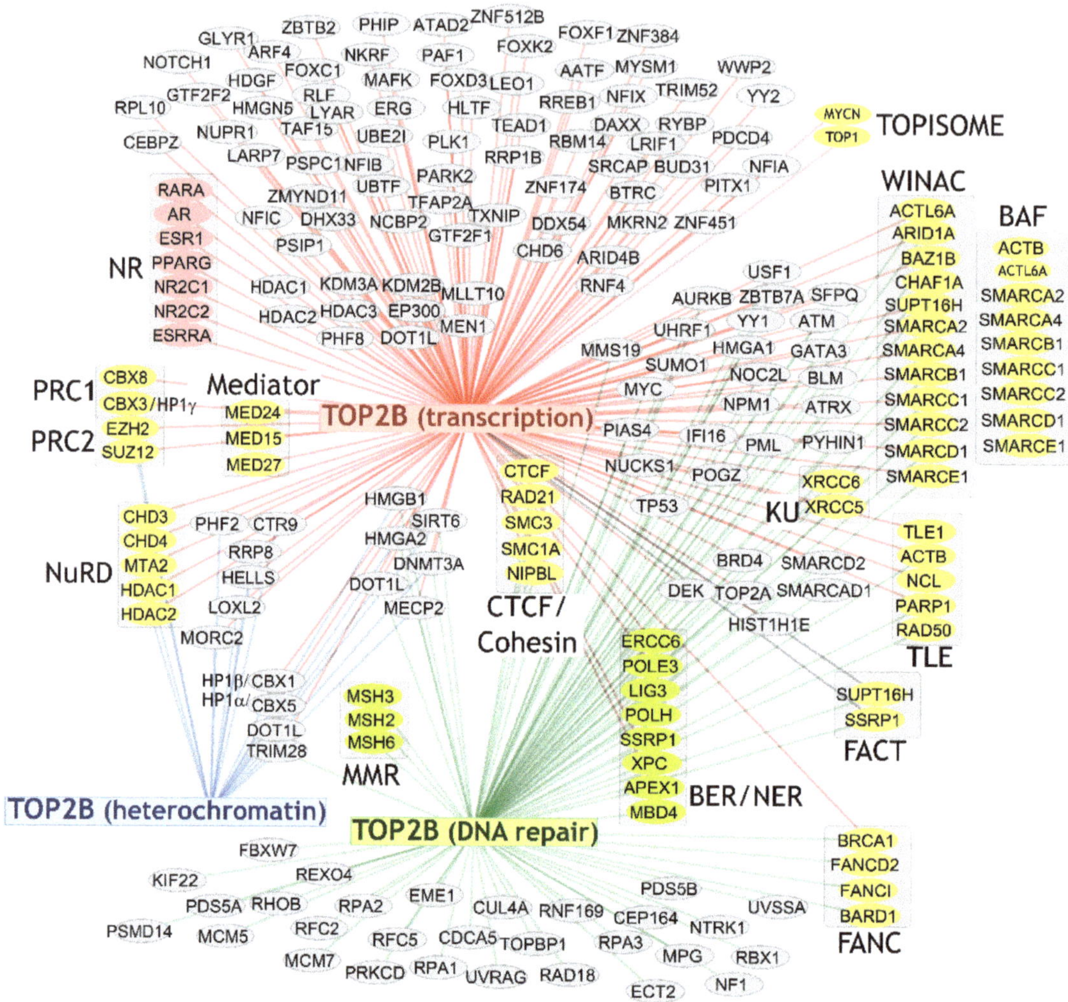

Figure 4. Transcription, DNA repair and heterochromatin—associated TOP2B interactors, complexes and pathways. TOP2B interactors contained in transcription, DNA repair or heterochromatin categories of Biological Processes (see Figure 3) are grouped according to their category, and further grouped (shaded boxes) according to membership of particular protein complexes (Topoisome, WINAC, BAF, TLE, FACT, NURD, PRC1/2, Mediator, KU, CTCF/Cohesin) or functional group/pathway (BER, NER, MMR, FANC, NR). Data were plotted using Cytoscape [55].

Table 1. TOP2B association with chromatin remodeling complexes.

Component	Synonyms	Reference
WINAC/WSTF/ISWI Complex (CORUM:1230)—ATP Dependent Chromatin Remodeling		
ACTL6A	BAF53A, INO80K, Arp4, ACTL6	[48]
ARID1A	BAF250A, SMARCF1,	[48]
BAZ1B	WSTF (n.b. H2AX Tyr142-kinase)	[48]
CHAF1A	CAF1P150	[22,48]
SMARCA2	BRM, SNF2L2, STH1P, BAF190, SNF2	[48]
SMARCA4	BRG1, BAF190A, SNF2L4	[21,48]
SMARCB1	INI1, BAF47, SNF5L1, SWNTS1	[48]
SMARCC1	Rsc8, CRACC1, SWI3, BAF155, SRG3	[48]
SMARCC2	CRACC2, BAF170	[48]
SMARCD1	Rsc6p, BAF60A, CRACD1	[48]
SMARCE1	BAF57, CSS5	[48]
SUPT16H	CDC68, SPT16, FACTP140	[22,56]
BAF Complex (CORUM:18)—ATP Dependent Chromatin Remodeling		
ACTB	β-actin	
ACTL6A	BAF53A, INO80K, Arp4	
ARID1A	BAF250A, SMARCF1,	[48]
SMARCA2	BRM, SNF2L2, STH1P, BAF190, SNF2	[48]
SMARCA4	BRG1, BAF190A, SNF2L4	[21,48]
SMARCB1	INI1, BAF47, SNF5L1, SWNTS1	[48]
SMARCC1	Rsc8, CRACC1, SWI3, BAF155, SRG3	[48]
SMARCC2	CRACC2, BAF170	[48]
SMARCD1	Rsc6p, BAF60A, CRACD1	[48]
SMARCE1	BAF57, CSS5	[48]
FACT Complex (CORUM:936)—Histone Chaperone		
SUPT16H	CDC68, SPT16, FACTP140	[22,56]
SSRP1	FACT, FACT80, T160	[22,47]
NuRD Complex (CORUM: 587)—Chromatin Remodeling and Histone Deacetylation		
CHD3	MI-2A	[57]
CHD4	MI-2B	[57]
MTA2	PID	[58]
HDAC1		[58,59]
HDAC2		[58,59]

4.3. TOP2B Transcription Factor Associations

The protein interaction studies described above identify a number of transcription factors that can associate with TOP2B (see Figure 4), this includes CTCF, MYCN and several nuclear hormone receptors such as RARA, AR, ESR1, ESR2 and PPARG. However, in an alternative approach to determine whether TOP2B colocalizes with specific transcription factors in chromatin we previously compared sequences contained under TOP2B ChIP-seq peaks from human MCF7 cells with motifs contained in the JASPAR database of transcription factor DNA binding sites [60,61]. Motifs corresponding to the following transcription factors were significantly enriched under TOP2B ChIP-seq peaks: CTCF, ESR1, ESR2, PPARG, TFAP2, MYF, REST, TFCP2l1, PAX5, INSM1 and AP1. The association with CTCF was later confirmed by others in mouse cells as alluded to above, where about 50% of CTCF genomic binding sites were shown to coincide with peaks of TOP2B [22,23]. Furthermore, CTCF and factors representing some of the other motifs (ESR1 and PPARG) were also present amongst the previously described TOP2B protein interactors. However, it should be noted that colocalization on chromatin does not necessarily require direct protein interaction. We subsequently carried out a similar analysis of motifs enriched under TOP2B peaks in high coverage mouse ChIP-seq data [22,62] from B cells, embryonic fibroblasts and liver and using the JASPAR enrichment tool (https://jaspar.genereg.net/enrichment/ (accessed on 29 March 2022)) [63]. Taking motifs with the highest odds ratio, present in all three sets, but not present in data from TOP2B null MEF cells yielded 158 motifs. These included motifs coinciding with CTCF, TFAP2, ESR2, REST that were also observed in the MCF7 set.

5. A Model for TOP2B Function in Transcription

Previous studies have demonstrated that TOP2B is required for efficient execution of certain developmental transcriptional programs and signal-induced transcription (see Sections 2 and 3 above). While there is evidence that TOP2B-mediated DNA strand breaks may be involved (particularly in the later scenario), the precise role of TOP2 in these gene regulatory events is poorly understood. Protein–protein interaction studies, evidence for promoter recruitment of TOP2B from ChIP and ChIP-seq experiments and TOP2B-mediated promoter DSB induction (see above) together point to multiple stages in the RNA pol II transcription initiation pathway where TOP2B could play a key role. Importantly, these are not mutually exclusive possibilities. Figure 5 shows a simplified model for transcriptional activation by RNA pol II, highlighting points where TOP2B could potentially have a functional role based on available data (indicated by numerals inside circles). Firstly, the early stage of gene activation is characterized by the replacement of co-repressors by coactivators such as P300 and recruitment of chromatin-remodeling factors (Figure 5B). Protein interaction studies have demonstrated physical association between TOP2B and P300 (EP300) and between TOP2B and multiple components of ATP-dependent chromatin remodeling complexes including BAF and WINAC (Figure 4, Table 1) and with the histone chaperone complex FACT. These interactions may reflect co-recruitment of TOP2B with these complexes during transcriptional activation. In addition, the histone acetylase and nucleosome remodeling activities (Figure 5B, purple arrows) which make chromatin structure more accessible to other factors including the basal transcription machinery can also generate torsional stress resulting from the displacement or translocation of nucleosomes [64,65]. Thus, it is possible that the association of TOP2B with these chromatin remodeling factors helps put TOP2B in the correct place to manage this remodeling-linked torsion. Recruitment of Mediator and chromatin looping subsequently enables promoter–enhancer contacts, facilitating the recruitment of general transcription factors (GTFs) and pre-initiation complex (PIC) assembly (Figure 5C). Intriguingly, TOP2 has been identified as an interactor for three components of the Mediator complex (Figure 4) [47]. Although the mechanistic significance of this interaction is unknown, it is possible that it reflects stabilization or mutual stimulation of chromatin binding. Alternatively, the association with TOP2B could aid promoter–enhancer loop formation stimulated by Mediator [66], either by removing inhibitory torsional stress or via formation of a pf-DSB that facilitates the necessary rapid chromatin reconfiguration as suggested previously [35]. Transcriptional activation following a signal such as nuclear hormone ligand binding or neural activity in mouse primary neurons results in the appearance of apparently TOP2B-mediated DNA strand beaks in the promoters of target genes [15,35,41] (Figure 5 red asterisks, see Section 3). Neither the mechanistic significance nor any precise role for these DNA breaks are currently understood, but they may provide a way of quickly remodeling promoter regions to enable rapid and large magnitude changes in gene expression level. Phosphorylation of the CTD of the largest subunit of RNA pol II by CDK7 allows RNA pol promoter escape (Figure 5D) and the transition to the elongation phase of transcription. This results in the generation of negative helical torsion behind the polymerase, which is propagated into the promoter region and managed by topoisomerases including TOP2. This may be particularly significant for high-output genes [7,10]. Recently, TOP2A and TOP2B have been shown to work in concert with TOP1 and either MYC or MYCN respectively to manage transcription-associated topological stress [50]. After promoter escape, RNA Pol II often generates a short nascent RNA before pausing and awaiting further signals to enter productive transcription into the gene body (Figure 5E). This phenomenon of promoter–proximal pausing is a key transcriptional regulatory point [42]. Notably, DSB formation, at least in some cases mediated by TOP2, has been implicated in the pause release mechanism [38,39,67]. Thus, in addition to modulating elongation-associate superhelical torsion in gene bodies and promoter regions, the evidence suggests that TOP2 has a role in promoter proximal pausing and release, through a mechanism that may involve production of a pf-DSB.

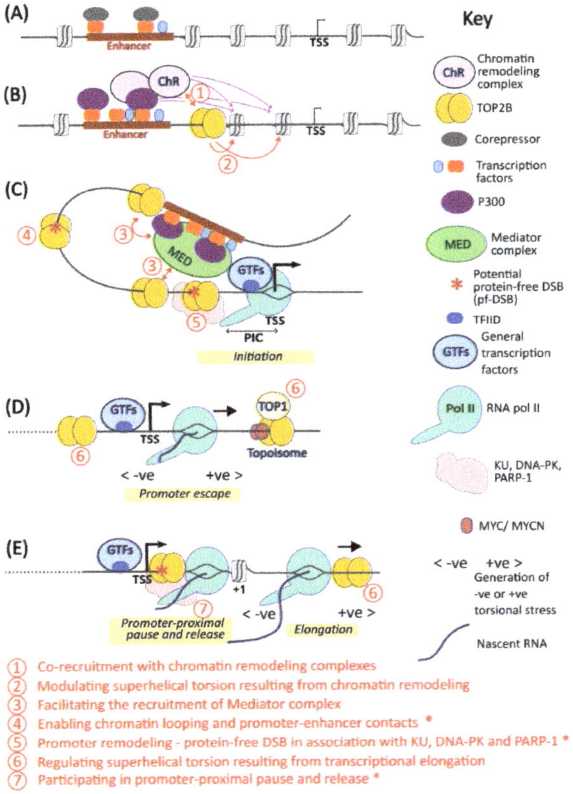

Figure 5. Model depicting potential modes of action of TOP2B in transcription. A simplified model for transcriptional activation by RNA pol II, highlighting points where TOP2B could potentially have a functional role (indicated by red numerals inside circles defined at the bottom of the diagram). (**A**) Gene in an "off" state, some transcription factor binding in an upstream enhancer, but these are associated with corepressors. (**B**) Early stage of transcriptional activation. Corepressors replaced by coactivators/s such as P300 along with binding of further TFs to the enhancer. The resulting histone acetylase and nucleosome remodeling activities (purple arrows) make chromatin structure more available to other factors including the basal transcription machinery but may also generate or be hindered by superhelical stress [64,65], requiring the action of topoisomerases. (**C**) Recruitment of Mediator and chromatin looping facilitates promoter–enhancer contacts, the recruitment of GTFs and PIC assembly (including RNA pol II). (**D**) Transcription initiation and promoter escape. The elongating polymerase generates +ve torsional stress ahead and -ve torsional stress behind [7,8,10]. (**E**) The initiation of transcription may be followed by transcriptional pausing a short distance downstream from the TSS [38,42]. Transcription, particularly of high output genes, generates DNA torsion and supercoiling that if not relieved by topoisomerase action could impede elongation by RNA pol II. Similarly, topoisomerases are required to manage (-) torsional stress propagating behind the early elongating polymerase into the promoter region [8,10]. The combined action of TOP1, TOP2 and MYC/MYCN in a protein complex coined the Topoisome may contribute to controlling this torsional stress [50]. Points where a pf-DSB could be envisaged to play a mechanistic role are highlighted with an asterisk.

High-coverage ChIP-seq analyses of TOP2B genomic distribution are limited, but from examination of available data, TOP2B is present in the 5′-regions of active genes, often in a broad area of enrichment superimposed on a sharp peak or peaks corresponding to

CTCF binding (see Figure 2). In addition, End-seq, focused on DNA breaks induced by etoposide and thus measuring sites of enzymatically active TOP2, gives a broadly similar distribution [23] (Figure 2A). This is consistent with TOP2B performing an enzymatic role in this setting. This broad distribution of TOP2B chromatin occupancy and TOP2 activity does not rule out any of the possible transcriptional modes of action of TOP2B discussed above, but rather suggests that TOP2B functions at multiple points in transcriptional activation.

6. Conclusions and Perspective

The evidence described above supports multiple ways in which TOP2B could be required for transcriptional regulation to ensure correct gene expression. ChIP-seq analysis shows that the enzyme is enriched in gene regulatory regions [22,23] and this makes it available to solve DNA topological challenges as they arise as part of the normal processes of gene expression. This is likely to be particularly important in non-dividing cells, where TOP2A is at a low level or absent, and in situations such as nuclear hormone signaling where a rapid and large induction of transcription is required.

In addition, activation of signal-induced genes is accompanied by the appearance of TOPB-dependent DSBs. However, it is not fully understood whether DSBs themselves (as opposed to standard TOP2 strand passage activity which does not result in pf-DSBs) are required for activation under physiological conditions. Evidence supporting a central role for DSBs includes the fact that etoposide treatment can mimic aspects of signal-induced transcription and that gene activation can be accompanied by the appearance of γH2AX in promoter regions and throughout the gene body of regulated genes [35]. Although it is unclear what mechanistic role these pf-DSBs may have in transcriptional activation, it is possible to envisage models such as DSB-facilitated chromatin remodeling or enhancer–promoter looping. Furthermore, the H2AX phosphorylation associated with transcriptional activation has some curious features. Classically, induction of genomic DSBs by agents such as etoposide or ionizing radiation results in histone H2AX ser-139 phosphorylation in megabase domains flanking each DSB [40,68]. However, in the context of signal-induced transcription, the zone of ser-139 phosphorylation appears to be tightly limited to the gene body and a short distance upstream [35,39]. This possibly reflects very transient pf-DSB formation or specific features of signal-induced transcription that limit spread of H2AX-139 along the chromatin. Furthermore, DSBs arising in transcription units are generally thought to be repressive to transcription [69] rather than associated with gene activation.

It is also unclear how such TOP2B-dependent DSBs might arise since the standard strand passage activity does not produce free DNA ends (Figure 1). The appearance of DSBs that lead to H2AX phosphorylation (or can be detected by other means) points to TOP2 poisoning (i.e., a block to cleavage complex religation, requiring processing to a pf-DSB before repair (see Figure 1)). However, it is not known if this is a targeted process or a consequence of other events occurring during gene activation. For example, histone and DNA demethylation associated with a change in transcription state following nuclear hormone action generate reactive oxygen species that lead to DNA damage (oxidized bases and SSBs), requiring base excision repair [18,70]. Notably, BER-intermediates act as TOP2 poisons [4]. Thus, via the formation and processing of stabilized cleavage complexes, TOP2 can convert BER intermediates to pf-DSBs.

In most studies that report strand breaks upon stimulation of signal-induced genes the methodology used has not allowed precise location of the break. Thus, it is not known if breaks are induced at specific locations or broadly in the vicinity of the regulatory region. Nor is it clear whether the observed breaks occur at most/every active allele, or just at a minority. The answers to these questions are likely to shed light on the mode of action of TOP2B at signal-induced promoters.

Thus, it is possible that evolution has co-opted DNA breaks and DNA repair proteins to facilitate rapid changes in transcriptional state required for signal-induced transcriptional activation. TOP2B colocalizes with DNA-PK and PARP at signal induced promoters, perhaps reflecting a requirement to rapidly repair transcription-associated DSBs which would otherwise inhibit transcription [71]. But alternatively, it has been suggested that the enzymatic activities of the repair proteins may themselves facilitate steps in transcriptional activation [16,39]. More work is required to fully understand the mechanistic role of TOP2B in transcriptional regulation, both in the context of developmental programs and in signal induced transcription.

Author Contributions: Writing—original draft preparation, I.G.C.; writing—review and editing, C.A.A.; data analysis, J.W.C. and I.G.C.; visualization, I.G.C. All authors have read and agreed to the published version of the manuscript.

Funding: This research received no external funding.

Institutional Review Board Statement: Not applicable.

Informed Consent Statement: Not applicable.

Data Availability Statement: Sources of previously published data are given in the figure legends including details of GEO accession numbers.

Conflicts of Interest: The authors declare no conflict of interest.

References

1. Pommier, Y.; Nussenzweig, A.; Takeda, S.; Austin, C. Human topoisomerases and their roles in genome stability and organization. *Nat. Rev. Mol. Cell Biol.* **2022**, *23*, 407–427. [CrossRef] [PubMed]
2. Austin, C.A.; Cowell, I.G.; Khazeem, M.M.; Lok, D.; Ng, H.T. TOP2B's contributions to transcription. *Biochem. Soc. Trans.* **2021**, *49*, 2483–2493. [CrossRef] [PubMed]
3. Austin, C.A.; Lee, K.C.; Swan, R.L.; Khazeem, M.M.; Manville, C.M.; Cridland, P.; Treumann, A.; Porter, A.; Morris, N.J.; Cowell, I.G. TOP2B: The First Thirty Years. *Int. J. Mol. Sci.* **2018**, *19*, 2765. [CrossRef]
4. Wilstermann, A.M.; Osheroff, N. Base excision repair intermediates as topoisomerase II poisons. *J. Biol. Chem.* **2001**, *276*, 46290–46296. [CrossRef] [PubMed]
5. Swan, R.L.; Cowell, I.G.; Austin, C.A. Mechanisms to Repair Stalled Topoisomerase II-DNA Covalent Complexes. *Mol. Pharmacol.* **2022**, *101*, 24–32. [CrossRef] [PubMed]
6. McKie, S.J.; Neuman, K.C.; Maxwell, A. DNA topoisomerases: Advances in understanding of cellular roles and multi-protein complexes via structure-function analysis. *Bioessays* **2021**, *43*, e2000286. [CrossRef] [PubMed]
7. Liu, L.F.; Wang, J.C. Supercoiling of the DNA template during transcription. *Proc. Natl. Acad. Sci. USA* **1987**, *84*, 7024–7027. [CrossRef]
8. Ma, J.; Wang, M.D. DNA supercoiling during transcription. *Biophys. Rev.* **2016**, *8*, 75–87. [CrossRef]
9. Khazeem, M.M.; Casement, J.W.; Schlossmacher, G.; Kenneth, N.S.; Sumbung, N.K.; Chan, J.Y.T.; McGow, J.F.; Cowell, I.G.; Austin, C.A. TOP2B Is Required to Maintain the Adrenergic Neural Phenotype and for ATRA-Induced Differentiation of SH-SY5Y Neuroblastoma Cells. *Mol. Neurobiol.* **2022**, *59*, 5987–6008. [CrossRef]
10. Kouzine, F.; Gupta, A.; Baranello, L.; Wojtowicz, D.; Ben-Aissa, K.; Liu, J.; Przytycka, T.M.; Levens, D. Transcription-dependent dynamic supercoiling is a short-range genomic force. *Nat. Struct. Mol. Biol.* **2013**, *20*, 396–403. [CrossRef]
11. Joshi, R.S.; Pina, B.; Roca, J. Topoisomerase II is required for the production of long Pol II gene transcripts in yeast. *Nucleic Acids Res.* **2012**, *40*, 7907–7915. [CrossRef] [PubMed]
12. Broderick, L.; Yost, S.; Li, D.; McGeough, M.D.; Booshehri, L.M.; Guaderrama, M.; Brydges, S.D.; Kucharova, K.; Patel, N.C.; Harr, M.; et al. Mutations in topoisomerase IIβ result in a B cell immunodeficiency. *Nat. Commun.* **2019**, *10*, 3644. [CrossRef] [PubMed]
13. Broderick, L.; Clay, G.M.; Blum, R.H.; Liu, Y.; McVicar, R.; Papes, F.; Booshehri, L.M.; Cowell, I.G.; Austin, C.A.; Putnam, C.D.; et al. Disease-associated mutations in topoisomerase IIβ result in defective NK cells. *J. Allergy Clin. Immunol.* **2022**, *149*, 2171–2176.e3. [CrossRef] [PubMed]
14. Yang, X.; Li, W.; Prescott, E.D.; Burden, S.J.; Wang, J.C. DNA topoisomerase IIβ and neural development. *Science* **2000**, *287*, 131–134. [CrossRef] [PubMed]
15. Ju, B.G.; Lunyak, V.V.; Perissi, V.; Garcia-Bassets, I.; Rose, D.W.; Glass, C.K.; Rosenfeld, M.G. A topoisomerase IIβ-mediated dsDNA break required for regulated transcription. *Science* **2006**, *312*, 1798–1802. [CrossRef] [PubMed]
16. Wong, R.H.F.; Chang, I.; Hudak, C.S.S.; Hyun, S.; Kwan, H.-Y.; Sul, H.S. A Role of DNA-PK for the Metabolic Gene Regulation in Response to Insulin. *Cell* **2009**, *136*, 1056–1072. [CrossRef] [PubMed]
17. Madabhushi, R.; Kim, T.K. Emerging themes in neuronal activity-dependent gene expression. *Mol. Cell. Neurosci.* **2018**, *87*, 27–34. [CrossRef] [PubMed]
18. Perillo, B.; Ombra, M.N.; Bertoni, A.; Cuozzo, C.; Sacchetti, S.; Sasso, A.; Chiariotti, L.; Malorni, A.; Abbondanza, C.; Avvedimento, E.V. DNA oxidation as triggered by H3K9me2 demethylation drives estrogen-induced gene expression. *Science* **2008**, *319*, 202–206. [CrossRef]

19. McNamara, S.; Wang, H.; Hanna, N.; Miller, W.H. Topoisomerase IIβ Negatively Modulates Retinoic Acid Receptor α Function: A Novel Mechanism of Retinoic Acid Resistance. *Mol. Cell. Biol.* **2008**, *28*, 2066–2077. [CrossRef]
20. Haffner, M.C.; Aryee, M.J.; Toubaji, A.; Esopi, D.M.; Albadine, R.; Gurel, B.; Isaacs, W.B.; Bova, G.S.; Liu, W.; Xu, J.; et al. Androgen-induced TOP2B-mediated double-strand breaks and prostate cancer gene rearrangements. *Nat. Genet.* **2010**, *42*, 668–675. [CrossRef]
21. Trotter, K.W.; King, H.A.; Archer, T.K. Glucocorticoid Receptor Transcriptional Activation via the BRG1-Dependent Recruitment of TOP2β and Ku70/86. *Mol. Cell. Biol.* **2015**, *35*, 2799–2817. [CrossRef]
22. Uuskula-Reimand, L.; Hou, H.; Samavarchi-Tehrani, P.; Rudan, M.V.; Liang, M.; Medina-Rivera, A.; Mohammed, H.; Schmidt, D.; Schwalie, P.; Young, E.J.; et al. Topoisomerase II beta interacts with cohesin and CTCF at topological domain borders. *Genome Biol.* **2016**, *17*, 182. [CrossRef] [PubMed]
23. Canela, A.; Maman, Y.; Jung, S.; Wong, N.; Callen, E.; Day, A.; Kieffer-Kwon, K.R.; Pekowska, A.; Zhang, H.; Rao, S.S.P.; et al. Genome Organization Drives Chromosome Fragility. *Cell* **2017**, *170*, 507–521.e518. [CrossRef] [PubMed]
24. Carpenter, A.J.; Porter, A.C. Construction, characterization, and complementation of a conditional-lethal DNA topoisomerase IIα mutant human cell line. *Mol. Biol. Cell* **2004**, *15*, 5700–5711. [CrossRef] [PubMed]
25. Akimitsu, N.; Adachi, N.; Hirai, H.; Hossain, M.S.; Hamamoto, H.; Kobayashi, M.; Aratani, Y.; Koyama, H.; Sekimizu, K. Enforced cytokinesis without complete nuclear division in embryonic cells depleting the activity of DNA topoisomerase IIα. *Genes Cells* **2003**, *8*, 393–402. [CrossRef]
26. King, I.F.; Yandava, C.N.; Mabb, A.M.; Hsiao, J.S.; Huang, H.S.; Pearson, B.L.; Calabrese, J.M.; Starmer, J.; Parker, J.S.; Magnuson, T.; et al. Topoisomerases facilitate transcription of long genes linked to autism. *Nature* **2013**, *501*, 58–62. [CrossRef] [PubMed]
27. Nevin, L.M.; Xiao, T.; Staub, W.; Baier, H. Topoisomerase IIβ is required for lamina-specific targeting of retinal ganglion cell axons and dendrites. *Development* **2011**, *138*, 2457–2465. [CrossRef] [PubMed]
28. Tsutsui, K.; Tsutsui, K.; Sano, K.; Kikuchi, A.; Tokunaga, A. Involvement of DNA topoisomerase IIβ in neuronal differentiation. *J. Biol. Chem.* **2001**, *276*, 5769–5778. [CrossRef]
29. Sano, K.; Miyaji-Yamaguchi, M.; Tsutsui, K.M.; Tsutsui, K. Topoisomerase IIβ activates a subset of neuronal genes that are repressed in AT-rich genomic environment. *PLoS ONE* **2008**, *3*, e4103. [CrossRef]
30. Li, Y.; Hao, H.; Tzatzalos, E.; Lin, R.-K.; Doh, S.; Liu, L.F.; Lyu, Y.L.; Cai, L. Topoisomerase IIbeta is required for proper retinal development and survival of postmitotic cells. *Biol. Open* **2014**, *3*, 172–184. [CrossRef]
31. Feng, W.; Kawauchi, D.; Korkel-Qu, H.; Deng, H.; Serger, E.; Sieber, L.; Lieberman, J.A.; Jimeno-Gonzalez, S.; Lambo, S.; Hanna, B.S.; et al. Chd7 is indispensable for mammalian brain development through activation of a neuronal differentiation programme. *Nat. Commun.* **2017**, *8*, 14758. [CrossRef]
32. Lopes, I.; Altab, G.; Raina, P.; de Magalhaes, J.P. Gene Size Matters: An Analysis of Gene Length in the Human Genome. *Front. Genet.* **2021**, *12*, 559998. [CrossRef] [PubMed]
33. Tiwari, V.K.; Burger, L.; Nikoletopoulou, V.; Deogracias, R.; Thakurela, S.; Wirbelauer, C.; Kaut, J.; Terranova, R.; Hoerner, L.; Mielke, C.; et al. Target genes of Topoisomerase IIβ regulate neuronal survival and are defined by their chromatin state. *Proc. Natl. Acad. Sci. USA* **2012**, *109*, E934–E943. [CrossRef] [PubMed]
34. Lyu, Y.L.; Lin, C.P.; Azarova, A.M.; Cai, L.; Wang, J.C.; Liu, L.F. Role of topoisomerase IIβ in the expression of developmentally regulated genes. *Mol. Cell Biol.* **2006**, *26*, 7929–7941. [CrossRef] [PubMed]
35. Madabhushi, R.; Gao, F.; Pfenning, A.R.; Pan, L.; Yamakawa, S.; Seo, J.; Rueda, R.; Phan, T.X.; Yamakawa, H.; Pao, P.C.; et al. Activity-Induced DNA Breaks Govern the Expression of Neuronal Early-Response Genes. *Cell* **2015**, *161*, 1592–1605. [CrossRef] [PubMed]
36. The ENCODE Project Consortium. An integrated encyclopedia of DNA elements in the human genome. *Nature* **2012**, *489*, 57–74. [CrossRef] [PubMed]
37. Cao, X.; Wang, J.; Zhang, T.; Liu, Z.; Liu, L.; Chen, Y.; Li, Z.; Zhao, Y.; Yu, Q.; Liu, T.; et al. Chromatin accessibility dynamics dictate renal tubular epithelial cell response to injury. *Nat. Commun.* **2022**, *13*, 7322. [CrossRef] [PubMed]
38. Bunch, H. DNA Breaks and Damage Response Signaling are Coupled with RNA Polymerase II Promoter-Proximal Pause Release and Required for Effective Transcriptional Elongation. *FASEB J.* **2016**, *30*, 589.581.
39. Bunch, H.; Lawney, B.P.; Lin, Y.-F.; Asaithamby, A.; Murshid, A.; Wang, Y.E.; Chen, B.P.C.; Calderwood, S.K. Transcriptional elongation requires DNA break-induced signalling. *Nat. Commun.* **2015**, *6*, 10191. [CrossRef]
40. Rogakou, E.P.; Pilch, D.R.; Orr, A.H.; Ivanova, V.S.; Bonner, W.M. DNA double-stranded breaks induce histone H2AX phosphorylation on serine 139. *J. Biol. Chem.* **1998**, *273*, 5858–5868. [CrossRef]
41. Delint-Ramirez, I.; Konada, L.; Heady, L.; Rueda, R.; Jacome, A.S.V.; Marlin, E.; Marchioni, C.; Segev, A.; Kritskiy, O.; Yamakawa, S.; et al. Calcineurin dephosphorylates topoisomerase IIβ and regulates the formation of neuronal-activity-induced DNA breaks. *Mol. Cell* **2022**, *82*, 3794–3809.e8. [CrossRef] [PubMed]
42. Core, L.; Adelman, K. Promoter-proximal pausing of RNA polymerase II: A nexus of gene regulation. *Genes Dev.* **2019**, *33*, 960–982. [CrossRef]
43. Abbasi, S.; Schild-Poulter, C. Mapping the Ku Interactome Using Proximity-Dependent Biotin Identification in Human Cells. *J. Proteome Res.* **2019**, *18*, 1064–1077. [CrossRef] [PubMed]
44. Singh, S.; Szlachta, K.; Manukyan, A.; Raimer, H.M.; Dinda, M.; Bekiranov, S.; Wang, Y.H. Pausing sites of RNA polymerase II on actively transcribed genes are enriched in DNA double-stranded breaks. *J. Biol. Chem.* **2020**, *295*, 3990–4000. [CrossRef] [PubMed]

45. Ju, B.G.; Solum, D.; Song, E.J.; Lee, K.J.; Rose, D.W.; Glass, C.K.; Rosenfeld, M.G. Activating the PARP-1 sensor component of the groucho/TLE1 corepressor complex mediates a CaMKinase IIδ-dependent neurogenic gene activation pathway. *Cell* **2004**, *119*, 815–829. [CrossRef] [PubMed]
46. Hossain, M.B.; Ji, P.; Anish, R.; Jacobson, R.H.; Takada, S. Poly(ADP-ribose) Polymerase 1 Interacts with Nuclear Respiratory Factor 1 (NRF-1) and Plays a Role in NRF-1 Transcriptional Regulation. *J. Biol. Chem.* **2009**, *284*, 8621–8632. [CrossRef] [PubMed]
47. Cho, N.H.; Cheveralls, K.C.; Brunner, A.D.; Kim, K.; Michaelis, A.C.; Raghavan, P.; Kobayashi, H.; Savy, L.; Li, J.Y.; Canaj, H.; et al. OpenCell: Endogenous tagging for the cartography of human cellular organization. *Science* **2022**, *375*, eabi6983. [CrossRef] [PubMed]
48. Havugimana, P.C.; Goel, R.K.; Phanse, S.; Youssef, A.; Padhorny, D.; Kotelnikov, S.; Kozakov, D.; Emili, A. Scalable multiplex co-fractionation/mass spectrometry platform for accelerated protein interactome discovery. *Nat. Commun.* **2022**, *13*, 4043. [CrossRef]
49. Buchel, G.; Carstensen, A.; Mak, K.Y.; Roeschert, I.; Leen, E.; Sumara, O.; Hofstetter, J.; Herold, S.; Kalb, J.; Baluapuri, A.; et al. Association with Aurora-A Controls N-MYC-Dependent Promoter Escape and Pause Release of RNA Polymerase II during the Cell Cycle. *Cell Rep.* **2017**, *21*, 3483–3497. [CrossRef]
50. Das, S.K.; Kuzin, V.; Cameron, D.P.; Sanford, S.; Jha, R.K.; Nie, Z.; Rosello, M.T.; Holewinski, R.; Andresson, T.; Wisniewski, J.; et al. MYC assembles and stimulates topoisomerases 1 and 2 in a "topoisome". *Mol. Cell* **2021**, *82*, 140–158.e12. [CrossRef]
51. Sears, R.M.; May, D.G.; Roux, K.J. BioID as a Tool for Protein-Proximity Labeling in Living Cells. *Methods Mol. Biol.* **2019**, *2012*, 299–313. [CrossRef] [PubMed]
52. Oughtred, R.; Rust, J.; Chang, C.; Breitkreutz, B.J.; Stark, C.; Willems, A.; Boucher, L.; Leung, G.; Kolas, N.; Zhang, F.; et al. The BioGRID database: A comprehensive biomedical resource of curated protein, genetic, and chemical interactions. *Protein Sci.* **2021**, *30*, 187–200. [CrossRef] [PubMed]
53. Tsitsiridis, G.; Steinkamp, R.; Giurgiu, M.; Brauner, B.; Fobo, G.; Frishman, G.; Montrone, C.; Ruepp, A. CORUM: The comprehensive resource of mammalian protein complexes—2022. *Nucleic Acids Res.* **2022**, *51*, D539–D545. [CrossRef] [PubMed]
54. Malik, S.; Roeder, R.G. The metazoan Mediator co-activator complex as an integrative hub for transcriptional regulation. *Nat. Rev. Genet.* **2010**, *11*, 761–772. [CrossRef] [PubMed]
55. Shannon, P.; Markiel, A.; Ozier, O.; Baliga, N.S.; Wang, J.T.; Ramage, D.; Amin, N.; Schwikowski, B.; Ideker, T. Cytoscape: A software environment for integrated models of biomolecular interaction networks. *Genome Res.* **2003**, *13*, 2498–2504. [CrossRef] [PubMed]
56. Wan, C.; Borgeson, B.; Phanse, S.; Tu, F.; Drew, K.; Clark, G.; Xiong, X.; Kagan, O.; Kwan, J.; Bezginov, A.; et al. Panorama of ancient metazoan macromolecular complexes. *Nature* **2015**, *525*, 339–344. [CrossRef] [PubMed]
57. Hoffmeister, H.; Fuchs, A.; Erdel, F.; Pinz, S.; Gröbner-Ferreira, R.; Bruckmann, A.; Deutzmann, R.; Schwartz, U.; Maldonado, R.; Huber, C.; et al. CHD3 and CHD4 form distinct NuRD complexes with different yet overlapping functionality. *Nucleic Acids Res.* **2017**, *45*, 10534–10554. [CrossRef]
58. Tsai, S.C.; Valkov, N.; Yang, W.M.; Gump, J.; Sullivan, D.; Seto, E. Histone deacetylase interacts directly with DNA topoisomerase II. *Nat. Genet.* **2000**, *26*, 349–353. [CrossRef]
59. Johnson, C.A.; Padget, K.; Austin, C.A.; Turner, B.M. Deacetylase activity associates with topoisomerase II and is necessary for etoposide-induced apoptosis. *J. Biol. Chem.* **2001**, *276*, 4539–4542. [CrossRef]
60. Mathelier, A.; Zhao, X.; Zhang, A.W.; Parcy, F.; Worsley-Hunt, R.; Arenillas, D.J.; Buchman, S.; Chen, C.-y.; Chou, A.; Ienasescu, H.; et al. JASPAR 2014: An extensively expanded and updated open-access database of transcription factor binding profiles. *Nucleic Acids Res.* **2013**, *42*, D142–D147. [CrossRef]
61. Manville, C.M.; Smith, K.; Sondka, Z.; Rance, H.; Cockell, S.; Cowell, I.G.; Lee, K.C.; Morris, N.J.; Padget, K.; Jackson, G.H.; et al. Genome-wide ChIP-seq analysis of human TOP2B occupancy in MCF7 breast cancer epithelial cells. *Biol. Open* **2015**, *4*, 1436–1447. [CrossRef]
62. Canela, A.; Maman, Y.; Huang, S.N.; Wutz, G.; Tang, W.; Zagnoli-Vieira, G.; Callen, E.; Wong, N.; Day, A.; Peters, J.M.; et al. Topoisomerase II-Induced Chromosome Breakage and Translocation Is Determined by Chromosome Architecture and Transcriptional Activity. *Mol. Cell* **2019**, *75*, 252–266.e8. [CrossRef] [PubMed]
63. Castro-Mondragon, J.A.; Riudavets-Puig, R.; Rauluseviciute, I.; Berhanu Lemma, R.; Turchi, L.; Blanc-Mathieu, R.; Lucas, J.; Boddie, P.; Khan, A.; Manosalva Pérez, N.; et al. JASPAR 2022: The 9th release of the open-access database of transcription factor binding profiles. *Nucleic Acids Res.* **2021**, *50*, D165–D173. [CrossRef] [PubMed]
64. Gavin, I.; Horn, P.J.; Peterson, C.L. SWI/SNF chromatin remodeling requires changes in DNA topology. *Mol. Cell* **2001**, *7*, 97–104. [CrossRef] [PubMed]
65. Jha, R.K.; Levens, D.; Kouzine, F. Mechanical determinants of chromatin topology and gene expression. *Nucleus* **2022**, *13*, 94–115. [CrossRef] [PubMed]
66. Kagey, M.H.; Newman, J.J.; Bilodeau, S.; Zhan, Y.; Orlando, D.A.; van Berkum, N.L.; Ebmeier, C.C.; Goossens, J.; Rahl, P.B.; Levine, S.S.; et al. Mediator and cohesin connect gene expression and chromatin architecture. *Nature* **2010**, *467*, 430–435. [CrossRef] [PubMed]
67. Dellino, G.I.; Palluzzi, F.; Chiariello, A.M.; Piccioni, R.; Bianco, S.; Furia, L.; Conti, G.D.; Bouwman, B.A.M.; Melloni, G.; Guido, D.; et al. Release of paused RNA polymerase II at specific loci favors DNA double-strand-break formation and promotes cancer translocations. *Nat. Genet.* **2019**, *51*, 1011–1023. [CrossRef] [PubMed]

68. Iacovoni, J.S.; Caron, P.; Lassadi, I.; Nicolas, E.; Massip, L.; Trouche, D.; Legube, G. High-resolution profiling of γH2AX around DNA double strand breaks in the mammalian genome. *EMBO J.* **2010**, *29*, 1446–1457. [CrossRef]
69. Caron, P.; van der Linden, J.; van Attikum, H. Bon voyage: A transcriptional journey around DNA breaks. *DNA Repair* **2019**, *82*, 102686. [CrossRef]
70. Sengupta, S.; Wang, H.; Yang, C.; Szczesny, B.; Hegde, M.L.; Mitra, S. Ligand-induced gene activation is associated with oxidative genome damage whose repair is required for transcription. *Proc. Natl. Acad. Sci. USA* **2020**, *117*, 22183–22192. [CrossRef]
71. Pankotai, T.; Bonhomme, C.; Chen, D.; Soutoglou, E. DNAPKcs-dependent arrest of RNA polymerase II transcription in the presence of DNA breaks. *Nat. Struct. Mol. Biol.* **2012**, *19*, 276–282. [CrossRef]

International Journal of
Molecular Sciences

MDPI

Article

Actions of a Novel Bacterial Topoisomerase Inhibitor against *Neisseria gonorrhoeae* Gyrase and Topoisomerase IV: Enhancement of Double-Stranded DNA Breaks

Soziema E. Dauda [1,†], Jessica A. Collins [1,†], Jo Ann W. Byl [1], Yanran Lu [2], Jack C. Yalowich [3], Mark J. Mitton-Fry [2] and Neil Osheroff [1,4,5,*]

1 Department of Biochemistry, Vanderbilt University School of Medicine, Nashville, TN 37232, USA
2 Division of Medicinal Chemistry and Pharmacognosy, College of Pharmacy, The Ohio State University, Columbus, OH 43210, USA
3 Division of Pharmaceutics and Pharmacology, College of Pharmacy, The Ohio State University, Columbus, OH 42310, USA
4 Department of Medicine (Hematology/Oncology), Vanderbilt University School of Medicine, Nashville, TN 37232, USA
5 VA Tennessee Valley Healthcare System, Nashville, TN 37212, USA
* Correspondence: neil.osheroff@vanderbilt.edu
† These authors contributed equally to this work.

Abstract: Novel bacterial topoisomerase inhibitors (NBTIs) are an emerging class of antibacterials that target gyrase and topoisomerase IV. A hallmark of NBTIs is their ability to induce gyrase/topoisomerase IV-mediated single-stranded DNA breaks and suppress the generation of double-stranded breaks. However, a previous study reported that some dioxane-linked amide NBTIs induced double-stranded DNA breaks mediated by *Staphylococcus aureus* gyrase. To further explore the ability of this NBTI subclass to increase double-stranded DNA breaks, we examined the effects of OSUAB-185 on DNA cleavage mediated by *Neisseria gonorrhoeae* gyrase and topoisomerase IV. OSUAB-185 induced single-stranded and suppressed double-stranded DNA breaks mediated by *N. gonorrhoeae* gyrase. However, the compound stabilized both single- and double-stranded DNA breaks mediated by topoisomerase IV. The induction of double-stranded breaks does not appear to correlate with the binding of a second OSUAB-185 molecule and extends to fluoroquinolone-resistant *N. gonorrhoeae* topoisomerase IV, as well as type II enzymes from other bacteria and humans. The double-stranded DNA cleavage activity of OSUAB-185 and other dioxane-linked NBTIs represents a paradigm shift in a hallmark characteristic of NBTIs and suggests that some members of this subclass may have alternative binding motifs in the cleavage complex.

Keywords: novel bacterial topoisomerase inhibitor; NBTI; type II topoisomerase; gyrase; topoisomerase IV; DNA cleavage; *Neisseria gonorrhoeae*

Citation: Dauda, S.E.; Collins, J.A.; Byl, J.A.W.; Lu, Y.; Yalowich, J.C.; Mitton-Fry, M.J.; Osheroff, N. Actions of a Novel Bacterial Topoisomerase Inhibitor against *Neisseria gonorrhoeae* Gyrase and Topoisomerase IV: Enhancement of Double-Stranded DNA Breaks. *Int. J. Mol. Sci.* **2023**, *24*, 12107. https://doi.org/10.3390/ijms241512107

Academic Editor: Philippe Pourquier

Received: 28 June 2023
Revised: 18 July 2023
Accepted: 25 July 2023
Published: 28 July 2023

1. Introduction

Fluoroquinolones, such as ciprofloxacin (Figure 1) and moxifloxacin, are among the most efficacious and broad-spectrum oral antibacterials in clinical use [1–3]. The World Health Organization (WHO) lists fluoroquinolones in their five "Highest Priority Critically Important Antimicrobials", and these drugs are heavily used worldwide [1,3].

There is a growing crisis in antibacterial resistance. Unfortunately, fluoroquinolone resistance is becoming prevalent and is impeding the clinical efficacy of this important drug class [2,4–7]. For example, ciprofloxacin was used as frontline treatment for gonorrheal infections in humans starting in 1993 [8]. By 2003, >40% of cases of gonorrhea were treated with fluoroquinolones [9]. However, their use as routine therapy was discontinued in 2006 due to the high incidence of resistance (>30% of gonorrhea cases in the United States are currently resistant to fluoroquinolones) [10,11]. Drug-resistant *Neisseria gonorrhoeae*, the causative agent of gonorrhea, is categorized by the Centers for Disease Control (CDC) as one of the five "urgent level" antibiotic resistance threats in the United States [12]. It is a major cause of pelvic inflammatory disease and infertility and appears to facilitate the transmission of HIV [13,14]. Current estimates suggest new cases of gonorrhea exceed 82 million annually, and the WHO has issued dire warnings that gonorrhea has the potential to join herpes and HIV/AIDS as the third "incurable" sexually transmitted disease [15,16].

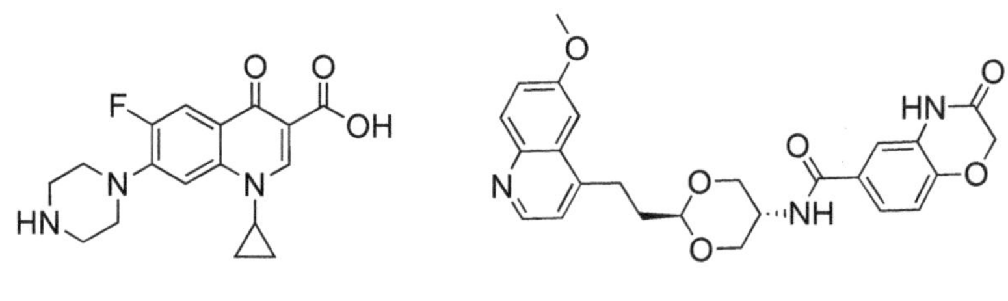

Figure 1. Structures of the fluoroquinolone ciprofloxacin and the NBTI OSUAB-185.

The targets for fluoroquinolones are the bacterial type II topoisomerases, gyrase and topoisomerase IV [4–7]. These enzymes modulate the superhelical state of the bacterial chromosome, remove positive supercoils that accumulate ahead of replication forks and transcription complexes, and remove tangles and knots from the genome [17–23]. Gyrase and topoisomerase IV regulate DNA topology by passing an intact double helix through a transient break that they generate in a second DNA segment [19,21,24–26]. To maintain genomic integrity while the DNA is cut, the enzymes form covalent bonds between active-site tyrosine residues and the newly generated 5′-DNA termini. These covalent enzyme-cleaved DNA complexes are known as "cleavage complexes" [4,19,21,23,25,26]. Fluoroquinolones kill bacterial cells by inserting into the cleaved DNA scissile bonds on both strands of the double helix (one drug molecule per strand), thereby stabilizing cleavage complexes, inhibiting enzyme function, and increasing levels of gyrase/topoisomerase IV-mediated double-stranded DNA breaks [2,4–7,27]. When stabilized cleavage complexes are approached by replication forks, transcription complexes, and other systems that require opening of the double helix, they disrupt the ability of the enzymes to ligate the DNA [6,25,28]. This requires the breaks to be resected via recombination/repair pathways [6,7,25,28]. When these breaks reach a critical level, they activate SOS pathways and ultimately lead to cell death [29–31].

The most important mechanism of fluoroquinolone resistance is target-mediated and is caused by mutations in a highly conserved serine residue (originally described as Ser83 in the GyrA subunit of *Escherichia coli* gyrase) and an acidic residue four amino acids away at position 87 [2,4–7,32]. These residues interact with two water molecules that are coordinated by a metal ion that is chelated to the fluoroquinolone C3-C4 keto acid group [4,6,27,33,34]. This "water-metal ion bridge" is the main conduit for interactions between fluoroquinolones and gyrase/topoisomerase IV [4,6,27,33,34].

To address fluoroquinolone resistance, new classes of compounds are under development that also target gyrase/topoisomerase IV but interact with different amino acid residues in the enzymes [2,6,35–38]. The most clinically advanced class of new compounds are Novel Bacterial Topoisomerase Inhibitors (NBTIs) [2,6,39–43]. Phase 3 clinical trials for the treatment of uncomplicated urinary tract infections (caused by uropathogens, including *E. coli*) with a member of this class were stopped early because of demonstrated efficacy [39,41,43,44]. An NBTI is also in phase 3 trials for uncomplicated urogenital gonorrhea (*N. gonorrhoeae*) [40,42]. In contrast to fluoroquinolones, only a single NBTI molecule binds per cleavage complex [35,37,45]. The left-hand side of the molecule sits in a pocket in the DNA on the two-fold axis of the complex, midway between the two DNA cleavage sites, and the right-hand side sits in a pocket on the two-fold axis between the two GyrA/ParC subunits [35,37,45]. While fluoroquinolones induce gyrase/topoisomerase IV-mediated double-stranded breaks [4,6,7,27,46,47], a hallmark of NBTIs is the generation of enzyme-mediated single-stranded DNA breaks and the suppression of double-stranded breaks [37,48,49]. However, some dioxane-linked amide members of the NBTI class have been shown to induce double-stranded and single-stranded DNA breaks mediated by *Staphylococcus aureus* gyrase [50].

To further explore the ability of NBTIs to induce double-stranded DNA breaks mediated by bacterial type II topoisomerases, we examined the effects of OSUAB-185 (Figure 1) on the DNA cleavage activity of *N. gonorrhoeae* gyrase and topoisomerase IV. Although a member of the NBTI class, the substituents on this dioxane-linked amide NBTI differ considerably from those of canonical members. Similar to other NBTIs, OSUAB-185 induced single-stranded and suppressed double-stranded DNA breaks mediated by *N. gonorrhoeae* gyrase. However, in contrast to NBTIs such as GSK126 [49], OSUAB-185 stabilized both single- and double-stranded DNA breaks mediated by topoisomerase IV. It does not appear that the induction of double-stranded breaks is due to the binding of a second NBTI molecule. The ability to induce double-stranded DNA breaks extends to fluoroquinolone-resistant *N. gonorrhoeae* topoisomerase IV, as well as type II enzymes from other species. Together with previous work on the dioxane-linked amide NBTIs [50], the double-stranded DNA cleavage activity of OSUAB-185 represents a paradigm shift in the hallmark characteristic of NBTIs and suggests that some members of this class may have alternative binding motifs in the cleavage complex.

2. Results

2.1. OSUAB-185 Induces Double-Stranded DNA Cleavage Mediated by *N. gonorrhoeae* Topoisomerase IV

A previous study demonstrated that some dioxane-linked NBTIs were able to induce double-stranded (in addition to single-stranded) DNA breaks mediated by *S. aureus* gyrase [50]. Therefore, to further explore the ability of this NBTI subclass to induce double-stranded DNA breaks generated by bacterial type II topoisomerases and to address the basis for this activity, we assessed the effects of OSUAB-185 on DNA cleavage mediated by *N. gonorrhoeae* gyrase and topoisomerase IV (Figure 2). The NBTI displayed moderate activity against gyrase, increasing single-stranded DNA breaks to ~12.0% at 5 μM compound (compared to a baseline level of ~3.1%) with a CC_{50} value (concentration of compound required to induce 50% of maximal DNA cleavage) of ~0.9 μM. As reported previously for other NBTIs [37,48,49], induction of double-stranded DNA cleavage was not observed. Although OSUAB-185 was less potent against topoisomerase IV ($CC_{50} \approx 4.8$ μM), it was considerably more efficacious, inducing ~40% single-stranded DNA breaks at 15 μM compound (compared to a baseline level of ~3.0%). Strikingly, OSUAB-185 also increased double-stranded DNA breaks mediated by topoisomerase IV. Approximately 10% double-stranded DNA breaks were observed at 15 μM compound (compared to a baseline level of ~1%) with a CC_{50} value of ~5.7 μM. The increase in double-stranded breaks is contrary to the hallmark of NBTIs, which is the induction of single-stranded DNA breaks by gyrase/topoisomerase IV and the suppression of enzyme-mediated double-stranded DNA breaks [37,48,49].

To ensure that the DNA breaks induced by OSUAB-185 were being generated by gyrase and topoisomerase IV, two controls were carried out (Figure 3). First, no single- or double-stranded DNA cleavage was observed in the presence of 10 μM compound in the absence of enzyme. Second, when DNA cleavage reactions were terminated by the addition of EDTA prior to SDS, DNA cleavage levels dropped for both enzymes. EDTA chelates the required catalytic Mg^{2+} ion, but only when the DNA is ligated [51]. Therefore, the decrease in cleaved DNA following EDTA treatment is inconsistent with a non-enzymatic reaction. These data provide strong evidence that the DNA breaks observed in the presence of OSUAB-185 were generated by gyrase and topoisomerase IV. It is notable that all of the NBTI-induced double-stranded breaks generated by topoisomerase IV disappeared in the presence of EDTA. This finding suggests that they may be less stable than the single-stranded DNA breaks generated in the presence of the compound.

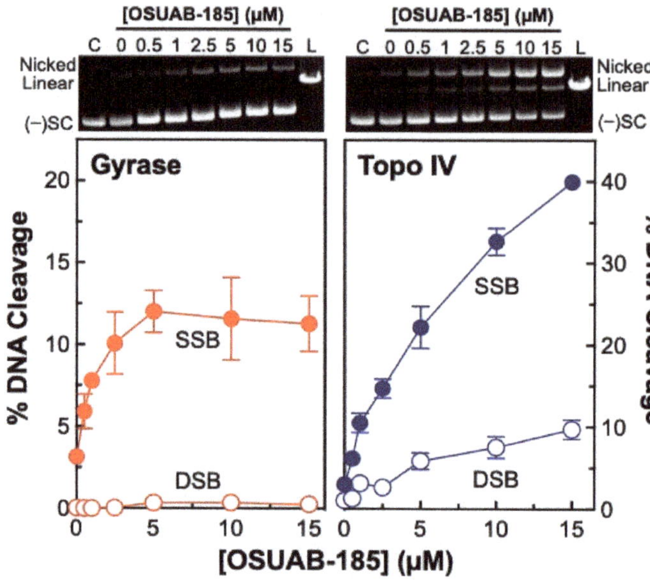

Figure 2. OSUAB-185 enhances DNA cleavage mediated by *N. gonorrhoeae* gyrase and topoisomerase IV. The graphs show the effects of OSUAB-185 on single-stranded (SSB, closed circle) and double-stranded (DSB, opened circle) DNA cleavage mediated by *N. gonorrhoeae* gyrase (red, (**left**)) and topoisomerase IV (Topo IV, blue, (**right**)). Note that the scaling of percent DNA cleavage on the y-axis differs between gyrase and topoisomerase IV. Error bars represent SDs for at least three independent experiments. Representative agarose gels for DNA cleavage assays with gyrase (**left**) and topoisomerase IV (**right**) are shown above the graphs. Control DNA (C) in the absence of enzyme and a linear DNA standard (L) are indicated. The positions of negatively supercoiled [(−)SC], nicked, and linear plasmid are shown.

To determine whether the relative levels of topoisomerase IV-mediated double- vs. single-stranded DNA breaks change with time, a time course for DNA scission with both enzymes was monitored (Figure 4). As with the NBTI titration, no appreciable double-stranded breaks were observed with gyrase (Figure 4, left panel). In contrast, double- and single-stranded DNA breaks were observed with topoisomerase IV and appeared to be generated coordinately (Figure 4, right panel). To analyze the formation of double-stranded DNA breaks by topoisomerase IV in greater detail, data from the [OSUAB-185] titration (Figure 2, as well as NBTI concentrations up to 200 μM) and the time course for cleavage (Figure 4) were converted to ratios of single-stranded:double-stranded DNA breaks (Figure 5, left and right panel, respectively). The single-stranded:double-stranded DNA break ratios remained constant over a wide range of NBTI concentrations and cleavage reaction times. These findings provide further evidence that the single- and double-stranded DNA breaks generated by *N. gonorrhoeae* topoisomerase IV in the presence of OSUAB-185 were generated coordinately. This result argues against a second NBTI molecule entering the active site of topoisomerase IV at higher concentrations of OSUAB-185 or over longer reaction times. It also argues against a change in the binding conformation of the NBTI over time.

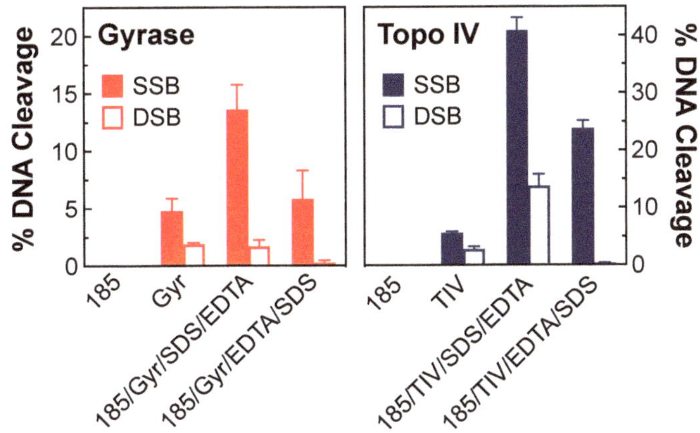

Figure 3. DNA cleavage induced by OSUAB-185 is medicated by *N. gonorrhoeae* gyrase and topoisomerase IV. Bar graphs show single-stranded (SSB, filled bars) and double-stranded (DSB, open bars) DNA cleavage mediated by *N. gonorrhoeae* gyrase (red, (**left**)) and topoisomerase IV (Topo IV, blue, (**right**)). Reactions contained negatively supercoiled DNA, with 10 μM OSUAB-185 in the absence of enzyme (185), gyrase (Gyr) or topoisomerase IV (TIV) in the absence of OSUAB-185, or complete reaction mixtures that contained enzyme, DNA, and the NBTI (185/Gyr/SDS/EDTA and 185/TIV/SDS/EDTA). All of these reactions were stopped with SDS prior to the addition of EDTA. Alternatively, reactions that contained enzyme, DNA, and the NBTI were treated with 2 μL of 250 mM EDTA for 10 min prior to the addition of SDS (185/Gyr/EDTA/SDS and 185/TIV/EDTA/SDS). Note that the scaling of percent DNA cleavage on the y-axis differs between gyrase and topoisomerase IV and that gyrase reactions contained 200 nM enzyme, which is twice the concentration used in other cleavage assays, to increase baseline levels of DNA cleavage. Error bars represent SDs for at least three independent experiments.

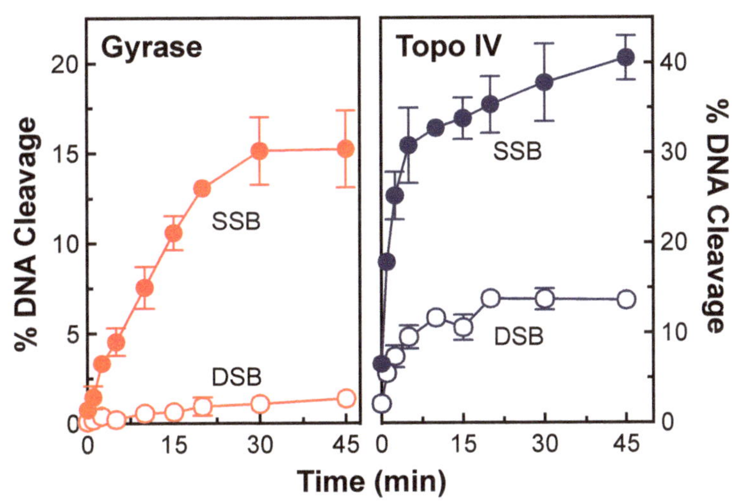

Figure 4. Time courses for DNA cleavage mediated by *N. gonorrhoeae* gyrase and topoisomerase IV. Time courses for single-stranded (SSB, closed circle) and double-stranded (DSB, opened circle) DNA cleavage mediated by gyrase (red, (**left**)) and topoisomerase IV (Topo IV, blue, (**right**)) in the presence of 10 μM OSUAB-185 are shown. Error bars represent SDs for at least three independent experiments.

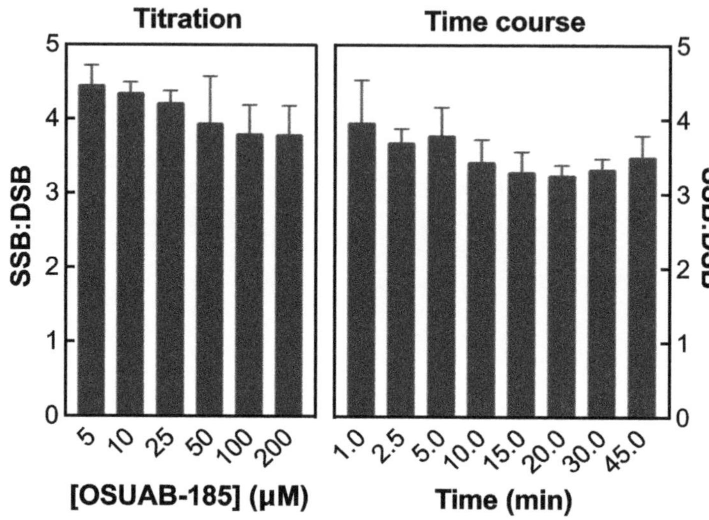

Figure 5. Ratios of NBTI-induced single-stranded to double-stranded DNA cleavage mediated by *N. gonorrhoeae* topoisomerase IV are maintained over OSUAB-185 concentrations and reaction time. The ratios of single-stranded to double-stranded (SSB:DSB) DNA breaks over a concentration range of OSUAB-185 ((**left**) panel) or a time course of DNA cleavage ((**right**) panel) induced by 10 μM NBTI with topoisomerase IV are shown. Error bars represent SDs for at least three independent experiments.

As a final demonstration that the induction of topoisomerase IV-mediated double-stranded DNA breaks by OSUAB-185 represents a novel mechanism of action, a titration of the NBTI was carried out in cleavage assays that replaced $MgCl_2$ with $CaCl_2$ (Figure 6). With some enzymes, Ca^{2+} dramatically raises baseline levels of single- and double-stranded DNA breaks, which affords a more ready assessment of NBTI effects on DNA cleavage [34,48,49,52]. Similar to previous results with other NBTIs, OSUAB-185 induced single-stranded DNA breaks and suppressed double-stranded breaks mediated by *N. gonorrhoeae* gyrase (Figure 6, left panel). However, in marked contrast, the NBTI enhanced both single- and double-stranded DNA cleavage mediated by topoisomerase IV (Figure 6, right panel). This finding provides strong evidence that OSUAB-185 induces double-stranded DNA breaks mediated by *N. gonorrhoeae* topoisomerase IV.

2.2. Stability of Single- and Double-Stranded DNA Breaks Induced by OSUAB-185

Other things being equal, drugs that generate the most stable cleavage complexes appear to be the most lethal in cells [53]. Therefore, two approaches were utilized to assess the effects of OSUAB-185 on the stability of cleavage complexes formed by *N. gonorrhoeae* gyrase and topoisomerase IV. In the first approach, the effects of OSUAB-185 on the persistence of cleavage complexes were determined. In this assay, cleavage complexes were formed in the presence of high concentrations of enzyme and DNA, and the lifetimes of cleavage complexes were monitored following 20-fold dilution into reaction buffer that lacked the catalytic divalent metal ion. While the shift in condition does not alter the DNA cleavage–ligation equilibrium in established cleavage complexes, complexes that disassociate are unlikely to reform. As seen in Figure 7, cleavage complexes formed with gyrase (monitoring single-stranded DNA cleavage, left panel) or topoisomerase IV (monitoring single-stranded DNA cleavage, middle panel, or double-stranded DNA cleavage, right panel) in the absence of the NBTI were highly unstable and rapidly disassociated following dilution. Complexes became considerably more stable in the presence of OSUAB-185. Consistent with the data from Figure 3, NBTI-induced double-stranded DNA breaks ($t_{1/2} \approx 4$ min) were less stable than single-stranded DNA breaks ($t_{1/2} \approx 8$ min) with topoisomerase IV.

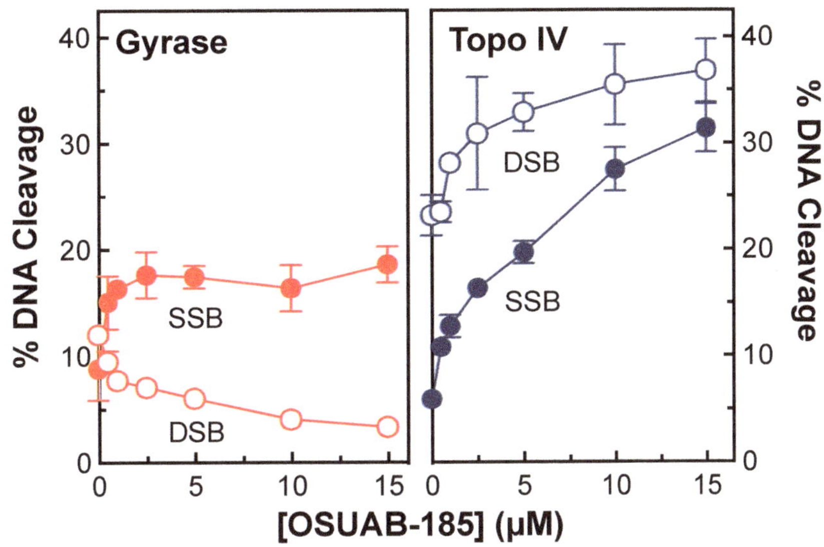

Figure 6. OSUAB-185 suppresses levels of double-stranded DNA cleavage mediated by *N. gonorrhoeae* gyrase and increases levels of double-stranded DNA breaks generated by topoisomerase IV. DNA cleavage reactions contained $CaCl_2$ in place of $MgCl_2$ to raise baseline levels of DNA scission. Graphs show the effects of OSUAB-185 on single-stranded (SSB, closed circle) and double-stranded (DSB, opened circle) DNA cleavage mediated by gyrase (red, (**left**)) and topoisomerase IV (Topo IV, blue, (**right**)). Error bars represent SDs for at least three independent experiments.

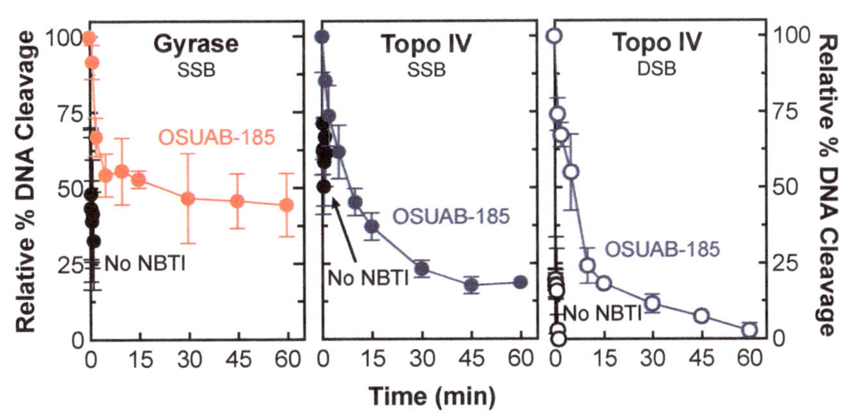

Figure 7. OSUAB-185 induces stable DNA cleavage complexes with *N. gonorrhoeae* gyrase and topoisomerase IV. Persistence reactions were allowed to reach cleavage–ligation equilibrium before dilution in reaction buffer that lacked $MgCl_2$. The subsequent stability of cleavage complexes was monitored. Persistence of single-stranded DNA (SSB, closed circle) cleavage complexes by gyrase (red, (**left**) panel) and topoisomerase IV (blue, (**middle**) panel) and double-stranded DNA (DSB, open circle) cleavage complexes by topoisomerase IV (blue, (**right**) panel) formed in the presence of 10 μM OSUAB-185 are shown. Persistence reactions carried out in the absence of the NBTI (No NBTI) are shown in black (SSB, closed circle; DSB, open circle). DNA cleavage prior to dilution of cleavage complexes was set to 100%. Error bars represent SDs of three independent experiments.

In the second approach, the effects of OSUAB-185 on the rate of gyrase/topoisomerase IV-mediated DNA ligation were monitored by shifting cleavage complexes from 37 °C to 65 °C (a temperature that allows ligation but not cleavage of the DNA) [54]. As seen in Figure 8, OSUAB-185 had a modest effect on rates of ligation for single-stranded DNA breaks with gyrase (left panel) and topoisomerase IV (middle panel) and double-stranded breaks with topoisomerase IV (right panel). The NBTI generated slightly more stable single-stranded ($t_{1/2} \approx 35$ s) DNA cleavage complexes than double-stranded ($t_{1/2} \approx 27$ s) DNA cleavage complexes with topoisomerase IV.

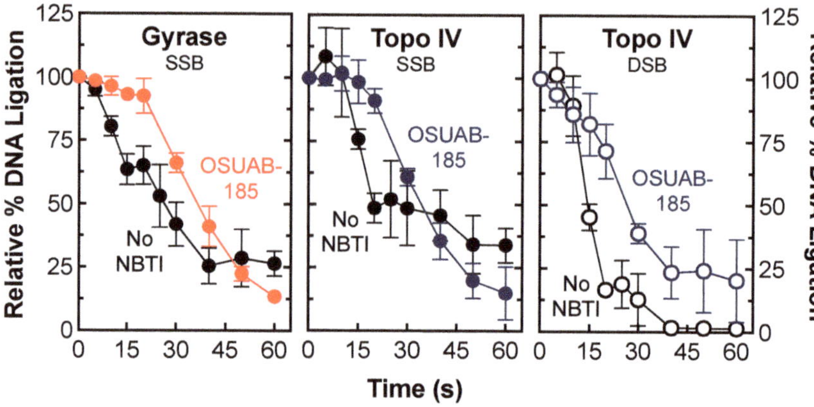

Figure 8. OSUAB-185 inhibits DNA ligation mediated by *N. gonorrhoeae* gyrase and topoisomerase IV. Ligation of single-stranded DNA (SSB, closed circle) cleavage complexes by gyrase (red, (**left**) panel) and topoisomerase IV (blue, (**middle**) panel) and double-stranded DNA (DSB, open circle) cleavage complexes by topoisomerase IV (blue, (**right**) panel) formed in the presence of 10 μM OSUAB-185 are shown. DNA ligation was also monitored in the absence of the NBTI (No NBTI, black, SSB, closed circle; DSB, open circle). Levels of single- and double-stranded DNA cleavage prior to the induction of ligation were set to 100%. Error bars represent SDs for at least three independent experiments.

2.3. Effects of OSUAB-185 on Fluoroquinolone-Resistant N. gonorrhoeae Gyrase and Topoisomerase IV

To determine whether OSUAB-185 is able to overcome fluoroquinolone resistance in *N. gonorrhoeae* type II enzymes, the effects of the NBTI on gyrase that contained $GyrA^{S91F}$ and topoisomerase IV that contained $ParC^{S87N}$ were examined. These two amino acid substitutions represent the most prevalent fluoroquinolone-resistant mutations in *N. gonorrhoeae* gyrase and topoisomerase IV, respectively (Figure 9) [55,56]. Levels of DNA cleavage with $GyrA^{S91F}$ gyrase (top left panel) were similar to those obtained with the wild-type enzyme and only a single-stranded DNA cleavage was induced by OSUAB-185. However, the NBTI was ~4-fold less potent against the $GyrA^{S91F}$ mutant. In contrast, levels of NBTI-induced DNA scission with $ParC^{S87N}$ topoisomerase IV (top right panel) were ~3 to 4-fold lower than observed for the wild-type enzyme, and OSUAB-185 maintained its potency. Once again, the NBTI induced single- and double-stranded DNA scission with fluoroquinolone-resistant topoisomerase IV. These findings predict that the NBTI would retain at least some activity against fluoroquinolone-resistant *N. gonorrhoeae* cells harboring common mutations in gyrase/topoisomerase IV. They also suggest that the fluoroquinolone-resistant mutations do not alter the mechanism of action of OSUAB-185 against bacterial type II topoisomerases.

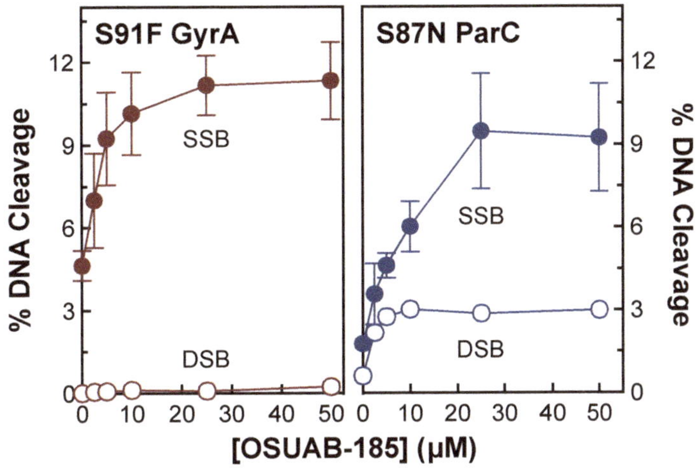

Type II Enzyme		CC_{50} SSB (μM)	CC_{50} DSB (μM)	Max % SSB	Max % DSB	SSB:DSB
Gyrase	WT	0.9	–	12.0	–	–
	S91F	3.4	–	11.3	–	–
Topo IV	WT	4.8	5.7	39.9	9.7	4.1
	S87N	7.0	2.0	9.5	3.0	3.2

Figure 9. Activity of OSUAB-185 against *N. gonorrhoeae* gyrase and topoisomerase IV that contain fluoroquinolone-resistance mutations. The effects of OSUAB-185 on single-stranded (SSB, closed circle) and double-stranded (DSB, open circle) DNA cleavage mediated by fluoroquinolone-resistant GyrA^{S91F} gyrase (S91F GyrA; (**top left**), red) and ParC^{S87N} topoisomerase IV (S87N ParC; (**top right**), blue) are shown. Error bars represent SDs of three independent experiments. A summary table of DNA cleavage mediated by WT and mutant *N. gonorrhoeae* gyrase and topoisomerase IV is shown at the bottom. CC_{50} values are indicated for the enhancement of single-stranded and double-stranded DNA cleavage. Maximal levels of NBTI-induced DNA scission (Max %) are also shown. Significant levels of double-stranded DNA breaks were not observed with WT or GyrA^{S91F} gyrase. The ratio of single-stranded to double-stranded DNA breaks (SSB:DSB) were calculated at maximal levels of DNA cleavage for WT and ParC^{S87N} topoisomerase IV.

2.4. *Effects of OSUAB-185 on Gyrase and Topoisomerase IV from E. coli and S. aureus*

To determine whether OSUAB-185 is able to induce double-stranded DNA breaks with type II topoisomerases from other species (Figure 10), the effects of the NBTI on DNA cleavage mediated by gyrase (left panels) and topoisomerase IV (right panels) from *E. coli* (Figure 10A) and *S. aureus* (Figure 10B) were assessed. Note that the induction of double-stranded DNA breaks by OSUAB-185 and *S. aureus* gyrase has been reported previously [50]. To at least some extent, OSUAB-185 induced double-stranded breaks with all four of the enzymes. Double-stranded DNA breaks were especially prominent with *E. coli* topoisomerase IV. These data indicate that the ability of the NBTI to generate enzyme-mediated double-stranded DNA breaks is not confined to *N. gonorrhoeae* or either type II topoisomerase.

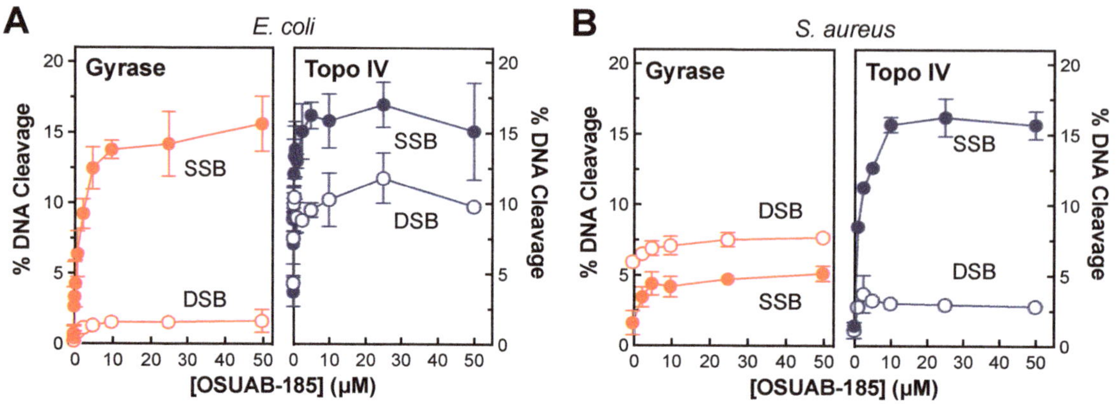

Figure 10. OSUAB-185 induces double-stranded DNA breaks mediated by gyrase and topoisomerase IV from *E. coli* and *S. aureus*. The effects of OSUAB-185 on single-stranded (SSB, closed circles) and double-stranded (DSB, open circles) DNA cleavage mediated by gyrase (red, left panels) and topoisomerase IV (blue, right panels) from *E. coli* (**A**) and *S. aureus* (**B**) are shown. Error bars represent SDs of three independent experiments.

2.5. Effects of OSUAB-185 on Human Topoisomerase IIα

Very little is known about the interaction of NBTIs with human type II topoisomerases [57–59]. Therefore, the effects of OSUAB-185 on DNA cleavage mediated by human topoisomerase IIα was determined (Figure 11). The NBTI displayed reasonable activity against the human type II enzyme, with CC_{50} values in the low μM range. In addition, similar to the results with some of the bacterial enzymes, OSUAB-185 induced moderate levels of double-stranded DNA breaks with the human enzyme (maximal levels of cleavage of ~11% and ~8.5% for single- and double-stranded DNA breaks, respectively). This last finding provides further evidence that some members of the NBTI class are capable of generating enzyme-mediated double-stranded DNA breaks.

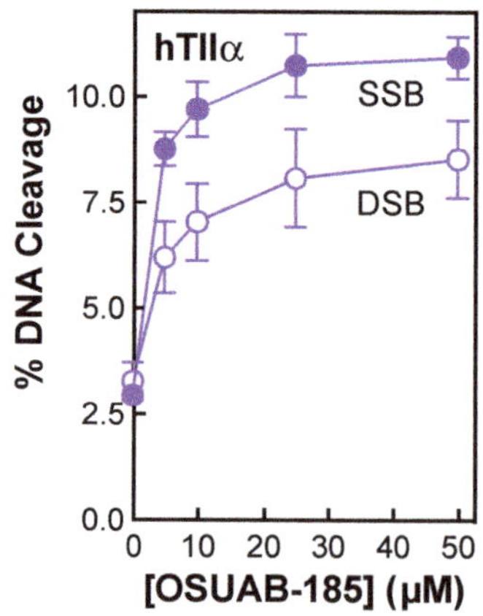

Figure 11. Effects of OSUAB-185 on DNA cleavage mediated by human topoisomerase IIα. The graph shows the effects of OSUAB-185 on single-stranded (SSB, purple, closed circle) and double-stranded (DSB, purple, open circle) DNA cleavage mediated by human topoisomerase IIα. Error bars represent standard error of the mean of two independent experiments.

3. Discussion

NBTIs are an emerging class of compounds with antibacterial activity. In contrast to fluoroquinolones, a hallmark of most NBTIs is their ability to induce gyrase/topoisomerase IV-mediated single-stranded breaks in DNA and suppress the formation of double-stranded breaks [37,48,49]. Together with a previous study, the present work provides strong evidence that select dioxane-linked amide NBTIs are also capable of generating double-stranded DNA breaks mediated by type II topoisomerases from a variety of bacterial species, as well as humans [50].

Fluoroquinolones induce double-stranded DNA breaks because two drug molecules bind in the active site of gyrase and topoisomerase IV, with one molecule stabilizing a cleaved scissile bond on each strand of the double helix [27,46,47]. In contrast, only a single NBTI molecule binds in the active site of the bacterial type II topoisomerases, with the molecule binding midway between the two scissile bonds [35,37,45]. It is believed that canonical NBTIs stabilize single-stranded and suppress double-stranded DNA breaks by distorting the active site of gyrase/topoisomerase IV in a manner that inhibits ligation of the first strand but does not allow cleavage of the second [35,37]. It is not known how OSUAB-185 and potentially other dioxane-linked amide NBTIs generate double-stranded breaks. Results from OSUAB-185 titrations (Figures 2 and 5) and time course experiments (Figures 4 and 5) strongly suggest that it is not due to the binding of a second molecule of compound. Potentially, this NBTI has two mutually exclusive binding configurations that distort the active site of the enzyme differently. In the first, the NBTI acts as a canonical member of this class and induces distortion after one strand is cut to inhibit ligation and prevent cleavage of the second strand (thus enhancing single-stranded DNA breaks). In the second, the NBTI acts in a non-canonical manner and only induces the dramatic active site distortion after both DNA strands are cleaved (thus enhancing double-stranded DNA breaks). Future structural studies are needed to determine if this is the case.

The activity of OSUAB-185 against topoisomerase IIα suggests that this member of this NBTI subclass might cross over to human systems if used to treat infections. However, our results raise an interesting possibility. The double-stranded DNA breaks generated by human type II topoisomerases during transcription have the potential to trigger secondary leukemias in a small percentage of patients treated with topoisomerase II-targeted anticancer drugs [25,60–63]. If a drug was developed that induced only single-stranded DNA breaks with human type II topoisomerases, it is possible that it could overcome the secondary leukemias observed with current drugs. Although OSUAB-185 induced both single- and double-stranded DNA breaks mediated by human topoisomerase IIα, our findings open the door for future work on NBTIs and human type II topoisomerases.

4. Materials and Methods

4.1. Enzymes and Materials

Wild-type *N. gonorrhoeae* gyrase (GyrA, GyrB) and topoisomerase IV (ParC, ParE) subunits were prepared as described previously [35,46]. Fluoroquinolone-resistant *N. gonorrhoeae* GyrA^{S91F} was a gift from Dr. Pan Chan (GlaxoSmithKline, Brentford, UK) and fluoroquinolone-resistant *N. gonorrhoeae* ParC^{S87N} was generated using a QuickChange II XL site-directed mutagenesis kit (Agilent Technologies, Santa Clara, CA, USA) with custom primers for the desired mutations. Mutant *N. gonorrhoeae* ParC^{S87N} was expressed and purified as described by Ashley et al. [64] with the following modifications to optimize protein expression and lysis: (1) ParC^{S87N} was expressed for 3 h before harvesting, and (2) cells were lysed by sonication using a digital sonifier (Branson, Danbury, CT, USA). Wild-type *E. coli* gyrase (GyrA, GyrB) subunits were expressed and purified as described by Chan et al. [46], and wild-type *E. coli* topoisomerase IV (ParC, ParE, a gift from Dr. Keir Neumann, NHLBI), subunits were expressed and purified as described by Peng and Marians [65]. Wild-type *S. aureus* gyrase (GyrA, GyrB) and topoisomerase IV (GrlA and GrlB) subunits were ordered from GenScript and expressed and purified as described previously [37,46]. Recombinant human wild-type topoisomerase IIα was expressed in

Saccharomyces cerevisiae and purified as described previously [66,67]. The identities of all constructs were confirmed by DNA sequencing, and all enzymes were stored at −80 °C.

Negatively supercoiled pBR322 DNA was prepared from *E. coli* using a Plasmid Mega Kit (Qiagen, Hilden, Germany) as described by the manufacturer. OSUAB-185 was synthesized as described previously [50] (shown as compound 23) and stored at 4 °C as a 20 mM stock solution in 100% DMSO. All other chemicals were of analytical reagent grade.

4.2. DNA Cleavage

DNA cleavage reactions were performed according to the procedure of Aldred et al. [33] (gyrase and topoisomerase IV) or Fortune and Osheroff [68] (human topoisomerase IIα). Reactions were performed in the absence of an NBTI or in the presence of increasing concentrations of OSUAB-185. Unless stated otherwise, assay mixtures contained 10 nM pBR322 and 100 nM wild-type or mutant ($GyrA^{S91F}$) *N. gonorrhoeae* gyrase, 100 nM wild-type or mutant ($ParC^{S87N}$) *N. gonorrhoeae* topoisomerase IV, 100 nM wild-type *E. coli* gyrase, 20 nM wild-type *E. coli* topoisomerase IV, 100 nM wild-type *S. aureus* gyrase, 20 nM *S. aureus* topoisomerase IV, or 300 nM wild-type human topoisomerase IIα in a total volume of 20 μL of cleavage buffer: 40 mM Tris-HCl (pH 7.9), 50 mM NaCl, 2.5% (*w*/*v*) glycerol, and 5 mM or 10 mM $MgCl_2$ for *E. coli* or *N. gonorrhoeae* type II topoisomerases, respectively; 50 mM Tris-HCl (pH 7.5), 100 mM KGlu, 5 mM $MgCl_2$, 5 mM dithiothreitol (DTT), and 50 μg/mL BSA for *S. aureus* gyrase; 40 mM Tris-HCl (pH 7.5), 6 mM $MgCl_2$, 20 mM NaCl, 10 mM DTT, and 50 μg/mL BSA for *S. aureus* topoisomerase IV; 50 mM Tris-HCl (pH 7.9), 0.5 mM EDTA (pH 8.0), 500 mM KCl, 25 mM $MgCl_2$, and 12.5% (*w*/*v*) glycerol for human topoisomerase IIα. In some cases, $MgCl_2$ in the cleavage buffer was replaced with an equivalent concentration of $CaCl_2$. Reactions were incubated at 37 °C for 30 min for wild-type and mutant ($GyrA^{S91F}$) *N. gonorrhoeae* gyrase, mutant ($ParC^{S87N}$) *N. gonorrhoeae* topoisomerase IV, and wild-type *S. aureus* gyrase; 20 min with wild-type *E. coli* gyrase; and 10 min for all other enzymes. Enzyme–DNA cleavage complexes were trapped by adding 2 μL of 4% SDS, followed by 2 μL of 250 mM EDTA (pH 8.0). Proteinase K was added (2 μL of a 0.8 mg/mL solution), and reaction mixtures were incubated for 30 min at 45 °C to digest the enzyme. Samples were mixed with 2 μL of agarose loading buffer [60% sucrose, 10 mM Tris-HCl (pH 7.9), 0.5% bromophenol blue, and 0.5% xylene cyanol FF] and heated for 2 min at 45 °C prior to electrophoresis in 1% agarose gels in 40 mM Tris-acetate (pH 8.3), and 2 mM EDTA containing 0.5 μg/mL ethidium bromide. DNA bands were visualized by mid-range ultraviolet light and quantified using an Alpha Innotech digital imaging system (Protein Simple, San Jose, CA, USA). DNA cleavage was monitored by the conversion of negatively supercoiled plasmid DNA to nicked (single-stranded break) or linear (double-stranded break) molecules. CC_{50} values (the concentration of compound that induced 50% maximal DNA cleavage complex formation) were calculated on GraphPad Prism Version 9.5.1 using a non-linear regression analysis with 95% confidence intervals.

4.3. Persistence of Cleavage Complexes

The persistence of gyrase/topoisomerase IV–DNA cleavage complexes was determined as described previously [33]. Cleavage complexes were formed by combining 50 nM pBR322 and 200 nM *N. gonorrhoeae* gyrase or 100 nM *N. gonorrhoeae* topoisomerase IV in the presence of 10 μM OSUAB-185 in a total volume of 20 μL of a cleavage buffer. Parallel control experiments were conducted to assess cleavage complexes formed in the absence of the NBTI by combining 50 nM pBR322 and 500 nM *N. gonorrhoeae* gyrase/topoisomerase IV in a 20 μL cleavage buffer. Reactions were incubated at 37 °C until the DNA cleavage/ligation equilibria were reached (30 min with gyrase and 10 min with topoisomerase IV) and diluted 20-fold in a DNA cleavage buffer lacking Mg^{2+}. Samples (20 μL) were removed at time points ranging from 0 to 60 min, and reactions were stopped with 2 μL of 4% SDS, followed by 2 μL of 250 mM EDTA (pH 8.0). Proteinase K was added (2 μL of a 0.8 mg/mL solution), and the reactions mixtures were incubated for 30 min at 45 °C. Samples were mixed with 2 μL of agarose gel loading buffer and processed and analyzed as described above. Levels of DNA cleavage were set to 100% at time zero, and the persistence of cleavage complexes were determined by the decay of nicked (single-stranded breaks) or linear (double-stranded breaks) reaction products over time. Cleavage complex stability (half-life, $t_{1/2}$) was calculated on GraphPad Prism Version 9.5.1 using non-linear regression analysis with 95% confidence intervals.

4.4. DNA Ligation

DNA ligation mediated by *N. gonorrhoeae* gyrase and topoisomerase IV was monitored using the procedure of Aldred et al. [33]. Initial reactions contained 10 nM pBR322 and 100 nM wild-type *N. gonorrhoeae* gyrase or topoisomerase IV in the absence or presence of 10 μM OSUAB-185. In order to better visualize levels of DNA breaks mediated by gyrase or topoisomerase IV in the absence of the NBTI, 10 mM $MgCl_2$ was replaced by 10 mM $CaCl_2$. DNA cleavage/ligation equilibria were established for 30 min with gyrase and 10 min with topoisomerase IV at 37 °C. DNA ligation was initiated by shifting samples from 37 °C to 65 °C, which allows enzyme-mediated ligation but prevents new rounds of DNA cleavage from occurring [54]. This results in a unidirectional sealing of the cleaved DNA. Reactions were stopped at time points ranging from 0 to 60 s by the addition of 4% SDS, followed by 2 μL 250 mM EDTA (pH 8.0). Proteinase K was added (2 μL of a 0.8 mg/mL solution), and reaction mixtures were incubated for 30 min at 45 °C. Samples were mixed with 2 μL of agarose loading buffer and processed and analyzed as described above. Nicked and linear DNA cleavage products at time zero were set to 100% to allow direct comparison between different conditions, and DNA ligation of single- or double-stranded breaks was monitored by the loss of nicked or linear DNA, respectively. The rate of DNA ligation ($t_{1/2}$) was calculated on GraphPad Prism Version 9.5.1 using non-linear regression analysis with 95% confidence intervals.

4.5. Accession Codes

N. gonorrhoeae Gyrase, GyrA: UniProtKB P48371, GyrB: UniProtKB P22118; *N. gonorrhoeae* Topoisomerase IV, ParC: UniProtKB P48374, ParE: UniProtKB A0A8D9YA86; *E. coli* Gyrase, GyrA: UniProtKB P0AES4, GyrB: UniProtKB P0AES6; *E. coli* Topoisomerase IV, ParC: UniProtKB P0AFI2, ParE: UniProtKB P20083; *B. anthracis* Gyrase, GyrA: UniProtKB A0A0F7R8R3, GyrB: UniProtKB Q9X3Y6; *B. anthracis* Topoisomerase IV, GrlA: UniProtKB A0A6H3AG07, GrlB: UniProtKB A0A2B0YAF3; *S. aureus* Gyrase, GyrA: UniProtKB Q2FKQ0, GyrB: UniProtKB P0A0K8; *S. aureus* Topoisomerase IV, GrlA: UniProtKB P0C1U9, GrlB: UniProtKB P0C1S7; Human Topoisomerase IIα: UniProtKB P11388.

Author Contributions: Conceptualization, J.C.Y., M.J.M.-F. and N.O.; methodology, investigation, and analysis, S.E.D., J.A.C. and J.A.W.B.; visualization, S.E.D. and J.A.C.; resources, Y.L., J.C.Y., M.J.M.-F. and N.O.; funding acquisition, M.J.M.-F. and N.O.; writing—original draft preparation, S.E.D., J.A.C. and N.O.; writing—review and editing, all authors. All authors have read and agreed to the published version of the manuscript.

Funding: This research was funded by US Veterans Administration Merit Review Award I01 Bx002198 (N.O.), Dr. Ralph and Marian Fak Medical Research Trust (Transformational Award) (M.J.M-F.), and National Institutes of Health grants R01 GM126363 (N.O.), R01 AI170546 (N.O.), and R01 AI173072 (M.J.M-F.). S.E.D. was supported in part by the Aspirnaut Undergraduate Discovery Science Experience in Renal Biology and Disease (R25 DK096999).

Institutional Review Board Statement: Not applicable.

Informed Consent Statement: Not applicable.

Data Availability Statement: All data are contained within the article.

Acknowledgments: We thank Jeffrey Y. Jian, Jillian F. Armenia, and Samika S. Joshi for their critical reading of the manuscript.

Conflicts of Interest: The authors declare no conflict of interest.

References

1. Centers for Disease Control and Prevention. *Outpatient Antibiotic Prescriptions—United States, 2017*; Centers for Disease Control and Prevention: Atlanta, GA, USA, 2017.
2. Basarab, G.S. Four ways to skin a cat: Inhibition of bacterial topoisomerases leading to the clinic. In *Antibacterials: Volume I*; Fisher, J.F., Mobashery, S., Miller, M.J., Eds.; Springer International Publishing: Cham, Switzerland, 2018; pp. 165–188.
3. World Health Organization. *Critically Important Antimicrobials for Human Medicine*; World Health Organization: Geneva, Switzerland, 2019.
4. Aldred, K.J.; Kerns, R.J.; Osheroff, N. Mechanism of quinolone action and resistance. *Biochemistry* **2014**, *53*, 1565–1574. [CrossRef] [PubMed]
5. Hooper, D.C.; Jacoby, G.A. Mechanisms of drug resistance: Quinolone resistance. *Ann. N. Y. Acad. Sci.* **2015**, *1354*, 12–31. [PubMed]
6. Gibson, E.G.; Ashley, R.E.; Kerns, R.J.; Osheroff, N. Fluoroquinolone interactions with bacterial type II topoisomerases and target-mediated drug resistance. In *Antimicrobial Resistance and Implications for the 21st Century*; Drlica, K., Shlaes, D., Fong, I.W., Eds.; Springer: New York, NY, USA, 2018; pp. 507–529.
7. Bush, N.G.; Diez-Santos, I.; Abbott, L.R.; Maxwell, A. Quinolones: Mechanism, lethality and their contributions to antibiotic resistance. *Molecules* **2020**, *25*, 5662. [CrossRef]
8. Unemo, M.; Del Rio, C.; Shafer, W.M. Antimicrobial resistance expressed by *Neisseria gonorrhoeae*: A major global public health problem in the 21st century. *Microbiol. Spectr.* **2016**, *4*, 213–237.
9. Kirkcaldy, R.D.; Harvey, A.; Papp, J.R.; Del Rio, C.; Soge, O.O.; Holmes, K.K.; Hook, E.W., 3rd; Kubin, G.; Riedel, S.; Zenilman, J.; et al. *Neisseria gonorrhoeae* antimicrobial susceptibility surveillance—The gonococcal isolate surveillance project, 27 sites, United States, 2014. *MMWR Surveill. Summ.* **2016**, *65*, 1–19. [CrossRef]
10. CDC. Update to CDC's sexually transmitted diseases treatment guidelines, 2006: Fluoroquinolones no longer recommended for treatment of gonococcal infections. *MMWR Morb. Mortal. Wkly. Rep.* **2007**, *56*, 332–336.
11. *Sexually Transmitted Disease Surveillance 2021*; Centers for Disease Control and Prevention: Atlanta, GA, USA, 2021.
12. CDC. *Antibiotic Resistance Threats in the United States, 2019*; U.S. Department of Health and Human Services, CDC: Atlanta, GA, USA, 2019.
13. Galvin, S.R.; Cohen, M.S. The role of sexually transmitted diseases in HIV transmission. *Nat. Rev. Microbiol.* **2004**, *2*, 33–42. [CrossRef]
14. Morgan, M.K.; Decker, C.F. Gonorrhea. *Dis. Mon.* **2016**, *62*, 260–268. [CrossRef]
15. Koebler, J. World Health Organization Warns Gonorrhea Could Join HIV as 'Uncurable'. U.S. News and World Report. 2012. Available online: https://www.usnews.com/news/articles/2012/06/06/world-health-organization-warns-gonorrhea-could-join-hiv-as-uncurable- (accessed on 20 June 2023).
16. World Health Organization. *Multi-Drug Resistant Gonorrhoea*; World Health Organization: Geneva, Switzerland, 2022.
17. Liu, Z.; Deibler, R.W.; Chan, H.S.; Zechiedrich, L. The why and how of DNA unlinking. *Nucleic Acids Res.* **2009**, *37*, 661–671. [CrossRef]
18. Vos, S.M.; Tretter, E.M.; Schmidt, B.H.; Berger, J.M. All tangled up: How cells direct, manage and exploit topoisomerase function. *Nat. Rev. Mol. Cell Biol.* **2011**, *12*, 827–841.
19. Chen, S.H.; Chan, N.L.; Hsieh, T.S. New mechanistic and functional insights into DNA topoisomerases. *Annu. Rev. Biochem.* **2013**, *82*, 139–170. [CrossRef] [PubMed]
20. Bush, N.G.; Evans-Roberts, K.; Maxwell, A. DNA topoisomerases. *EcoSal Plus* **2015**, *6*, 1–34. [CrossRef]
21. Dalvie, E.D.; Osheroff, N. DNA topoisomerases: Type II. In *The Encyclopedia of Biological Chemistry*, 3rd ed.; Allewell, N.M., Ed.; Academic Press: Waltham, MA, USA, 2021; Volume 3, pp. 479–486.
22. McKie, S.J.; Neuman, K.C.; Maxwell, A. DNA topoisomerases: Advances in understanding of cellular roles and multi-protein complexes via structure-function analysis. *Bioessays* **2021**, *43*, 2000286. [CrossRef] [PubMed]
23. Ashley, R.E.; Osheroff, N. Regulation of DNA by topoisomerases: Mathematics at the molecular level. In *Knots, Low-Dimensional Topology and Applications*; Adams, C., Gordon, C.M., Jones, V., Kauffman, L., Lambropoulou, S., Millet, K., Przytycki, J., Ricca, R., Sazdanovic, R., Eds.; Springer International Publishing: Cham, Switzerland, 2019.
24. Dong, K.C.; Berger, J.M. Structural basis for gate-DNA recognition and bending by type IIA topoisomerases. *Nature* **2007**, *450*, 1201–1205. [CrossRef] [PubMed]
25. Deweese, J.E.; Osheroff, N. The DNA cleavage reaction of topoisomerase II: Wolf in sheep's clothing. *Nucleic Acids Res.* **2009**, *37*, 738–749. [CrossRef]
26. Laponogov, I.; Pan, X.S.; Veselkov, D.A.; McAuley, K.E.; Fisher, L.M.; Sanderson, M.R. Structural basis of gate-DNA breakage and resealing by type II topoisomerases. *PLoS ONE* **2010**, *5*, e11338. [CrossRef]
27. Wohlkonig, A.; Chan, P.F.; Fosberry, A.P.; Homes, P.; Huang, J.; Kranz, M.; Leydon, V.R.; Miles, T.J.; Pearson, N.D.; Perera, R.L.; et al. Structural basis of quinolone inhibition of type IIA topoisomerases and target-mediated resistance. *Nat. Struct. Mol. Biol.* **2010**, *17*, 1152–1153. [CrossRef] [PubMed]
28. Vann, K.R.; Oviatt, A.A.; Osheroff, N. Topoisomerase II poisons: Converting essential enzymes into molecular scissors. *Biochemistry* **2021**, *60*, 1630–1641. [CrossRef]
29. Michel, B. After 30 years of study, the bacterial SOS response still surprises us. *PLoS Biol.* **2005**, *3*, e255. [CrossRef]

30. Drlica, K.; Malik, M.; Kerns, R.J.; Zhao, X. Quinolone-mediated bacterial death. *Antimicrob. Agents Chemother.* **2008**, *52*, 385–392. [CrossRef]
31. Schroder, W.; Goerke, C.; Wolz, C. Opposing effects of aminocoumarins and fluoroquinolones on the SOS response and adaptability in *Staphylococcus aureus*. *J. Antimicrob. Chemother.* **2013**, *68*, 529–538. [CrossRef]
32. Heisig, P.; Schedletzky, H.; Falkensteinpaul, H. Mutations in the gyrA gene of a highly fluoroquinolone-resistant clinical isolate of *Escherichia coli*. *Antimicrob. Agents Chemother.* **1993**, *37*, 696–701. [CrossRef]
33. Aldred, K.J.; McPherson, S.A.; Wang, P.; Kerns, R.J.; Graves, D.E.; Turnbough, C.L.; Osheroff, N. Drug interactions with *Bacillus anthracis* topoisomerase IV: Biochemical basis for quinolone action and resistance. *Biochemistry* **2012**, *51*, 370–381.
34. Aldred, K.J.; McPherson, S.A.; Turnbough, C.L.; Kerns, R.J.; Osheroff, N. Topoisomerase IV-quinolone interactions are mediated through a water-metal ion bridge: Mechanistic basis of quinolone resistance. *Nucleic Acids Res.* **2013**, *41*, 4628–4639. [CrossRef] [PubMed]
35. Bax, B.D.; Chan, P.F.; Eggleston, D.S.; Fosberry, A.; Gentry, D.R.; Gorrec, F.; Giordano, I.; Hann, M.M.; Hennessy, A.; Hibbs, M.; et al. Type IIA topoisomerase inhibition by a new class of antibacterial agents. *Nature* **2010**, *466*, 935–940. [CrossRef]
36. Basarab, G.S.; Kern, G.H.; McNulty, J.; Mueller, J.P.; Lawrence, K.; Vishwanathan, K.; Alm, R.A.; Barvian, K.; Doig, P.; Galullo, V.; et al. Responding to the challenge of untreatable gonorrhea: ETX0914, a first-in-class agent with a distinct mechanism-of-action against bacterial type II topoisomerases. *Sci. Rep.* **2015**, *5*, 11827. [CrossRef] [PubMed]
37. Gibson, E.G.; Bax, B.; Chan, P.F.; Osheroff, N. Mechanistic and structural basis for the actions of the antibacterial gepotidacin against *Staphylococcus aureus* gyrase. *ACS Infect. Dis.* **2019**, *5*, 570–581. [CrossRef] [PubMed]
38. Bax, B.D.; Murshudov, G.; Maxwell, A.; Germe, T. DNA topoisomerase inhibitors: Trapping a DNA-cleaving machine in motion. *J. Mol. Biol.* **2019**, *431*, 3427–3449. [CrossRef] [PubMed]
39. GlaxoSmithKline. *A Phase III, Randomized, Multicenter, Parallel-Group, Double-Blind, Double-Dummy Study in Adolescent and Adult Female Participants Comparing the Efficacy and Safety of Gepotidacin to Nitrofurantoin in the Treatment of Uncomplicated Urinary Tract Infection (Acute Cystitis)*; National Library of Medicine: Bethesda, MD, USA, 2021.
40. GlaxoSmithKline. *A Phase III, Randomized, Multicenter, Open-Label Study in Adolescent and Adult Participants Comparing the Efficacy and Safety of Gepotidacin to Ceftriaxone Plus Azithromycin in the Treatment of Uncomplicated Urogenital Gonorrhea Caused by Neisseria gonorrhoeae*; National Library of Medicine: Bethesda, MD, USA, 2022.
41. Perry, C.; Hossain, M.; Powell, M.; Raychaudhuri, A.; Scangarella-Oman, N.; Tiffany, C.; Xu, S.; Dumont, E.; Janmohamed, S. Design of two phase III, randomized, multicenter studies comparing gepotidacin with nitrofurantoin for the treatment of uncomplicated urinary tract infection in female participants. *Infect. Dis. Ther.* **2022**, *11*, 2297–2310. [CrossRef]
42. Scangarella-Oman, N.E.; Hossain, M.; Perry, C.R.; Tiffany, C.; Powell, M.; Swift, B.; Dumont, E.F. Dose selection for a phase III study evaluating gepotidacin (GSK2140944) in the treatment of uncomplicated urogenital gonorrhoea. *Sex. Transm. Infect.* **2023**, *99*, 64–69. [CrossRef]
43. Watkins, R.R.; Thapaliya, D.; Lemonovich, T.L.; Bonomo, R.A. Gepotidacin: A novel, oral, 'first-in-class' triazaacenaphthylene antibiotic for the treatment of uncomplicated urinary tract infections and urogenital gonorrhoea. *J. Antimicrob. Chemother.* **2023**, *78*, 1137–1142. [CrossRef] [PubMed]
44. GlaxoSmithKline. *EAGLE-2 and EAGLE-3 Phase III Trials for Gepotidacin Stopped Early for Efficacy Following Pre-Planned Interim Analysis by Independent Data Monitoring Committee*; GlaxoSmithKline: Brentford, UK, 2022.
45. Vanden Broeck, A.; Lotz, C.; Ortiz, J.; Lamour, V. Cryo-EM structure of the complete *E. coli* DNA gyrase nucleoprotein complex. *Nat. Commun.* **2019**, *10*, 4935. [CrossRef] [PubMed]
46. Chan, P.F.; Srikannathasan, V.; Huang, J.; Cui, H.; Fosberry, A.P.; Gu, M.; Hann, M.M.; Hibbs, M.; Homes, P.; Ingraham, K.; et al. Structural basis of DNA gyrase inhibition by antibacterial QPT-1, anticancer drug etoposide and moxifloxacin. *Nat. Commun.* **2015**, *6*, 10048. [CrossRef] [PubMed]
47. Blower, T.R.; Williamson, B.H.; Kerns, R.J.; Berger, J.M. Crystal structure and stability of gyrase-fluoroquinolone cleaved complexes from *Mycobacterium tuberculosis*. *Proc. Natl. Acad. Sci. USA* **2016**, *113*, 1706–1713. [CrossRef] [PubMed]
48. Gibson, E.G.; Blower, T.R.; Cacho, M.; Bax, B.; Berger, J.M.; Osheroff, N. Mechanism of action of *Mycobacterium tuberculosis* gyrase inhibitors: A novel class of gyrase poisons. *ACS Infect. Dis.* **2018**, *4*, 1211–1222. [CrossRef]
49. Gibson, E.G.; Oviatt, A.A.; Cacho, M.; Neuman, K.C.; Chan, P.F.; Osheroff, N. Bimodal actions of a naphthyridone/aminopiperidine-based antibacterial that targets gyrase and topoisomerase IV. *Biochemistry* **2019**, *58*, 4447–4455. [CrossRef]
50. Lu, Y.; Papa, J.L.; Nolan, S.; English, A.; Seffernick, J.T.; Shkolnikov, N.; Powell, J.; Lindert, S.; Wozniak, D.J.; Yalowich, J.; et al. Dioxane-linked amide derivatives as novel bacterial topoisomerase inhibitors against gram-positive *Staphylococcus aureus*. *ACS Med. Chem. Lett.* **2020**, *11*, 2446–2454. [CrossRef]
51. Osheroff, N. Role of the divalent cation in topoisomerase II mediated reactions. *Biochemistry* **1987**, *26*, 6402–6406. [CrossRef]
52. Osheroff, N.; Zechiedrich, E.L. Calcium-promoted DNA cleavage by eukaryotic topoisomerase II: Trapping the covalent enzyme-DNA complex in an active form. *Biochemistry* **1987**, *26*, 4303–4309. [CrossRef]
53. Bandele, O.J.; Osheroff, N. The efficacy of topoisomerase II-targeted anticancer agents reflects the persistence of drug-induced cleavage complexes in cells. *Biochemistry* **2008**, *47*, 11900–11908. [CrossRef]
54. Anderson, V.E.; Zaniewski, R.P.; Kaczmarek, F.S.; Gootz, T.D.; Osheroff, N. Quinolones inhibit DNA religation mediated by *Staphylococcus aureus* topoisomerase IV: Changes in drug mechanism across evolutionary boundaries. *J. Biol. Chem.* **1999**, *274*, 35927–35932. [PubMed]

55. Hall, C.L.; Harrison, M.A.; Pond, M.J.; Chow, C.; Harding-Esch, E.M.; Sadiq, S.T. Genotypic determinants of fluoroquinolone and macrolide resistance in *Neisseria gonorrhoeae*. *Sex. Health* **2019**, *16*, 479–487. [CrossRef] [PubMed]
56. Dan, M. The use of fluoroquinolones in gonorrhoea: The increasing problem of resistance. *Expert Opin. Pharmacother.* **2004**, *5*, 829–854. [CrossRef] [PubMed]
57. Mitton-Fry, M.J.; Brickner, S.J.; Hamel, J.C.; Brennan, L.; Casavant, J.M.; Chen, M.; Chen, T.; Ding, X.Y.; Driscoll, J.; Hardink, J.; et al. Novel quinoline derivatives as inhibitors of bacterial DNA gyrase and topoisomerase IV. *Bioorg. Med. Chem. Lett.* **2013**, *23*, 2955–2961. [CrossRef] [PubMed]
58. Li, L.S.; Okumu, A.; Dellos-Nolan, S.; Li, Z.; Karmahapatra, S.; English, A.; Yalowich, J.C.; Wozniak, D.J.; Mitton-Fry, M.J. Synthesis and anti-staphylococcal activity of novel bacterial topoisomerase inhibitors with a 5-amino-1,3-dioxane linker moiety. *Bioorg. Med. Chem. Lett.* **2018**, *28*, 2477–2480. [CrossRef] [PubMed]
59. Li, L.; Okumu, A.A.; Nolan, S.; English, A.; Vibhute, S.; Lu, Y.; Hervert-Thomas, K.; Seffernick, J.T.; Azap, L.; Cole, S.L.; et al. 1,3-dioxane-linked bacterial topoisomerase inhibitors with enhanced antibacterial activity and reduced hERG inhibition. *ACS Infect. Dis.* **2019**, *5*, 1115–1128. [CrossRef]
60. Felix, C.A.; Kolaris, C.P.; Osheroff, N. Topoisomerase II and the etiology of chromosomal translocations. *DNA Repair* **2006**, *5*, 1093–1108. [CrossRef]
61. Joannides, M.; Grimwade, D. Molecular biology of therapy-related leukaemias. *Clin. Transl. Oncol.* **2010**, *12*, 8–14. [CrossRef]
62. Cowell, I.G.; Austin, C.A. Mechanism of generation of therapy related leukemia in response to anti-topoisomerase II agents. *Int. J. Environ. Res. Public Health* **2012**, *9*, 2075–2091. [CrossRef]
63. Pendleton, M.; Lindsey, R.H., Jr.; Felix, C.A.; Grimwade, D.; Osheroff, N. Topoisomerase II and leukemia. *Ann. N. Y. Acad. Sci.* **2014**, *1310*, 98–110. [CrossRef]
64. Ashley, R.E.; Dittmore, A.; McPherson, S.A.; Turnbough, C.L., Jr.; Neuman, K.C.; Osheroff, N. Activities of gyrase and topoisomerase IV on positively supercoiled DNA. *Nucleic Acids Res.* **2017**, *45*, 9611–9624. [CrossRef]
65. Peng, H.; Marians, K.J. *Escherichia coli* topoisomerase IV. Purification, characterization, subunit structure, and subunit interactions. *J. Biol. Chem.* **1993**, *268*, 24481–24490. [CrossRef] [PubMed]
66. Worland, S.T.; Wang, J.C. Inducible overexpression, purification, and active site mapping of DNA topoisomerase II from the yeast *Saccharomyces cerevisiae*. *J. Biol. Chem.* **1989**, *264*, 4412–4416. [CrossRef] [PubMed]
67. Kingma, P.S.; Greider, C.A.; Osheroff, N. Spontaneous DNA lesions poison human topoisomerase IIα and stimulate cleavage proximal to leukemic 11q23 chromosomal breakpoints. *Biochemistry* **1997**, *36*, 5934–5939. [CrossRef] [PubMed]
68. Fortune, J.M.; Osheroff, N. Merbarone inhibits the catalytic activity of human topoisomerase IIα by blocking DNA cleavage. *J. Biol. Chem.* **1998**, *273*, 17643–17650. [CrossRef] [PubMed]

International Journal of
Molecular Sciences

MDPI

Review

Inhibition of Topoisomerases by Metal Thiosemicarbazone Complexes

Xiaohua Jiang [1,*], Lauren A. Fielding [2], Hunter Davis [3], William Carroll [3], Edward C. Lisic [3] and Joseph E. Deweese [2,4,*]

1 Department of Chemistry, Vanderbilt University, Nashville, TN 37240, USA
2 Department of Biological, Physical and Human Sciences, Freed Hardeman University, Henderson, TN 38340, USA
3 Department of Chemistry, Tennessee Tech University, Cookeville, TN 38505, USA
4 Department of Biochemistry, Vanderbilt University School of Medicine, Nashville, TN 37240, USA
* Correspondence: xiaohua.jiang@vanderbilt.edu (X.J.); jdeweese@fhu.edu (J.E.D.); Tel.: +1-931-372-3814 (X.J.); +1-731-989-6669 (J.E.D.)

Abstract: Topoisomerases, common targets for anti-cancer therapeutics, are crucial enzymes for DNA replication, transcription, and many other aspects of DNA metabolism. The potential anti-cancer effects of thiosemicarbazones (TSC) and metal–TSC complexes have been demonstrated to target several biological processes, including DNA metabolism. Human topoisomerases were discovered among the molecular targets for TSCs, and metal-chelated TSCs specifically displayed significant inhibition of topoisomerase II. The processes by which metal–TSCs or TSCs inhibit topoisomerases are still being studied. In this brief review, we summarize the TSCs and metal–TSCs that inhibit various types of human topoisomerases, and we note some of the key unanswered questions regarding this interesting class of diverse compounds.

Keywords: thiosemicarbazone; topoisomerase; antitumor; bis-thiosemicarbazone

Citation: Jiang, X.; Fielding, L.A.; Davis, H.; Carroll, W.; Lisic, E.C.; Deweese, J.E. Inhibition of Topoisomerases by Metal Thiosemicarbazone Complexes. *Int. J. Mol. Sci.* **2023**, *24*, 12010. https://doi.org/10.3390/ijms241512010

Academic Editor: Ana Cristina Gonçalves

Received: 9 June 2023
Revised: 20 July 2023
Accepted: 25 July 2023
Published: 27 July 2023

1. Introduction

Topoisomerases (Tops) are essential enzymes for genome stability that are involved in DNA metabolism through the maintenance of DNA topology. All cells maintain DNA supercoiling in a dynamic process that is required for transcription, replication, and cell division to take place. Tops are found across all domains of life and several viruses encode topoisomerases. Tops are divided into two families: Type I and Type II.

Type I Tops cut and religate one strand of the double helix coupled with either a strand passage or controlled rotation mechanism to remove DNA positive (overwinding) and negative (underwinding) supercoiling. To modulate supercoiling, remove DNA knots, and unlink catenated DNA (intertwined DNA), Type II Tops form a double-stranded DNA break and passes a double-strand DNA section through the break before ligating the DNA back together. In both mechanisms, the DNA break is stabilized through a covalent intermediate with an active site tyrosine residue on the Top. There are six human Tops including Top1B, mitochondrial Top1B, Top2α, Top2β, Top3α, Top3β [1].

There are two main Type I subfamilies IA (Top3α/β in humans) and IB (Top1B/mitochondrial Top1B in humans) [1,2]. The subfamilies differ both structurally and mechanistically, and these differences have been reviewed elsewhere [1,2]. There is a Type IC that is only found in an archaeal species.

There are two main Type II subfamilies, which include Type IIA and Type IIB depending on either a 4- or 2-base stagger between the cleaved positions on the DNA, respectively [1,3–5]. Human Top2α and Top2β both belong to the Type IIA subfamily [1,5]. While the names and structures for the archaeal and bacterial counterparts differ, this review will focus on the mammalian forms of the eukaryotic enzymes.

During the catalytic cycle, Tops form temporary single- or double-stranded DNA breaks [1,5]. These momentary DNA breaks may become permanent, leading to DNA damage and cell death [1,2,4]. Several widely-used anti-cancer therapeutics target either Type I or Type II Tops, and in a similar manner, several antibacterial agents are used to target bacterial Tops to fight off infections. Interfacial poisons and catalytic inhibitors are the two primary groups of inhibitors for Top enzymes [2,4,6]. Whereas catalytic inhibitors prevent Tops from completing its catalytic cycle, interfacial poisons stabilize single-stranded or double-stranded breaks leading to further damage [2,4,6]. The general mechanism for catalytic inhibitors is interaction with the N-terminal ATPase domain, which is a clamp-like region at the "top" of the enzyme. Top poisons are thought to act by a "doorstop" mechanism where the drug molecule prevents ligation by slipping between the cleaved ends of DNA during catalysis. In general, poisons tend to lead to accumulation of cleaved DNA while catalytic inhibitors do not. In addition, some agents are reactive and lead to inactivation of the enzyme and/or covalent adduction, which could occur either in the active site or at or near to the ATPase domain [6,7].

Thiosemicarbazones (TSCs) are a varied class of compounds with common N-N-S coordinates. They were discovered in the 1950s with anti-bacterial, anti-fungal, and anti-tumor activities [8,9]. TSCs are multi-target drugs and the molecular mechanisms involved metal chelation, DNA interference, topoisomerase inhibition and ribonucleotide reductase inhibition [10]. Many metal–TSC complexes have been synthesized and were found to be more effective in cell toxicity experiments than the ligand form of the TSC [10,11]. The cell toxicity of metal–TSCs included a decrease in de novo purine synthesis and inhibition of IMP dehydrogenase, DNA polymerase activity, and topoisomerase II activity [10–12]. Recent studies in yeast have uncovered other cellular mechanisms for metal-bis(TSC), which includes chromatin remodeling, cytoskeleton organization, mitochondrial function, and iron metabolism [13]. Due to the varied structures of TSCs, they impact multiple cellular targets. In this paper, we summarize the TSCs that are targeting Tops with a particular emphasis on metal–TSCs.

Several examples of ligand TSCs that have been studied either clinically or in animal models are shown in Figure 1. Triapine (3-aminopyridine-2-carboxaldehyde thiosemicarbazone; 3-AP) is the first thiosemicarbazone to be approved for clinical trials. It inhibits ribonucleotide reductase and chelates iron to kill cancer cells. There are more than 30 Phase I and Phase II clinical trials with Triapine, but it has disadvantages of adverse events, such as methemoglobinemia, and a short plasma half-life [14]. In recent years, TSCs gained more attention as potential anticancer drugs since they impact multiple potential targets. Polypharmacology is particularly useful in the metastasis stage of cancer treatment. Dp44mT, (di-2-pyridylketone-4,4-dimethyl-3-thiosemicarbazone) has potent and broad antitumor activities in a panel of human xenografts in nude mice [15]. It not only chelates iron but also has redox properties, similar to Triapine [15]. Dp44mT showed conflicting results in Top inhibition, which is discussed later in this article. DpC, (di-2-pyridylketone 4-cyclohexyl-4-methyl-3-thiosemi-carbazone), optimized based on the structure of Dp44mT, has superior activity against human pancreatic cancer xenografts in nude mice [16].

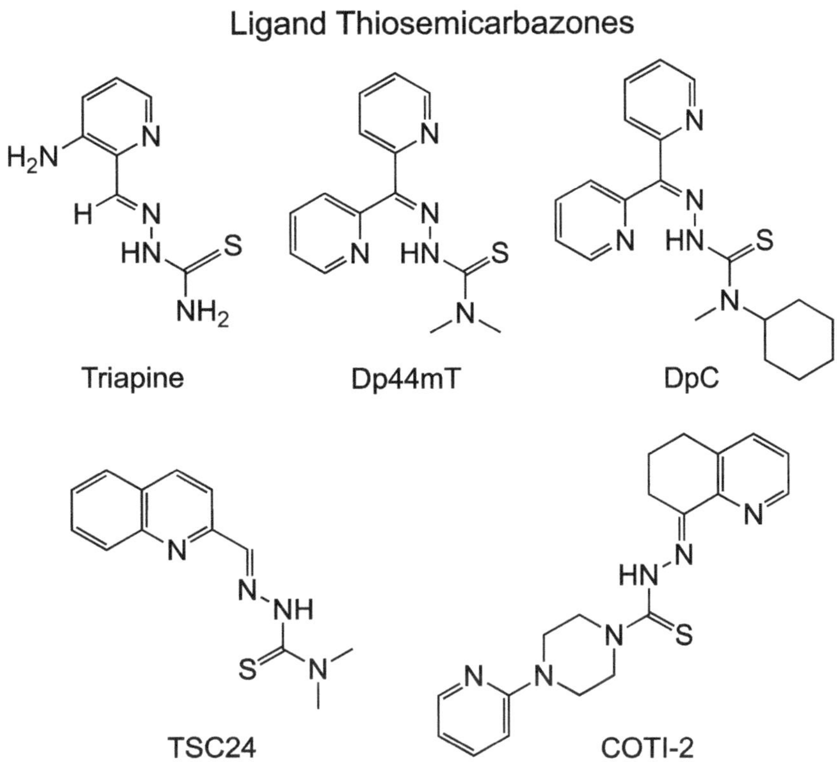

Figure 1. Ligand TSC compounds. Examples of several ligand TSCs are shown including some that have undergone clinical or pre-clinical trials. Some of these are known to chelate metals in cellular contexts. Structures prepared using ChemDraw 20.1.

COTI-2 [4-(2-pyridinyl)-2-(6,7-dihydro-8(5H)-quinolinylidene)-hydrazide], a TSC ligand, was identified through in silico screening and was found to inhibit many human cell lines in vivo [17]. COTI-2 has been shown to have anticancer activities through p53-dependent and p53-independent mechanisms [18]. Bis(thiosemicarbazone) metal complexes have been recently applied to neurodegenerative disease since they are able to restore metal balance in neurons. They showed promising results in animal models for Alzheimer's disease, Parkinson disease, and amyotrophic lateral sclerosis [19].

2. Thiosemicarbazones as Inhibitors of Topoisomerases

2.1. TSC Ligand or bis(TSC) Ligand Inhibition of Topoisomerases

TSCs have been studied for decades for antiviral, antifungal and antiproliferation activity (Table 1). In recent years, Triapine (Figure 1), a member of TSC family, has been studied in phase I and II clinical trials [20–23]. The major molecular target of Triapine was identified as ribonucleotide reductase [23,24]. Triapine showed some inhibition of Top2A but only in the presence of Cu(II) [25]. Previous work on Triapine did not show any inhibition of relaxation or poisoning of plasmid DNA cleavage with Top2α [26].

In a series of TSC compounds, Top2 was discovered as one of the molecular targets for the ligand TSC24 (Figure 1) [27]. TSC24 showed high potent activity with an IC_{50} of 0.02 μM against the HT-29 cell line [27]. TSC24 inhibited tumor growth of S-180 sarcoma-bearing mice in a dose-dependent manner, with inhibitory rates of 17.6%, 35.8%, and 76.7% at doses of 12.5, 25, and 50 (mg/kg)/day, respectively, after it was administered intraperitoneal (IP) for 7 days [21].

The pattern of TSC antiproliferative effects against the human cancer cell line (NCI-60) is similar to those found with recognized anticancer treatments [27]. These investigations revealed TSC24's profile is comparable to known Top2 agents. TSC24 was further investigated and it was found that it hindered DNA relaxing and decatenation by inhibiting the Top2α ATPase domain [27].

Another TSC ligand that has been well studied is Dp44mT (Figure 1). It induces G1 cell cycle arrest and reduces cancer cell clonogenic growth in the breast cancer line MDA-MB-231 at nanomolar concentrations [22]. In relaxation and cleavage experiments, Dp44mT was shown to preferentially target Top2α, with minimal effect on Top2β and little inhibition of Top1 [28]. But this result is controversial as an additional study found that the ligand Dp44mT did not inhibit Top2α or increase cellular cleavage complexes [26]. A more recent study on the ligand Dp44mT also found little effect of the compound alone, but they found inhibition when combined with Cu(II) [25]. The cell toxicity result is consistent with previous studies [22] and IC_{50} against the cancer cell line is in a nanomolar concentration. The Cu(II) complex of Dp44mT showed similar or lower IC_{50} compared with Dp44mT [25]. It is unclear whether it was the presence of Cu(II) alone or if the Cu(II) formed a complex with the ligand leading to the effect in this case. As is discussed below, it is possible that the ligand forms alone show little activity, while metal-chelated forms have varying levels of activity depending on the metal ion.

Computational docking and surface plasmon resonance studies support the ability of TSC24 to bind near the ATP binding pocket, but it is unclear if this is generalizable to other TSCs and whether this has been biochemically validated. Both TSC24 and Dp44mT appear to act as catalytic inhibitors, and there is an increase in DNA cleavage seen with Dp44mT [27,28]. TSC24 does not appear to increase strand breaks in cells, nor block the effects of VP16 [27].

A series of thiosemicarbazones and 4-thiazolindinones have been synthesized and some were identified with activity against Top2. The thiosemicarbazone (E)-2-(1H-indol-3ylmethylene)-*N*-(naphthalen-1-yl) hydrazinecarbothioamide (compound **2b**) has a IC_{50} of 0.01 μM against colorectal adenocarcinoma (HT-29) and leukemia (K562) cells and appear to weakly inhibit plasmid DNA relaxation by Top2α [29]. Compound **3a** [(Z)-2-(acridin-9-ylmethylene)-*N*-phenyl-hydrazinecarbothioamide)] and (compound **3h**) [(Z)-2-(acridin-9-ylmethylene)-*N*-(naphtalen-1-yl) hydrazinecarbothioamide] also had a limited ability to inhibit relaxation [30]. Detailed studies on the structure–activity relationship of TSCs against Top2α showed that most TSC ligands inhibit Top2α very weakly or not at all (Table 1) [31–34].

Despite the fact that TSC ligands were initially found to target Top2, the majority of the ligands produced had little or no effect on Top1 [35] or Top2 [32–34,36,37]. Metal–TSCs, on the other hand, showed profound inhibition on Tops, as discussed below.

Table 1. TSC ligand inhibition on Tops.

Name	Inhibition of Top	Reference
TSC24	Inhibits Top2α DNA relaxation at 25 μM Inhibits Top2 at decatenation at 100 μM	[27]
Dp44mT	Inhibits human Top2α in relaxation assay with 5′ labeled 161-bp fragment from pBluescript SK phagemid DNA Does not inhibit human Top2β or human TopoI	[28]
Dp44mT Triapine	Does not inhibit Top2α in decatenation Does not increase cleavage complex by human Top2α	[26]
Compound **2b**	Inhibits human Top2α weakly at 100 μM	[29]

Table 1. *Cont.*

Name	Inhibition of Top	Reference
Compound **3a**	Inhibit human Top2α-mediated DNA relaxation at 100 μM	[30]
Compound **3h**		
Triapine	Do not inhibit Top2α-mediated DNA relaxation at 50 μM	[25]
Dp44mT		
Compound **24**	Do not inhibit isolated Top2 from L1210	[12]
Compound **36**		
NQTS	Does not inhibit Top2α	[36]
HFp4mT, HFp4pyrrT	Inhibit Top2α at 100 μM	[37]
HFp4eT, HFp4ipT, HFp4alT, HAp4mT, HAp4-eT, HFp4bzT, HFpT, HAp4alT, and HApz4mT	Do not inhibit Top2α at 100 μM	[37]
BZP–TSC ligands series	Inhibits Top2α slightly at 50 μM	[20]
ATZ ligand series	Do not inhibit Top2α at 10 μM, do not increase cleavage complex	[33]
BZP ligands series		
APY	Does not inhibit Top2α at 100 μM	[32]
APY	Does not inhibit Top2β at 200 μM	[34]
APZ	Inhibits Top2α at 100 μM	[32]
BZP	Inhibits Top2β at 200 μM weakly	[34]
HPyCT4BrPh	Does not inhibit human Top1B at 50 μM	[35]

2.2. Metal–TSC or Metal–Bis(TSC) Inhibition of Topoisomerases

TSCs can chelate with various metal ions. Metal–TSC complexes have been produced from copper (Cu), nickel (Ni), palladium (Pd), ruthenium (Ru), tin (Sn), gallium (Ga), gold (Au), and cobalt (Co) [31,36,38–42]. Metal–TSCs inhibit Tops more strongly than their TSC ligands [25,31–34,37]. As will be discussed more below, metal–TSCs appear to be active against both Type I (Table 2) and Type II Tops (Table 3).

Table 2. Metal–TSC inhibition of TopI.

Name	Inhibition of TopI	Reference
Cu(PyCT4BrPh)Cl	Inhibits TopI The inhibition is severe with pre-incubation of the compound with TopI Inhibited the cleavage step and partially inhibited religation	[35]
Pd–pyrene–TSC	Inhibits human Top1B at 12.5 μM	[43]
Ga(III)–TSC complex (C4)	Inhibits TopI	[42]
Au(III)(TSC)Cl (complex 1)	Inhibits human Top1B activity starting at 1.5 μM Pre-incubation of Top1B with Complex 1 increased the inhibition	[31]
Ni-bis(TSC)	No inhibition of *E. coli* TopI	[44]
Nine copper complexes	Inhibits TopI	[45]
Cobalt (III)–TSC (Complex 4)	Inhibits TopI-induced DNA relaxation	[41]

2.2.1. Inhibition of Type I Top

Human Top1B belongs to the Type I Top family. It relaxes the DNA supercoils during DNA replication, recombination, and transcription by cutting one strand of DNA and performing a controlled rotation/swivel mechanism [1,2]. Top1B inhibitors are a class of compounds that target the enzyme and prevent it from relaxing DNA, leading to the accumulation of DNA damage. These inhibitors have potential as antitumor agents since tumor cells are under fast proliferation and DNA replication, making them more susceptible to DNA damage. Several classes of Top1B inhibitors have been developed, including camptothecin analogs and indolocarbazoles [1,2,46]. There are several metal–TSC compounds that have been studied against Top1B activity (Table 2 and Figure 2).

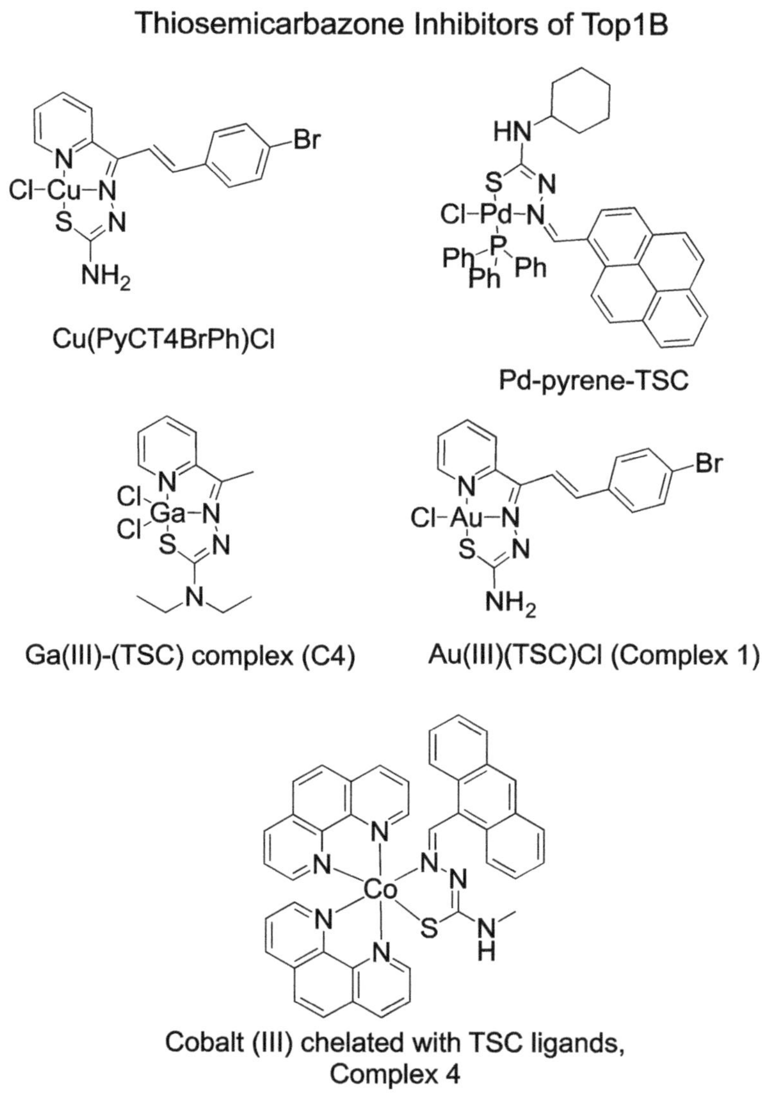

Figure 2. Examples of metal–TSC complexes studied with Top1B. These compounds were found to display varying activity against Top1B and include various metal ions.

Cu(PyCT4BrPh)Cl [Cu(3-(4-bromophenyl)-1-pyridin-2-ylprop-2-en-1-one-thiosemicarbazone)Cl] was studied against human Top1B [35]. It inhibited Topo1B by partially blocking ligation of the cleaved DNA [35]. The complex also reduced enzyme-DNA binding according to an EMSA assay [35]. Additional studies are needed to clarify the exact mechanism. The cytotoxicity of Cu(HPyCT4BrPh) increased 6-fold against THP-1 (IC_{50} = 0.20 μM) and 8-fold against MCF-7 cells (IC_{50} = 0.16 μM) compared with ligand HpyCT4BrPh [29].

In another study, pyrene TSCs were complexed with Pd (Complex 1) and examined for inhibition of human Top1B [43]. It has an IC_{50} of 7.59 μM in A2780 human ovarian carcinoma cells [38]. Its IC_{50} for A2780 cisplatin resistant human ovarian carcinoma cells is even lower with a value of 3.16 μM [43]. Pd–pyrene–TSC complexes inhibited relaxation of supercoiled plasmid by human Top1B at 12.5 μM [43]. Additionally, the Pd–pyrene–TSC complex displayed the ability to inhibit ligation of cleaved DNA with Top1B, similar to Cu(PyCT4BrPh)Cl [43].

One group reported the use of a Ga(III)–TSC complex, [*N*,*N*-diethyl-2-[1-(2-pyridinyl) ethylidene]hydrazinecarbothioamide-*N*,*N*,*S*-gallium(III)]bis(chloride), referred to as C4 in the study [42]. Based upon their results, human Top1B cleavage activity was inhibited by the Ga(III)–TSC complex while the ligand alone did not show significant inhibition [42]. C4 showed selective activities against tumor cells. It exhibited an IC_{50} of 0.30μM for lung cancer cell line NCI-H460 cells, 0.35 μM for T24 cells of the urinary bladder cancer cell line, 0.55 μM for BEL-7402 human liver cancer, and 0.76 μM for MSTO-211H, human mesothelioma cell line, while it showed low cell toxicity to the normal cell line—human fetal lung fibroblast cells with an IC_{50} higher than 28.65 μM [36].

The Au(III)–TSC complex [(3-(4-bromophenyl)-1-pyridin-2-ylprop-2-en-1-enone thiosemicarbazonato)chlorogold(III)] chloride, [Au(PyCT4BrPh)Cl]Cl, was studied with human Top1B and found to inhibit relaxation at 1.5 μM [31]. In contrast, $HAuCl_4 \cdot 3H_2O$ did not inhibit until 200 μM. Pre-incubation of Top1B with this compound increased the inhibition, which suggests gold(III)–TSC binds and inhibits the activity of Topo1B [31]. The gold(III)–TSC complex showed a high potency in cytotoxicity, with an IC_{50} of 0.26 μM for HL60 (human promyelocytic leukemia), 0.62 μM for THP-1 (human monocytic leukemia), 0.09 μM for MDA-MB 231 (human breast adenocarcinoma), and 0.42 μM for MCF-7 (human breast adenocarcinoma) [31].

In another study, Ni chelated with testosterone TSC to form a distorted square planar with ligand as a bidentate NS donor—Ni–bisTSC [44]. Ni–bis(TSC) did not inhibit *E. coli* TopI, but it showed DNA binding affinity similar to ethidium bromide, which results in selective activity against human prostate cancer cells [44].

In summary, the research on metal–TSCs inhibiting Top1B is limited. Some compounds displayed catalytic inhibition, such as [Au(PyCT4BrPh)Cl]Cl, others are interfacial poisons by inhibiting ligation, including Cu(PyCT4BrPh)Cl, Pd–Pyrene–TSC, and Ga(III)–TSC. The cell toxicity results are similar and the IC_{50} is between 0.1 and 10 μM, and some of the metal complexes showed selective activities towards tumor cells.

2.2.2. Inhibition of Type II Top

Type II Tops are the primary targets for studies of TSC antitumor activity. Multiple metal–TSC complexes showed higher inhibition compared with their ligand counterparts (Figure 3 and Table 3). Cu–TSCs are the most studied and have demonstrated the highest inhibition of Top2.

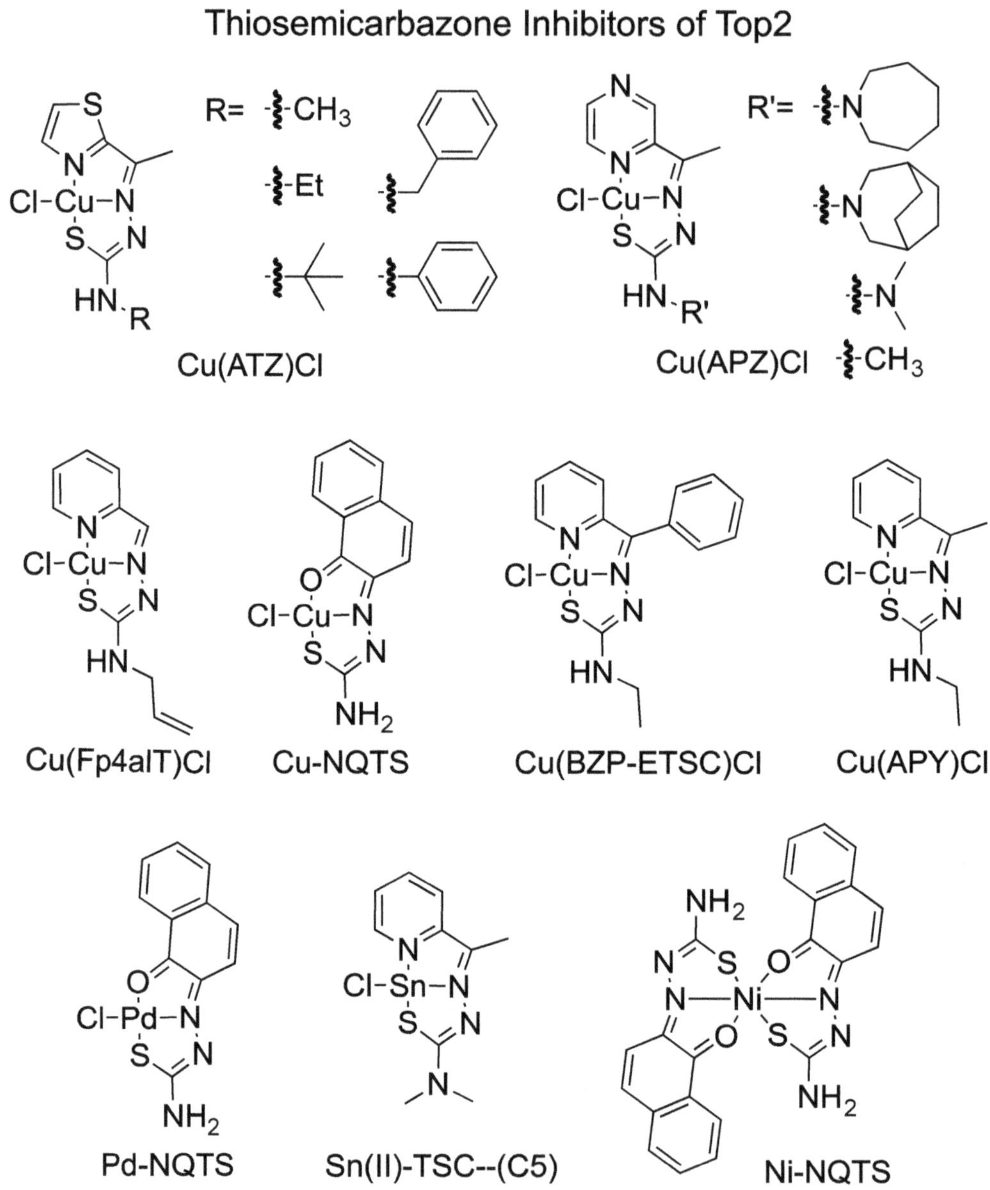

Figure 3. Examples of metal–TSC complexes studied with Top2. Several metal–TSC complexes are shown along with some varying side-chain examples.

1. Ni chelated with bis(TSC)

Ni–TSCs were discovered to block a variety of metabolic pathways, including purine synthesis, DNA polymerase, PRPP-amino transferase, IMP dehydrogenase, dihydrofolate reductase, TMP-kinase, and thymidylate synthetase activities, against the L1210 cell line in 1997 [38]. Despite the fact that Ni–TSCs demonstrated several cellular pathways for inhibition, research suggests that Ni–TSCs do not efficiently inhibit Top enzymes [13,38]. Ni(II) coordinates with two TSC ligands, Ni–bis(TSC), which lack the essential square planar structure for Top inhibition [32,38]. The ED_{50} is between 1 and 4 μg/mL against the growth of murine or human leukemias, human HeLa uterine suspended carcinoma, colon adenocarcinoma SW480, KB nasopharynx, lung MB 9812 bronchogenic carcinomas, solid HeLa uterine carcinoma, and rat osteosarcoma [38].

There are some controversies on the inhibition of Ni–TSCs against Top2. Ni-NQTS has very effective antiproliferation activity against the MCF-7 breast cancer cell line with an IC50 of 2.25 μM, better than its copper and palladium counterparts, and it showed inhibition of a Top2α-mediated DNA relaxation assay using a TopoGEN topoisomerase assay kit (Buena Vista, CO) [36]. However, the data are inconsistent with other reported results. For example, a yeast screen did not show that Ni–bisTSC interferes with Tops [13]. Our unpublished results showed that when bisTSCs chelate with metal ions, the metal–bisTSC compounds do not inhibit Top2α (Beckett and Jiang, unpublished). Ni-NQTS were also tested in DNA cleavage assays with Top2α [36]. The results seem to show that Ni-NQTS does not stabilize double-stranded DNA cleavage, but there was a low amount of nicking observed, though it was not quantified [36]. In another study, several Ni–TSC complexes were examined alongside Cu analogs discussed below [47]. Interestingly, Ni(L1)(HL1)Cl, $Ni(HL2)_2Cl_2$, $Ni(L3)_2$, $Ni(L4)_2$, and $Ni(L5)_2Cl_2$ did not appreciably inhibit Top2α from TopoGEN [47]. Although Ni–TSCs performed profound inhibition against cell proliferation, Top2 may not be the target (or the primary target) for Ni–TSCs [8–10].

2. Cu-chelated TSCs

In cell toxicity studies, copper (Cu^{2+})-chelated TSCs are one of the most active groups of metal–TSCs (Table 3) [8,10]. When copper chelates with TSC, it forms a square planar structure with Cu in the middle, which seems to be the crucial structure element for Top2 inhibition [32–34], Cu(TSC)s demonstrated greater inhibition compared with their ligands. For example, Cu(TSC)Cl (Compounds **1** and **2**) inhibited Top2 while the corresponding TSC ligands (Compounds **24** and **36**) did not [12]. In general, Cu(TSC) complexes act on Top2 as catalytic inhibitors through inhibiting the ATPase function and inhibiting relaxation.

Another study showed that Cu(TSC)Cl complexes (Compounds **1–3**) reduced the DNA cleavage observed in the presence of etoposide, and these compounds alone did not show any stabilization of cleavage complexes [48]. Cu-NQTS inhibited Top2α-mediated DNA relaxation assays and it showed comparable IC_{50} with etoposide in cytotoxicity [36].

Cu(Fp4alT)Cl completely inhibits Top2α without promoting the formation of linear DNA products [37]. Similar results were observed with the other Cu(TSC)Cl complexes in the study [37]. Thus, Cu(Fp4alT)Cl and its family of Cu(TSC)Cl complexes are catalytic inhibitors of Top2α rather than poisons of the enzyme [37].The IC_{50} of Top2α inhibition of Cu(Fp4alT)Cl is 0.3 μM and between 0.6 and 7.2 μM for the rest of the Cu–TSC complexes, while the IC_{50} of Top2α is between 50 and 90 μM for etoposide and 1 and 5 μM for doxorubicin [37]. The cell toxicity for Cu(Fp4alT)Cl is 0.8 μM for the SK-BR-3 cell line and 4.6 μM for MCF-7 cells [37]. The cytotoxicity data for other Cu–TSC complexes are between 0.4 and 12 μM[37]. Cu(L1)Cl, Cu(L2)Cl, Cu(L3)Cl, Cu(L4)Cl, and $Cu(L5)Cl_2$ all showed inhibition of Top2α from TopoGEN [47]. Cu(TSC) cation (Complex 1) increased DNA cleavage complexes and inhibited DNA relaxation [49]. It had better efficacy in inhibiting cell growth of the colorectal cancer cells when compared to etoposide [49]. In another study, the complexes [Cu(S,R)-L] and [Cu(R,S)-L] showed inhibition of Top2α relaxation at 300 μM [50]. However, the concentration of inhibition is similar to the ligand TSC and much higher (10–100+-fold) than other Cu(TSC)Cl [50].

Our collaboration worked on a series of Cu(TSC)Cl complexes that demonstrated their inhibition of both human Top2α and Top2β [32–34]. The structure–activity relationship of metal–TSCs showed that Cu(II) played a predominant role in the inhibition of Top2 [32–34]. The mechanism of Cu(TSC)Cl inhibition on Top2 is complicated. Cu(TSC)Cl inhibited ATP hydrolysis and plasmid DNA relaxation by Top2α and Top2β, which is consistent with these compounds acting as catalytic inhibitors. However, unlike other catalytic inhibitors, Cu(TSC)Cl complexes stabilize the DNA cleavage complexes and increase levels of DNA cleavage, which is the characteristic of interfacial poisons [32–34]. In addition, the complexes we tested led to higher levels of double-stranded breaks implying an increase in coordination between the two active sites [32]. The increase in DNA cleavage was not seen in a mutant lacking the ATPase domain [32]. Further, incubation of Cu(TSC)Cl complexes with Top2α or Top2β prior to DNA leads to a progressive inactivation of the enzyme [32,34]. Consistent with this data is the observation that Cu(TSC)Cl stabilizes a closed N-terminal region (ATPase domain) of Top2α or Top2β [34]. The significance of this particular aspect is that the ATPase domains of each half of the homodimer close around DNA in the presence of ATP. Our results demonstrate that the Cu(TSC)Cl complexes that were studied were able to induce closure of this N-terminal gate in a way that stabilized the gate, similar to what is seen with a non-hydrolyzable ATP analog (AMP-PNP) [34].

Although Top2α has been widely used as the molecular target to study Cu(TSC)Cl inhibition, our research found that Cu(TSC)Cl complexes inhibited ATPase and relaxation activity of both Top2α and Top2β [34]. Taken together, the data support the idea that these Cu(TSC)Cl complexes act on or near to the ATPase domain, which is highly similar between both isoforms. Using N-terminally and C-terminally truncated versions of Top2α or Top2β, both resulted in a lack of increased DNA cleavage [32,34]. Interestingly, some Cu(TSC)Cl showed inhibition both of Top1 and Top2, as will be discussed below [45].

Other metal–TSCs also showed inhibition of Top2. Pd-NQTS had an IC_{50} of 13 μM for MCF-7 and inhibited a Top2α-mediated DNA relaxation assay [36]. Cu-NQST is four times more efficient in cytotoxicity compared with Pd-NQST. When chelated with the same ligand, Pd(TSC)Cl seemed to be less active compared with its copper counterpart [33]. Ru(TSC)Cl {[(η-6-p-cymene)Ru(EtATSC)Cl]+ cation}, with a big substrate ring structure, inhibited human Top2α in a relaxation assay [39]. The ruthenium complex of TSC has been tested in a Top2α-mediated DNA relaxation assay and found to inhibit relaxation. The cell toxicity results showed that Ru–TSC complexes showed less or sometimes comparable anti-proliferation activities compared with cisplatin and etoposide [39]. Sn(II)-chelated TSC complexes (C5) inhibited Top2 at 20 μM [40].

Table 3. Metal–TSC inhibition of Top2.

Name	Inhibition of Top2	Reference
Nine compounds and their copper complexes	Inhibit human Top2α	[45]
Cobalt (III) chelated with TSC ligand Complex 4	Inhibits human Topo2α-induced DNA relaxation	[41]
Ni–bis(TSC) Complex 1	Does not inhibit isolated Top2 from L1210 cells at 100 μM	[38]
Cu–TSCs (Compound **1** and **2**)	Inhibits isolated Top2 from L1210 cells	[12]
Copper TSC	Inhibits Top2 of L1210 cells with IC_{50} value of 6.25–12.2 μM. Antagonizes the DNA break affect by etoposide.	[48]
Ni-NQTS Cu-NQTS Pd-NQTS	Inhibits DNA relaxation Inhibits DNA relaxation (TOPOGEN kit)	[36]
Cu(Fp4alT)Cl and its family of Cu(TSC)Cl	Inhibits relaxation by Top2 at 10 μM Completely inhibits Top2α without promoting the formation of linear DNA products	[37]
Ni–bis(TSC)	No effect in stabilizing DNA breaks	[51]
Cu(TSC)	Stabilizes DNA breaks	[51]
Ni(L1)(H1)Cl, $Ni(HL2)_2Cl_2$, $Ni(L3)_2$,$Ni(L4)_2$ $Ni(L5)_2Cl_2$	No inhibition of Top2 (TopoGEN)	[47]

Table 3. *Cont.*

Name	Inhibition of Top2	Reference
Cu(L1)Cl, Cu(L2)Cl, Cu(L3)Cl, Cu(L4)Cl, Cu(L5)Cl_2	Inhibit Top2 (TopoGEN)	[47]
Complex 1 (CuTSC cation)	Inhibits DNA relaxation and increase DNA cleavage	[49]
Sn(II)–TSC—(C5)	Inhibits Topo2α at 20 μM	[40]
Cu(S,R)L and Cu(R,S)L	Inhibit Top2α relaxation at 300 μM	[50]
[Cu(APY)Cl] and [Cu(APZ)Cl]	Inhibit Top2α from 0.5 μM Increase DNA cleavage Inhibit Top2α ATP hydrolysis No inhibition of ligation by Top2α Pre-incubating compounds with Top2α inactivated the enzyme	[32]
[Cu(APY)Cl] and [Cu(BZP)Cl]	Inhibit Top2β at 5 μM Increase DNA cleavage by Top2β Inhibit Top2β ATP hydrolysis Inhibit ligation by Top2β Pre-incubating compounds with Top2β inactivate the enzyme Stabilized closure of N-terminal Top2α and Top2β clamp	[34]
Cu(BZP)Cl series and Cu(ATZ)Cl series	Inhibit Top2α relaxation and increase DNA cleavage	[33]
Pd(BZP)Cl series	Inhibit Top2α relaxation and increase DNA cleavage	[33]

2.2.3. Inhibition of Type I and Type II Top

A few studies have compared the inhibition on Top1B and Top2α. Nine compounds and their copper complexes were investigated against human Top1B and Top2α from TopoGEN [45]. Relaxation assays were quantified to generate an IC_{50}. The Cu–TSC complexes were at least 10-fold more effective than the ligand alone [45]. They displayed greater inhibition of Top2α than Top1B [45]. Interestingly, the larger side chain substitutions generally displayed better inhibition of Top2α [45].

Complex 4 of Co(III)–TSC complexes inhibited Top1B-induced and Top2α-induced DNA relaxation, but neither the ligand nor its precursor was able to inhibit either enzyme [41]. Complex 4 did not cause a significant increase in DNA complexes with Top1B or Top2α, which suggests that Complex 4 is a catalytic inhibitor not a poison [41].

3. Discussion

TSCs are a broad group of compounds and thus many diverse TSCs have been synthesized and examined. Some of the TSC ligands have been tested in clinical trials such as Triapine and Dp44mT. Although metal–TSCs showed promising results in cytotoxicity and Top inhibition, none of these have been advanced to clinical studies. While TSCs display a broad range of possible mechanisms of action, direct studies on purified Top enzymes have been very helpful in identifying the mechanism of inhibition (Figure 4). Based upon the evidence in the literature, both Type IB (Top1B) and Type IIA (Top2) are affected by various TSCs. This review could not find any evidence of studies with Type IA (Top3) enzymes and TSCs. While Top1B is targeted by some TSCs, far fewer studies have examined Top1 than Top2 enzymes. Far more types of TSCs have been tested against Top2 than Top1B. While the main impact on Top1B appears to be inhibition of relaxation, there are a variety of impacts on Top2 depending on the specific TSC. This is likely due to the more complex reaction mechanism of Top2.

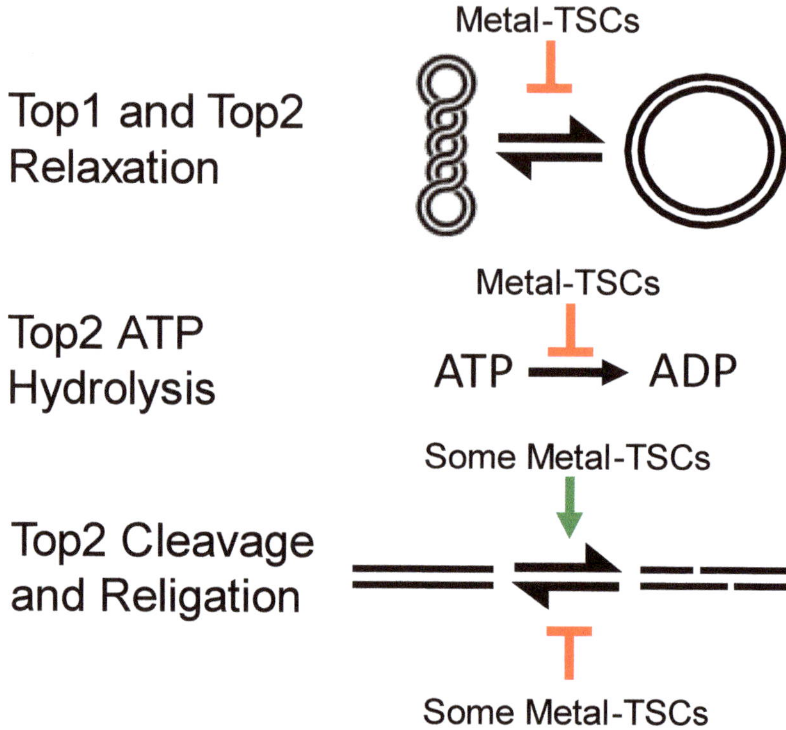

Figure 4. Mechanisms of TSCs against Top1B (Type IB) and Top2α/β (Type IIA). Metal–TSCs appear to inhibit plasmid DNA relaxation and ATP hydrolysis. Some metal TSCs have been shown to increase plasmid DNA cleavage and/or inhibit religation.

Cu–TSCs have been studied extensively against Top2α. The mechanism showed that Cu–TSCs are catalytic inhibitors that also exhibit some features of interfacial poisons (namely the increase in DNA cleavage levels). Some metal–TSCs inhibited DNA religation by Top1B, but religation of DNA by Top2 is not consistently inhibited by Cu–TSCs. The cleavage complexes created by Top2 are stabilized by the metal–TSC complexes may or may not involve inhibition of religation [32–34]. Further, Top2 enzymes generally are more impacted by metal–TSC compounds rather than the ligands alone. There remains a lack of a clear understanding of exactly where and how these compounds are acting. For example, biochemical evidence supports that these compounds can inhibit ATP hydrolysis, but some also increase DNA cleavage. These two mechanisms seem to contradict one another since the general thought is that catalytic inhibitors affect ATP hydrolysis but do not disturb DNA cleavage. Interestingly, some of these compounds appear to stabilize the N-terminal clamp of Top2α and Top2β, which may help clarify the increase in DNA cleavage [34]. For instance, 1,4-benzoquinone is known to increase DNA cleavage and stabilize the N-terminal clamp, likely through a covalent adduction mechanism [34,52–54]. Several computational studies indicated that TSC complexes can bind in or around the ATPase domain, which is similar to the mechanism observed with 1,4-benzoquinone. Again, these data are consistent with the observation that N-terminally and C-terminally truncated mutants of Top2α or Top2β are not affected by metal–TSC complexes [32,34]. Together, these data suggest that there may be an increase in coordination between the two active sites of Top2 when metal–TSCs are present, which could result in increased DNA cleavage without true poisoning.

Although there are many molecular modeling studies to predict where TSC binds on Tops, currently no NMR or crystal structure information is available on the exact location of where TSC or metal–TSC binds on Tops. Further structural studies are urgently needed to elucidate the molecular structural information of metal–TSC inhibition on Tops. The mode of interaction of the Cu(TSC)Cl complexes with Tops may be inferred from the literature of TSCs interacting with other molecules. It has been well established that Cu(TSC)+ complexes have been easily synthesized and crystal structures have been produced solved that demonstrate that the complexes can pick up water as a ligand and form five-coordinate complexes [44,45]. Many of the Cu(TSC)Cl complexes are often found in the solid state as dimers, $[Cu(TSC)Cl]_2$, which shows a weak bond, dissociable in solution, at the Cu center making it five-coordinate. This structure then forms the square planar four-coordinate complex Cu(TSC)Cl in solution. Also, Cu(TSC)Cl complexes can dissociate the chloride ion (Cl^-) in aqueous solution to replace Cl^- with other ligands and water [46]. These observations indicate that Cu(TSC)Cl complexes can shed Cl- and bond directly with Tops. The observation that they can become five-coordinate indicates that the complex possibly forms two bonds with Tops. This may suggest why structurally similar square planar Pd(TSC)Cl and Pt(TSC)Cl complexes may lose the Cl- ligand to bind to Tops, but they cannot form five-coordinate structures, and thus, are not as potent inhibitors of Tops as the Cu(TSC)Cl complexes.

The ability of metal–TSCs to inhibit Top2α and Top2β suggests that these compounds may have therapeutic potential. However, there are additional considerations that must be addressed before these compounds can be used clinically. First, it is unclear whether inhibition of cell growth can be attributed to Top2 or to other possible mechanisms, known or unknown. Given that the ATPase domain of Top2 is similar to some other enzymes (GHKL ATPase/kinase superfamily), it is possible that metal–TSCs may impact other enzymes as well [55]. Second, the reactivity of these compounds must be explored to determine whether these compounds can covalently interact with proteins and the consequences of this must be considered. Previous data indicate that Top2 can be inactivated through incubation with Cu(TSC)Cl complexes, and the mechanism of this needs to be explored to determine whether this is a specific action against Top2 or could occur more generally leading to unforeseen collateral damage. Also, the specific metal ions that interact with these compounds tend to be d-block metals that can participate in redox reactions. Thus, the extent of that reactivity needs to be considered as the mechanism(s) of action are studied. Third, there is significant structure–activity relationship data available to focus on specific families and classes of TSCs for further development. Fourth, the delivery, bioavailability, and metabolism of these compounds will need to be established and could vary widely depending on the structure of the complex and the metal ion chosen. Much work toward this area has been conducted with the few compounds that have reached clinical trials, and these efforts should help guide the development of additional compounds. In spite of the challenges that lie ahead, these compounds represent a promising area of further research.

Author Contributions: Writing—original draft preparation, X.J.; writing—review and editing, J.E.D., E.C.L., L.A.F. and X.J.; visualization, L.A.F., J.E.D., X.J., W.C. and H.D. All authors have read and agreed to the published version of the manuscript.

Funding: This research received no external funding.

Institutional Review Board Statement: Not applicable.

Informed Consent Statement: Not applicable.

Data Availability Statement: Not applicable.

Conflicts of Interest: The authors declare no conflict of interest.

References

1. Pommier, Y.; Sun, Y.; Huang, S.N.; Nitiss, J.L. Roles of eukaryotic topoisomerases in transcription, replication and genomic stability. *Nat. Rev. Mol. Cell Biol.* **2016**, *17*, 703–721. [CrossRef] [PubMed]
2. Thomas, A.; Pommier, Y. Targeting Topoisomerase I in the Era of Precision Medicine. *Clin. Cancer Res.* **2019**, *25*, 6581–6589. [CrossRef] [PubMed]
3. McKie, S.J.; Desai, P.R.; Seol, Y.; Allen, A.M.; Maxwell, A.; Neuman, K.C. Topoisomerase VI is a chirally-selective, preferential DNA decatenase. *Elife* **2022**, *11*, e67021. [CrossRef]
4. Vann, K.R.; Oviatt, A.A.; Osheroff, N. Topoisomerase II Poisons: Converting Essential Enzymes into Molecular Scissors. *Biochemistry* **2021**, *60*, 1630–1641. [CrossRef] [PubMed]
5. Deweese, J.E.; Osheroff, N. The DNA cleavage reaction of topoisomerase II: Wolf in sheep's clothing. *Nucleic Acids Res.* **2009**, *37*, 738–749. [CrossRef]
6. Murphy, M.B.; Mercer, S.L.; Deweese, J.E. Inhibitors and Poisons of Mammalian Type II Topoisomerases. In *Advances in Molecular Toxicology*; Fishbein, J.C., Heilman, J., Eds.; Academic Press: Cambridge, MA, USA, 2017; Volume 11, pp. 203–240.
7. Gibson, E.G.; Deweese, J.E. Covalent poisons of topoisomerase II. *Curr. Top. Pharm.* **2013**, *17*, 1–12.
8. Moorthy, N.S.; Cerqueira, N.M.; Ramos, M.J.; Fernandes, P.A. Aryl- and heteroaryl-thiosemicarbazone derivatives and their metal complexes: A pharmacological template. *Recent Pat. Anticancer Drug Discov.* **2013**, *8*, 168–182. [CrossRef]
9. Bai, X.G.; Zheng, Y.; Qi, J. Advances in thiosemicarbazone metal complexes as anti-lung cancer agents. *Front. Pharmacol.* **2022**, *13*, 1018951. [CrossRef]
10. Matesanz, A.I.; Herrero, J.M.; Quiroga, A.G. Chemical and Biological Evaluation of Thiosemicarbazone-Bearing Heterocyclic Metal Complexes. *Curr. Top. Med. Chem.* **2021**, *21*, 59–72. [CrossRef]
11. Liberta, A.E.; West, D.X. Antifungal and antitumor activity of heterocyclic thiosemicarbazones and their metal complexes: Current status. *Biometals* **1992**, *5*, 121–126. [CrossRef]
12. Miller, M.C., 3rd; Stineman, C.N.; Vance, J.R.; West, D.X.; Hall, I.H. The cytotoxicity of copper(II) complexes of 2-acetyl-pyridyl-4N-substituted thiosemicarbazones. *Anticancer Res.* **1998**, *18*, 4131–4139. [PubMed]
13. Baruffini, E.; Ruotolo, R.; Bisceglie, F.; Montalbano, S.; Ottonello, S.; Pelosi, G.; Buschini, A.; Lodi, T. Mechanistic insights on the mode of action of an antiproliferative thiosemicarbazone-nickel complex revealed by an integrated chemogenomic profiling study. *Sci. Rep.* **2020**, *10*, 10524. [CrossRef] [PubMed]
14. Merlot, A.M.; Kalinowski, D.S.; Richardson, D.R. Novel chelators for cancer treatment: Where are we now? *Antioxid. Redox Signal.* **2013**, *18*, 973–1006. [CrossRef] [PubMed]
15. Whitnall, M.; Howard, J.; Ponka, P.; Richardson, D.R. A class of iron chelators with a wide spectrum of potent antitumor activity that overcomes resistance to chemotherapeutics. *Proc. Natl. Acad. Sci. USA* **2006**, *103*, 14901–14906. [CrossRef]
16. Jansson, P.J.; Kalinowski, D.S.; Lane, D.J.; Kovacevic, Z.; Seebacher, N.A.; Fouani, L.; Sahni, S.; Merlot, A.M.; Richardson, D.R. The renaissance of polypharmacology in the development of anti-cancer therapeutics: Inhibition of the "Triad of Death" in cancer by Di-2-pyridylketone thiosemicarbazones. *Pharmacol. Res.* **2015**, *100*, 255–260. [CrossRef]
17. Salim, K.Y.; Maleki Vareki, S.; Danter, W.R.; Koropatnick, J. COTI-2, a novel small molecule that is active against multiple human cancer cell lines in vitro and in vivo. *Oncotarget* **2016**, *7*, 41363–41379. [CrossRef]
18. Lindemann, A.; Patel, A.A.; Silver, N.L.; Tang, L.; Liu, Z.; Wang, L.; Tanaka, N.; Rao, X.; Takahashi, H.; Maduka, N.K.; et al. COTI-2, A Novel Thiosemicarbazone Derivative, Exhibits Antitumor Activity in HNSCC through p53-dependent and -independent Mechanisms. *Clin. Cancer Res.* **2019**, *25*, 5650–5662. [CrossRef]
19. McKenzie-Nickson, S.; Bush, A.I.; Barnham, K.J. Bis(thiosemicarbazone) Metal Complexes as Therapeutics for Neurodegenerative Diseases. *Curr. Top. Med. Chem.* **2016**, *16*, 3058–3068. [CrossRef]
20. Karp, J.E.; Giles, F.J.; Gojo, I.; Morris, L.; Greer, J.; Johnson, B.; Thein, M.; Sznol, M.; Low, J. A phase I study of the novel ribonucleotide reductase inhibitor 3-aminopyridine-2-carboxaldehyde thiosemicarbazone (3-AP, Triapine) in combination with the nucleoside analog fludarabine for patients with refractory acute leukemias and aggressive myeloproliferative disorders. *Leuk. Res.* **2008**, *32*, 71–77. [CrossRef]
21. Ma, B.; Goh, B.C.; Tan, E.H.; Lam, K.C.; Soo, R.; Leong, S.S.; Wang, L.Z.; Mo, F.; Chan, A.T.; Zee, B.; et al. A multicenter phase II trial of 3-aminopyridine-2-carboxaldehyde thiosemicarbazone (3-AP, Triapine) and gemcitabine in advanced non-small-cell lung cancer with pharmacokinetic evaluation using peripheral blood mononuclear cells. *Investig. New Drugs* **2008**, *26*, 169–173. [CrossRef]
22. Nutting, C.M.; van Herpen, C.M.; Miah, A.B.; Bhide, S.A.; Machiels, J.P.; Buter, J.; Kelly, C.; de Raucourt, D.; Harrington, K.J. Phase II study of 3-AP Triapine in patients with recurrent or metastatic head and neck squamous cell carcinoma. *Ann. Oncol.* **2009**, *20*, 1275–1279. [CrossRef]
23. Finch, R.A.; Liu, M.C.; Cory, A.H.; Cory, J.G.; Sartorelli, A.C. Triapine (3-aminopyridine-2-carboxaldehyde thiosemicarbazone; 3-AP): An inhibitor of ribonucleotide reductase with antineoplastic activity. *Adv. Enzyme Regul.* **1999**, *39*, 3–12. [CrossRef] [PubMed]
24. Finch, R.A.; Liu, M.; Grill, S.P.; Rose, W.C.; Loomis, R.; Vasquez, K.M.; Cheng, Y.; Sartorelli, A.C. Triapine (3-aminopyridine-2-carboxaldehyde- thiosemicarbazone): A potent inhibitor of ribonucleotide reductase activity with broad spectrum antitumor activity. *Biochem. Pharmacol.* **2000**, *59*, 983–991. [CrossRef]

25. Posa, V.; Hajdu, B.; Toth, G.; Domotor, O.; Kowol, C.R.; Keppler, B.K.; Spengler, G.; Gyurcsik, B.; Enyedy, E.A. The coordination modes of (thio)semicarbazone copper(II) complexes strongly modulate the solution chemical properties and mechanism of anticancer activity. *J. Inorg. Biochem.* **2022**, *231*, 111786. [CrossRef] [PubMed]
26. Yalowich, J.C.; Wu, X.; Zhang, R.; Kanagasabai, R.; Hornbaker, M.; Hasinoff, B.B. The anticancer thiosemicarbazones Dp44mT and triapine lack inhibitory effects as catalytic inhibitors or poisons of DNA topoisomerase IIα. *Biochem. Pharmacol.* **2012**, *84*, 52–58. [CrossRef]
27. Huang, H.; Chen, Q.; Ku, X.; Meng, L.; Lin, L.; Wang, X.; Zhu, C.X.; Wang, Y.; Chen, Z.; Li, M.; et al. A series of α-heterocyclic carboxaldehyde thiosemicarbazones inhibit topoisomerase IIα catalytic activity. *J. Med. Chem.* **2010**, *53*, 3048–3064. [CrossRef] [PubMed]
28. Rao, V.A.; Klein, S.R.; Agama, K.K.; Toyoda, E.; Adachi, N.; Pommier, Y.; Shacter, E.B. The iron chelator Dp44mT causes DNA damage and selective inhibition of topoisomerase IIalpha in breast cancer cells. *Cancer Res.* **2009**, *69*, 948–957. [CrossRef]
29. de Oliveira, J.F.; Lima, T.S.; Vendramini-Costa, D.B.; de Lacerda Pedrosa, S.C.B.; Lafayette, E.A.; da Silva, R.M.F.; de Almeida, S.M.V.; de Moura, R.O.; Ruiz, A.; de Carvalho, J.E.; et al. Thiosemicarbazones and 4-thiazolidinones indole-based derivatives: Synthesis, evaluation of antiproliferative activity, cell death mechanisms and topoisomerase inhibition assay. *Eur. J. Med. Chem.* **2017**, *136*, 305–314. [CrossRef]
30. da Silva Filho, F.A.; de Freitas Souza, T.; Ribeiro, A.G.; Alves, J.E.F.; de Oliveira, J.F.; de Lima Souza, T.R.C.; de Moura, R.O.; do Carmo Alves de Lima, M.; de Carvalho Junior, L.B.; de Almeida, S.M.V. Topoisomerase inhibition and albumin interaction studies of acridine-thiosemicarbazone derivatives. *Int. J. Biol. Macromol.* **2019**, *138*, 582–589. [CrossRef]
31. Samia, L.B.; Parrilha, G.L.; Da Silva, J.G.; Ramos, J.P.; Souza-Fagundes, E.M.; Castelli, S.; Vutey, V.; Desideri, A.; Beraldo, H. Metal complexes of 3-(4-bromophenyl)-1-pyridin-2-ylprop-2-en-1-one thiosemicarbazone: Cytotoxic activity and investigation on the mode of action of the gold(III) complex. *Biometals* **2016**, *29*, 515–526. [CrossRef]
32. Wilson, J.T.; Jiang, X.; McGill, B.C.; Lisic, E.C.; Deweese, J.E. Examination of the Impact of Copper(II) alpha-(N)-Heterocyclic Thiosemicarbazone Complexes on DNA Topoisomerase IIalpha. *Chem. Res. Toxicol.* **2016**, *29*, 649–658. [CrossRef]
33. Morris, W.H.; Ngo, L.; Wilson, J.T.; Medawala, W.; Brown, A.R.; Conner, J.D.; Fabunmi, F.; Cashman, D.J.; Lisic, E.C.; Yu, T.; et al. Structural and Metal Ion Effects on Human Topoisomerase IIalpha Inhibition by alpha-(N)-Heterocyclic Thiosemicarbazones. *Chem. Res. Toxicol.* **2019**, *32*, 90–99. [CrossRef] [PubMed]
34. Keck, J.M.; Conner, J.D.; Wilson, J.T.; Jiang, X.; Lisic, E.C.; Deweese, J.E. Clarifying the Mechanism of Copper(II) alpha-(N)-Heterocyclic Thiosemicarbazone Complexes on DNA Topoisomerase IIalpha and IIbeta. *Chem. Res. Toxicol.* **2019**, *32*, 2135–2143. [CrossRef] [PubMed]
35. Vutey, V.; Castelli, S.; D'Annessa, I.; Samia, L.B.; Souza-Fagundes, E.M.; Beraldo, H.; Desideri, A. Human topoisomerase IB is a target of a thiosemicarbazone copper(II) complex. *Arch. Biochem. Biophys.* **2016**, *606*, 34–40. [CrossRef] [PubMed]
36. Chen, J.; Huang, Y.; Liu, G.; Afrasiabi, Z.; Sinn, E.; Padhye, S.; Ma, Y. The cytotoxicity and mechanisms of 1,2-naphthoquinone thiosemicarbazone and its metal derivatives against MCF-7 human breast cancer cells. *Tox. App. Pharm.* **2004**, *197*, 40–48. [CrossRef]
37. Zeglis, B.M.; Divilov, V.; Lewis, J.S. Role of metalation in the topoisomerase IIα inhibition and antiproliferation activity of a series of α-heterocyclic-N4-substituted thiosemicarbazones and their Cu(II) complexes. *J. Med. Chem.* **2011**, *54*, 2391–2398. [CrossRef]
38. Hall, I.H.; Miller, M.C.; West, D.X. Antineoplastic and Cytotoxic Activities of Nickel(II) Complexes of Thiosemicarbazones. *Met. Based Drugs* **1997**, *4*, 89–95. [CrossRef]
39. Beckford, F.; Thessing, J.; Woods, J.; Didion, J.; Gerasimchuk, N.; Gonzalez-Sarrias, A.; Seeram, N.P. Synthesis and structure of [(eta(6)-p-cymene)Ru(2-anthracen-9-ylmethylene-N-ethylhydrazinecarbothioamide)Cl]Cl; biological evaluation, topoisomerase II inhibition and reaction with DNA and human serum albumin. *Metallomics* **2011**, *3*, 491–502. [CrossRef]
40. Wu, J.; Yang, T.; Wang, X.; Li, W.; Pang, M.; Sun, H.; Liang, H.; Yang, F. Development of a multi-target anticancer Sn(ii) pyridine-2-carboxaldehyde thiosemicarbazone complex. *Dalton Trans.* **2021**, *50*, 10909–10921. [CrossRef] [PubMed]
41. Beebe, S.J.; Celestine, M.J.; Bullock, J.L.; Sandhaus, S.; Arca, J.F.; Cropek, D.M.; Ludvig, T.A.; Foster, S.R.; Clark, J.S.; Beckford, F.A.; et al. Synthesis, characterization, DNA binding, topoisomerase inhibition, and apoptosis induction studies of a novel cobalt(III) complex with a thiosemicarbazone ligand. *J. Inorg. Biochem.* **2020**, *203*, 110907. [CrossRef] [PubMed]
42. Qi, J.; Zheng, Y.; Qian, K.; Tian, L.; Zhang, G.X.; Cheng, Z.; Wang, Y. Synthesis, crystal structure and antiproliferative mechanisms of 2-acetylpyridine-thiosemicarbazones Ga(III) with a greater selectivity against tumor cells. *J. Inorg. Biochem.* **2017**, *177*, 110–117. [CrossRef]
43. Oliveira, C.G.; Romero-Canelon, I.; Silva, M.M.; Coverdale, J.P.C.; Maia, P.I.S.; Batista, A.A.; Castelli, S.; Desideri, A.; Sadler, P.J.; Deflon, V.M. Palladium(ii) complexes with thiosemicarbazones derived from pyrene as topoisomerase IB inhibitors. *Dalton Trans.* **2019**, *48*, 16509–16517. [CrossRef] [PubMed]
44. Heng, M.P.; Sinniah, S.K.; Teoh, W.Y.; Sim, K.S.; Ng, S.W.; Cheah, Y.K.; Tan, K.W. Synthesis of a DNA-targeting nickel (II) complex with testosterone thiosemicarbazone which exhibits selective cytotoxicity towards human prostate cancer cells (LNCaP). *Spectrochim. Acta A Mol. Biomol. Spectrosc.* **2015**, *150*, 360–372. [CrossRef] [PubMed]
45. Qi, J.; Zheng, Y.; Li, B.; Ai, Y.; Chen, M.; Zheng, X. Pyridoxal hydrochloride thiosemicarbazones with copper ions inhibit cell division via Topo-I and Topo-IIa. *J. Inorg. Biochem.* **2022**, *232*, 111816. [CrossRef] [PubMed]
46. Teicher, B.A. Next generation topoisomerase I inhibitors: Rationale and biomarker strategies. *Biochem. Pharmacol.* **2008**, *75*, 1262–1271. [CrossRef]

47. Bisceglie, F.; Musiari, A.; Pinelli, S.; Alinovi, R.; Menozzi, I.; Polverini, E.; Tarasconi, P.; Tavone, M.; Pelosi, G. Quinoline-2-carboxaldehyde thiosemicarbazones and their Cu(II) and Ni(II) complexes as topoisomerase IIα inhibitors. *J. Inorg. Biochem.* **2015**, *152*, 10–19. [CrossRef]
48. Miller III, M.C.; Stineman, C.N.; Vance, J.R.; West, D.X.; Hall, I.H. Multiple mechanisms for cytotoxicity induced by copper(II) complexes of 2-acetylpyrazine-N-substituted thiosemicarbazones. *Appl. Organomet. Chem.* **1999**, *13*, 9–19. [CrossRef]
49. Sandhaus, S.; Taylor, R.; Edwards, T.; Huddleston, A.; Wooten, Y.; Venkatraman, R.; Weber, R.T.; Gonzalez-Sarrias, A.; Martin, P.M.; Cagle, P.; et al. A novel copper(II) complex identified as a potent drug against colorectal and breast cancer cells and as a poison inhibitor for human topoisomerase IIalpha. *Inorg. Chem. Commun.* **2016**, *64*, 45–49. [CrossRef]
50. Bacher, F.; Enyedy, E.; Nagy, N.V.; Rockenbauer, A.; Bognar, G.M.; Trondl, R.; Novak, M.S.; Klapproth, E.; Kiss, T.; Arion, V.B. Copper(II) complexes with highly water-soluble L- and D-proline-thiosemicarbazone conjugates as potential inhibitors of Topoisomerase IIalpha. *Inorg. Chem.* **2013**, *52*, 8895–8908. [CrossRef]
51. Bisceglie, F.; Pinelli, S.; Alinovi, R.; Goldoni, M.; Mutti, A.; Camerini, A.; Piola, L.; Tarasconi, P.; Pelosi, G. Cinnamaldehyde and cuminaldehyde thiosemicarbazones and their copper(II) and nickel(II) complexes: A study to understand their biological activity. *J. Inorg. Biochem.* **2014**, *140*, 111–125. [CrossRef]
52. Lindsey, R.H.; Bender, R.P.; Osheroff, N. Stimulation of topoisomerase II-mediated DNA cleavage by benzene metabolites. *Chem. Biol. Interact.* **2005**, *153-154*, 197–205. [CrossRef] [PubMed]
53. Lindsey, R.H., Jr.; Bender, R.P.; Osheroff, N. Effects of benzene metabolites on DNA cleavage mediated by human topoisomerase IIα: 1,4-hydroquinone is a topoisomerase II poison. *Chem. Res. Toxicol.* **2005**, *18*, 761–770. [CrossRef]
54. Lindsey, R.H., Jr.; Bromberg, K.D.; Felix, C.A.; Osheroff, N. 1,4-Benzoquinone is a topoisomerase II poison. *Biochemistry* **2004**, *43*, 7563–7574. [CrossRef] [PubMed]
55. Dutta, R.; Inouye, M. GHKL, an emergent ATPase/kinase superfamily. *Trends Biochem. Sci.* **2000**, *25*, 24–28. [CrossRef] [PubMed]

International Journal of *Molecular Sciences*

MDPI

Review

Telling Your Right Hand from Your Left: The Effects of DNA Supercoil Handedness on the Actions of Type II Topoisomerases

Jeffrey Y. Jian [1] and Neil Osheroff [1,2,*]

1 Department of Biochemistry, Vanderbilt University School of Medicine, Nashville, TN 37232, USA; jeffrey.y.jian@vanderbilt.edu
2 Department of Medicine (Hematology/Oncology), Vanderbilt University School of Medicine, Nashville, TN 37232, USA
* Correspondence: neil.osheroff@vanderbilt.edu

Abstract: Type II topoisomerases are essential enzymes that modulate the topological state of DNA supercoiling in all living organisms. These enzymes alter DNA topology by performing double-stranded passage reactions on over- or underwound DNA substrates. This strand passage reaction generates a transient covalent enzyme–cleaved DNA structure known as the cleavage complex. Al-though the cleavage complex is a requisite catalytic intermediate, it is also intrinsically dangerous to genomic stability in biological systems. The potential threat of type II topoisomerase function can also vary based on the nature of the supercoiled DNA substrate. During essential processes such as DNA replication and transcription, cleavage complex formation can be inherently more dangerous on overwound versus underwound DNA substrates. As such, it is important to understand the profound effects that DNA topology can have on the cellular functions of type II topoisomerases. This review will provide a broad assessment of how human and bacterial type II topoisomerases recognize and act on their substrates of various topological states.

Keywords: DNA topoisomerase; DNA topology; supercoil handedness; type II topoisomerase; DNA relaxation; DNA cleavage

Citation: Jian, J.Y.; Osheroff, N. Telling Your Right Hand from Your Left: The Effects of DNA Supercoil Handedness on the Actions of Type II Topoisomerases. *Int. J. Mol. Sci.* **2023**, *24*, 11199. https://doi.org/10.3390/ijms241311199

Academic Editor: Yuri Lyubchenko

Received: 16 June 2023
Revised: 5 July 2023
Accepted: 5 July 2023
Published: 7 July 2023

1. Introduction

DNA is often visualized as a ladder. However, the stacking of the nucleotide base pairs upon one another introduces a twist in the structure, converting the ladder into a double helix in which one DNA strand is wrapped around the other [1]. Because of the double-stranded nature of DNA and its extreme compaction into a crowded cellular environment, this plectonemic coiling leads to a number of topological problems in DNA [2–5].

As long as the ends of DNA are fixed in space, topological properties are defined as those that cannot be changed without breaking one or both strands of the double helix [2–6]. For practical purposes, the ends of cellular DNA can be considered to be fixed in space; they are anchored and unable to rotate freely [3,6]. This is due to the high frictional energy associated with the extreme length of chromosomes in humans, the circular nature of plasmids and chromosomal DNA in bacteria, and the tethering of DNA to chromosomal scaffolds in humans and membranes in bacteria [3,5]. Although the genetic information is organized in a linear array of nucleotide bases, DNA topology plays an important role in facilitating access to this information [2,3,6–8].

The topology of DNA is described by three concepts: twist, writhe, and linking number [2–5,9,10]. Twist is the total number of double helical turns in a given DNA segment and represents the torsional stress that is present in the double helix at any time. By convention, positive twist (right-handed twist) is present in the normal right-handed Watson–Crick DNA structure (Figure 1). Writhe is defined as the number of times the double helix crosses itself if the DNA segment is projected in two dimensions and represents axial stress in the molecule. The directionality of the double helical crossover (i.e., node or juxtaposition) is assigned a positive or negative value based on its orientation (i.e., handedness). Positive supercoils [(+)SC] form left-handed crossovers in the double helix, while negative supercoils [(−)SC] form right-handed crossovers.

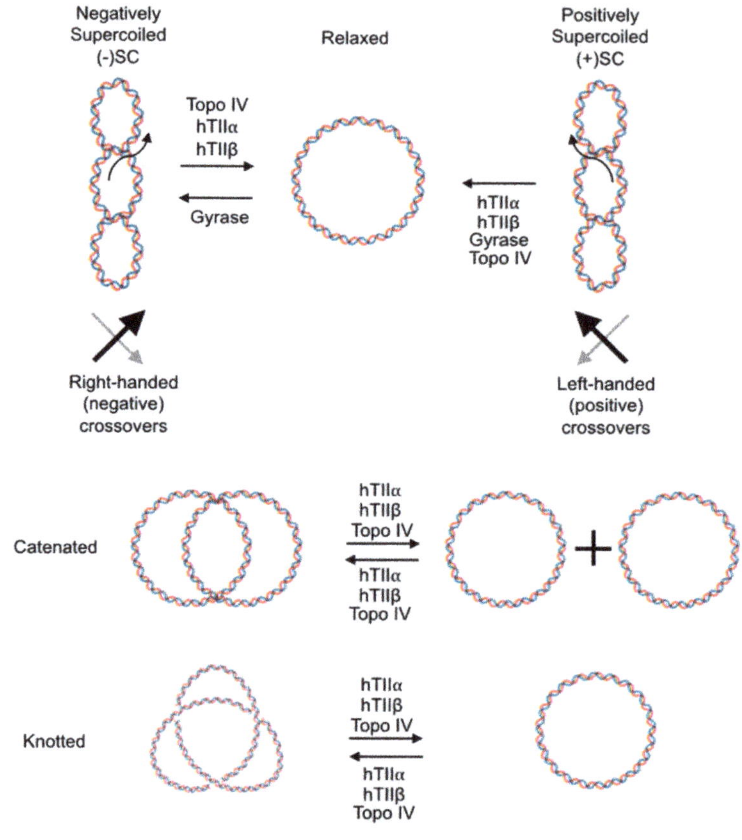

Figure 1. Topological states of DNA. DNA containing no torsional stress is considered "relaxed" (**top middle**). Underwinding or overwinding the DNA results in negatively supercoiled [(−)SC, **top left**] or positively supercoiled [(+)SC, **top right**] DNA. DNA supercoiling is depicted as writhe for visual clarity, but twist and writhe are interconvertible within these molecules. Intermolecular catenanes (**middle**) and intramolecular knots (**bottom**) can also form in DNA. In these cases, twist and writhe are not interconvertible. Type II enzymes (human topoisomerase IIα, hTIIα; human topoisomerase IIβ, hTIIβ; gyrase; topoisomerase IV, topo IV) that can perform each of the reactions to alter topological states are also listed. Created with BioRender.com.

Linking number represents the sum of twist and writhe. Assuming that the ends of DNA are "fixed" and the double helix is intact (i.e., unbroken), linking number is invariant. DNA duplexes that are not under torsional stress, such as that seen in the Watson–Crick structure, are denoted as "relaxed." In relaxed molecules, the two strands twist around the helical axis once every ~10.4 base pairs (Figure 1) [3,9,11]. DNA under- or overwinding induces torsional stress in the double helix. If this stress is unconstrained and allowed to freely distribute, it will be partially converted into axial stress (writhe) [2,3,5]. In this case, one portion of the DNA will form a "superhelical turn" around another portion of the molecule. Hence, DNA that is under torsional stress (either under- or overwound) is referred to as being "supercoiled".

DNA supercoiling is especially relevant to essential nucleic acid processes that require strand separation, such as replication and transcription [2,3,5,12–14]. In species ranging from bacteria to humans, the double helix is globally underwound (i.e., negatively supercoiled) ~6% [2,3,5]. This underwound nature of the genome reduces the energy required to separate complementary base pairs, facilitating the opening of the double helix to access the genetic information [15–18]. In contrast, once the movement of DNA tracking machinery begins, the deleterious effects of DNA topology manifest (Figure 2). Because helicases separate but do not unwind the two strands of DNA, they do not remove any turns of the double helix. Consequently, acutely overwound (i.e., positively supercoiled) DNA forms ahead of the tracking machinery, generating an increase in torsional stress that needs to be alleviated [19]. If unresolved, the accumulation of (+)SCs blocks replication and transcription, causing these processes to stall rapidly [17,18,20–22].

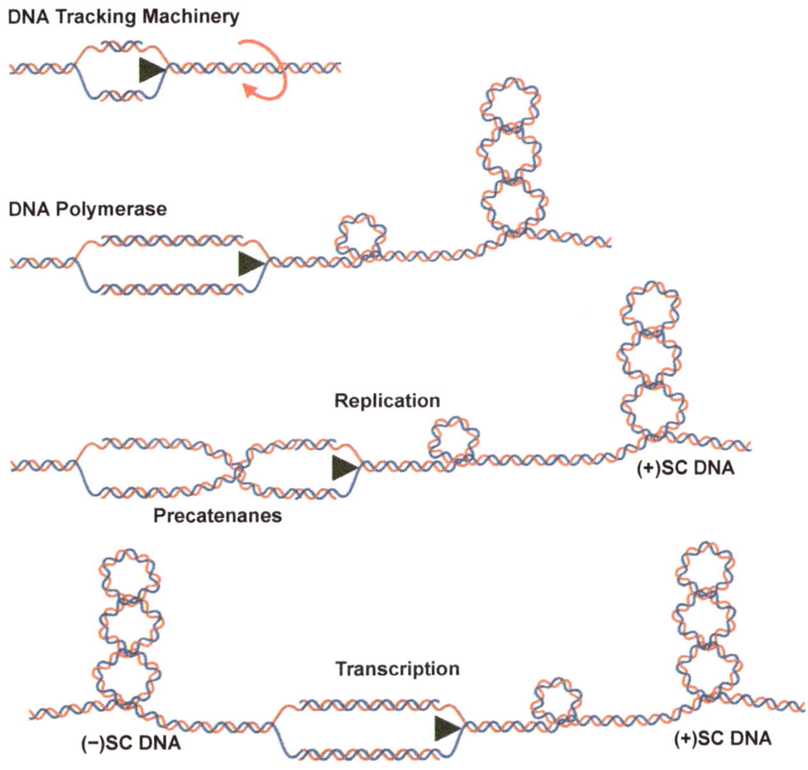

Figure 2. Movement of DNA tracking machinery causes topological problems as indicated by the overwinding shown by the circular red arrow. As DNA tracking systems move through the DNA, twists are pushed ahead of replication forks and transcription complexes, resulting in DNA overwinding and the formation of (+)SCs. In the case of replication, precatenanes also form behind the fork; during transcription, (−)SCs form behind the moving DNA tracking machinery. Created with BioRender.com.

Additional topological structures, such as tangles and knots, result from replication and recombination, respectively [5–7]. Tangles (i.e., catenanes) can form between distal segments of the same DNA molecule or separate molecules during processes such as replication [2,3,5]. Catenanes must be removed to allow for the proper separation of sister chromatids during mitosis [23–28]. Due to the long length of DNA, nucleic acid knots can form within a single DNA molecule during processes such as recombination. DNA knots prevent the separation of the two strands of DNA [2,3,5,7,13].

2. Topoisomerases

In order to maintain appropriate levels of DNA supercoiling and remove knots and tangles from the genome, cells encode enzymes known as topoisomerases [3–6,29,30]. These enzymes are ubiquitous to all domains of life and are necessary for cellular survival. All topoisomerases modulate the topological state of the genome through the creation of transient breaks in the DNA sugar–phosphate backbone. Broadly, there are two classes of topoisomerases, and they are both defined by the number of DNA strands they cleave per enzyme reaction cycle [2,3,5,6,30]. Type I topoisomerases generate a single-stranded break, or "nick", in the double helix [3,5,30]. In contrast, type II topoisomerases create a double-stranded break in the genetic material [2,3,5,6,29–31]. This review will focus on type II topoisomerases.

3. Type II Topoisomerases

There are two subclasses of type II topoisomerases: type IIA and type IIB. To date, functional type IIB enzymes have only been identified in plants and archaea and will not be discussed further [30,32,33].

The first type IIA enzyme, bacterial DNA gyrase, was discovered in 1976 [34]. Bacterial topoisomerase IV was later identified in 1990 [35]. Most bacterial species encode gyrase and topoisomerase IV [30,36]. However, a few species, such as *Mycobacterium tuberculosis*, encode only a single type II topoisomerase, gyrase, which can presumably perform the cellular functions of both type II enzymes [4,6,29,37,38].

The first eukaryotic type II enzyme was identified in *Drosophila* in 1980 [39]. *Drosophila* and other invertebrates, as well as lower eukaryotes, such as yeast, encode only one type II enzyme, topoisomerase II. In contrast, vertebrates, such as humans, express two forms of the type II enzyme: topoisomerase IIα and topoisomerase IIβ [3,5–7,30,31,40]. Human topoisomerase IIα and topoisomerase IIβ were identified in 1988 [41] and 1989 [42,43], respectively.

4. Type II Topoisomerase Domain Structures

Bacterial type II topoisomerases are heterotetrameric in structure (A_2B_2; Figure 3). The founding type II enzyme, gyrase, is comprised of two distinct subunits: GyrA and GyrB. Like gyrase, topoisomerase IV is a heterotetramer that is composed of two separate subunits: ParC and ParE (which are homologous to GyrA and GyrB, respectively) in Gram-negative species and the corresponding GrlA and GrlB subunits in Gram-positive species (Figure 3) [35,44].

Eukaryotic type II topoisomerases are homologous to the bacterial type II enzymes [3,5,6,29,30,45]. However, the two bacterial subunits have fused into a single polypeptide in the eukaryotic type II topoisomerases (Figure 3) [3,5,6,29,30,45].

All known type II topoisomerases share several common structural features across three regions. Using DNA gyrase as the model, the N-terminus is located in GyrB, the catalytic core spans portions of GyrB and GyrA, and the C-terminus is located in GyrA (Figure 3) [3,5,6,29,30,45].

The N-terminal region contains the N-gate, where the DNA enters the enzyme. This portion of the molecule includes the ATPase active site, also known as the GHKL (DNA gyrase, Hsp90, bacterial CheA-family histidine kinases, and MutL; Figure 3, blue) domain. The GHKL domain contains an ATP-binding region that is formed from an eight-stranded antiparallel beta sheet surrounded by alpha helices [46,47]. The N-terminal region also contains the transducer domain (Figure 3, green), which relays ATP binding/hydrolysis information to the catalytic core [48,49]. The binding of ATP induces the dimerization of the N-terminal region, which shifts the N-gate into a closed conformation. The bound ATP interacts with a lysine residue in the transducer domain and subsequently facilitates rotation between the GHKL and transducer domains [32,47].

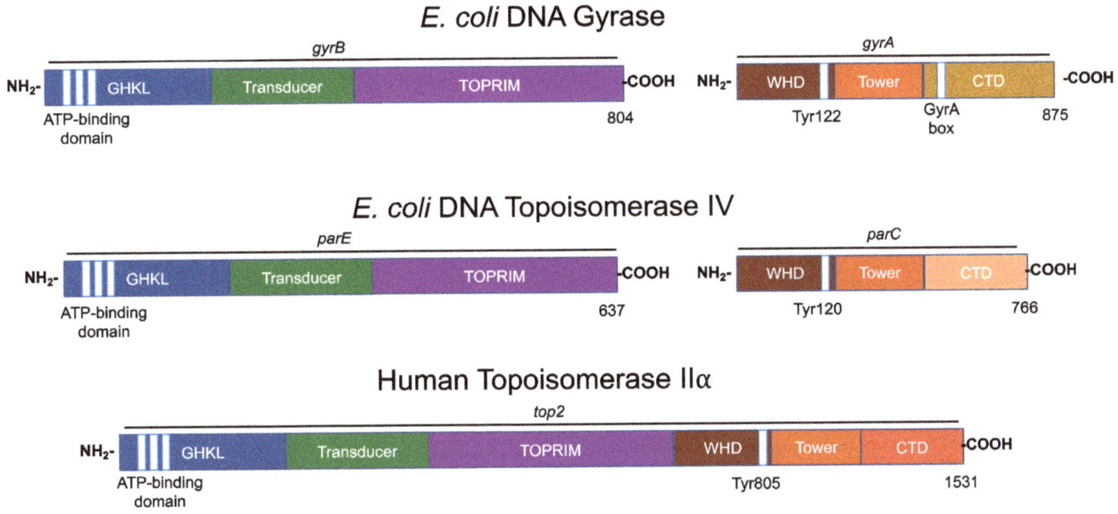

Figure 3. Domain structures of type II topoisomerases. The domain structures of three type II topoisomerases, bacterial (*Escherichia coli*) gyrase and topoisomerase IV, and human topoisomerase IIα are shown. Regions of homology among the enzymes are indicated by colors. The N-terminal (i.e., GyrB) homology domains contain the regions responsible for ATP binding and hydrolysis (GHKL, blue). The vertical white stripes represent the three conserved motifs that define the ATP-binding domain. The N-terminal domain also contains the binding site for divalent metal ions (TOPRIM, purple). The central (i.e., GyrA) region (WHD, brown) contains the active site tyrosyl residue that forms the covalent bond with DNA during scission. For bacterial gyrase, the variable C-terminal domain (gyrase, gold; topoisomerase IV, pink) contains the "GyrA box" that is necessary for the wrapping mechanism. For human topoisomerase IIα, the C-terminal homology domain (CTD, red) contains nuclear localization sequences (NLS) and phosphorylation sites (PO_4). The active site tyrosine residue is indicated for each enzyme.

The catalytic core contains the topoisomerase/primase (TOPRIM; Figure 3, purple) domain, which coordinates the active site divalent cations, the winged-helix domain (WHD; Figure 3, brown), which contains the active site tyrosine residue, and the tower domain (Figure 3; orange), which maintains polar and electrostatic interactions with the DNA substrate [48,50,51].

The TOPRIM domain is necessary for the transesterification reaction between the scissile phosphate of the DNA backbone and active site tyrosine residue [47,50]. The active site divalent cation is held by an aspartate-any residue-aspartate (DxD) motif and a glutamate residue that can act as a general acid–base moiety [52,53]. The DxD motif and its coordinate divalent cation in the TOPRIM domain, along with the active site tyrosine of the WHD, enable the formation of the two transient cuts of the DNA backbone via a non-canonical two-metal ion mechanism [6,51,54,55].

The WHD is able to bind DNA and also contains the active site tyrosine residue, which is responsible for the nucleophilic attack on the scissile phosphate of the DNA double helical backbone and the formation of the transient topoisomerase–DNA covalent bond [45,48].

The tower domain functions in DNA bending. This domain contains a beta sheet that can interact with one of the captured DNA double helices (the gate or G-segment, to be discussed later), bending the DNA segment to promote cleavage [56–58]. The presence of a conserved, invariant isoleucine residue has been found to intercalate between two base pairs of the G-segment, inducing a ~150° bend [56,59].The deletion or mutation of this isoleucine interferes with proper DNA bending, the subsequent cleavage, and the relaxation of supercoiled DNA [56,59].

The sequence of the C-terminal domain varies considerably between species, but it is characterized by the presence of charged amino acid residues [30,60]. In gyrase, this region contains a seven-amino acid motif known as the GyrA box (Figure 3, gold) [29,36,47,61,62]. The GyrA box is located within a six-blade beta pinwheel in the C-terminal domain, and it uniquely allows for the wrapping of the DNA substrate to introduce (−)SCs [29,36,47,61–63].

In comparison to gyrase, the C-terminal domain of topoisomerase IV does not contain the structure necessary to wrap and supercoil DNA (Figure 3, pink). Rather, topoisomerase IV contains a "broken" five- (not six) blade beta pinwheel and lacks a GyrA box [46,64–66]. Remnants of the canonical GyrA motif have been found in each of its pinwheel "blades" [64,65]. Nonetheless, the C-terminal domain of topoisomerase IV contains positively charged moieties on its outer surface, suggesting a role in binding DNA [46].

The C-termini of eukaryotic type II topoisomerases also contain the remnants of highly charged pinwheel blades but are inherently disordered in the absence of DNA (Figure 3, red) [67,68]. This portion of the eukaryotic enzyme also contains nuclear localization sequences and sites for posttranslational modifications such as phosphorylation and SUMOylation [5,45,47]. For the IIα isoform, these modifications enable the enzyme to be concentrated at centromeres during mitosis [67,69].

5. Catalytic Cycle of Type II Topoisomerases

All type II topoisomerases undergo similar catalytic cycles. These enzymes function by forming a transient double-stranded DNA break and modulate the topological state of DNA by a double-stranded passage reaction (Figure 4) [3,5,6,29–31,47]. The enzyme begins its catalytic cycle by capturing a segment of intact DNA through the opened N-terminal region (N-gate, gray) of the enzyme (Step 1). This first segment will be cut by the enzyme and is known as the "gate" or G-segment. The segment that is captured second and eventually transported through the transiently cleaved G-segment is known as the "transport" or T-segment. In the presence of two divalent cations, such as Mg^{2+}, and in coordination with the TOPRIM domain, the G-segment is assessed for bendability (Step 2) [57]. DNA sequences that can be bent are distorted to an angle of ~150° and can be used as the site for scission [56–59].

Both strands of the bent G-segment are then cleaved via a nucleophilic attack by the two active site tyrosine residues on the phosphate backbone of the double helix (Step 3). DNA cleavage is initiated when a general base, which is believed to be a conserved histidine residue, deprotonates the hydroxyl group of the active site tyrosine, allowing the oxyanion to attack the scissile phosphate. Two divalent cation molecules, such as magnesium (i.e., Mg^{2+}), are necessary for this nucleophilic attack [5,6,29,30,70]. Type II topoisomerases use a non-canonical two-metal ion mechanism [51,70]. The presence of one divalent cation enables interaction with the bridging 5′-oxygen molecule of the scissile bond and speeds up rates of enzyme-mediated cleavage at the first cut site. Once the first DNA strand is cut, the second strand is cleaved ~20-fold faster [71]. The resulting enzyme-cleaved DNA complex is a transient structure that has the enzyme covalently bound to the scissile 5′-phosphate of the double helical backbone.

To maintain the bond energy of the sugar–phosphate backbone as well as genomic integrity during the double-stranded DNA cleavage process, the type II enzyme forms covalent bonds between the two active site tyrosine residues and the newly generated 5′-phosphate groups of the DNA backbone, generating a phosphotyrosyl linkage and a four-base DNA overhang [5,6,29,31,47]. The transiently cleaved, covalently linked enzyme–DNA structure that is formed is known as the pre-strand passage "cleavage complex" [5,6,29,31]. The formation of the cleavage complex during enzyme catalysis is tightly regulated to prevent the generation of permanent DNA breaks or the disruption of genomic integrity [5,6,29,30].

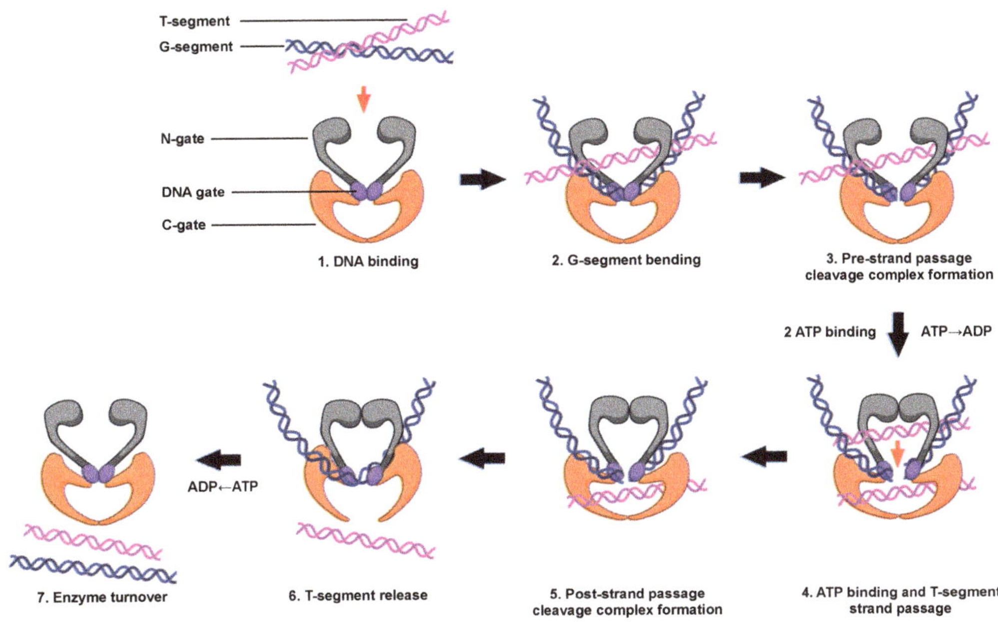

Figure 4. The catalytic cycle of type II topoisomerases. The double-stranded DNA passage reaction of type II topoisomerases can be separated into discrete steps. (1) Type II enzyme binding to two intact segments of DNA: the gate, or G-segment (blue, the first segment bound) and transport, or T-segment (purple, the second segment bound). (2) Bending of the G-segment to assess for sites of DNA cleavage. (3) Double-stranded DNA cleavage of the G-segment (i.e., formation of the pre-strand cleavage complex). (4) Binding of two ATP molecules, which triggers the closing of the N-gate, opening of the DNA gate, and the passage of the T-segment through the DNA gate. Strand passage occurs more rapidly if one of the two ATP molecules is hydrolyzed. (5) Formation of the post-strand passage cleavage complex. (6) Religation of the cleaved G-segment and release of the T-segment through the C-gate of the protein. (7) ATP hydrolysis, which triggers enzyme turnover and the regeneration of the enzyme to initiate a new round of catalysis. Created with BioRender.com.

When ATP enters the enzyme–DNA complex is not precisely known. This high-energy cofactor is not required for either DNA cleavage or religation. However, upon the binding of two ATP molecules, the N-gate is closed, triggering a conformational change in the enzyme that translocates the T-segment through the transient opening in the DNA (i.e., DNA "gate", Step 4). Although hydrolysis of the high-energy cofactor is not necessary for this strand passage event to occur, this step proceeds faster if one of the two bound ATP molecules is hydrolyzed [72].

After strand passage, a second, post-strand passage, cleavage complex is formed (Step 5). The type II enzyme then religates the cleaved DNA to regenerate the intact DNA double helix. DNA religation is initiated when a general acid removes the hydrogen from the 3′-terminal hydroxyl group [6,48]. Another nucleophilic attack is then initiated on the phosphotyrosyl bond, regenerating the intact DNA double helical backbone and the enzyme active site. The T-segment is then released from the protein (Step 6). The hydrolysis of a second ATP molecule occurs, resetting the type II enzyme conformation and allowing for enzyme turnover during the next cycle of catalysis (Step 7).

6. Cellular Functions of Gyrase and Topoisomerase IV

The main function of gyrase is to maintain the proper superhelical density of the bacterial genome (acting in conjunction with the ω protein, a type IA topoisomerase) and to remove

(+)SCs that accumulate ahead of DNA tracking machinery (i.e., polymerases and helicases) during essential nucleic acid processes such as replication and transcription [5,7,29,30,36,73–76]. Because of its ability to wrap DNA during catalysis (discussed below), gyrase functions primarily to generate (−)SCs, which also allows it to remove (+)SCs in a highly efficient manner [63,76–79].

Because topoisomerase IV is unable to wrap DNA, it functions as a canonical type II topoisomerase [36,63,76,79–81]. As such, the enzyme primarily acts to resolve precatenanes formed between daughter chromosomes during DNA replication and remove DNA knots that form during recombination [64,81]. Topoisomerase IV may also play a role ahead of DNA tracking systems, but the precise nature of this process is poorly understood [81,82].

7. Cellular Functions of Human Type II Topoisomerases

As discussed earlier, humans encode two isoforms of topoisomerase II: α and β. Topoisomerase IIα and topoisomerase IIβ are distinct in their expression patterns [3,6,29,30,41–43,83–85]. The levels of topoisomerase IIα are at their lowest during the G_1 phase and rise throughout the S phase, eventually peaking at the G_2/M phase boundary [84–86]. The enzyme is found almost exclusively in actively proliferating tissues, localizes predominantly in the nucleus, is associated with replication forks and transcription machinery, and is tightly bound to chromosomes and sister chromatids throughout mitosis [28,67,85–90]. Topoisomerase IIα is required for the survival of proliferating cells and is believed to be the main isoform that functions in growth-related processes such as replication and chromosomal segregation [4,30,31,47,83,84,88,91,92]. One of the primary functions of the enzyme is to resolve precatenanes that form behind replication forks [4,31,47,83,84,88,91,92]. However, genomic evidence also suggests a role for topoisomerase IIα during transcription [93–95].

In contrast to the α isoform, topoisomerase IIβ is not required for survival at the cellular level [40,88,91,96]. The concentration of topoisomerase IIβ is independent of the stage of the cell cycle, and the levels of the isoform are generally consistent irrespective of cell proliferation status [40,47,97]. Although cells can survive in the absence of topoisomerase IIβ, the enzyme is required for proper neural development in mice [31,40,88,98]. It also plays a role in the transcription of hormonally regulated genes [31,40,98,99].

8. When Good Enzymes Go Bad

Because type II topoisomerases generate transient double-stranded DNA breaks as requisite intermediates during their catalytic cycles, these enzymes also have the capacity to fragment the genome [6,17,31,100]. Thus, the type II enzymes are dualistic in nature; although essential for survival, they pose an eminent danger to the cell every time they act [6,17,31,100]. Consequently, the equilibrium between the forward cleavage reaction (enabling the subsequent strand passage) and the reverse religation reaction (resealing the DNA break) heavily favors religation to maintain genomic integrity during catalysis. As a result, under normal equilibrium conditions, covalent enzyme-cleaved DNA complexes generated by type II topoisomerases are tightly regulated [3,5,6,17,31,100]. These complexes are present at low steady-state levels, short-lived, and tolerated by the cell [3,5,6,17,31,100]. Cleavage complexes become more lethal when they are formed ahead of DNA tracking systems, such as replication forks and transcription complexes. When polymerases or helicases attempt to traverse the covalently bound topoisomerase "roadblock" in the genetic material, cleavage complexes can become disrupted, leaving the enzyme unable to religate the double-stranded DNA breaks [3,5,6,17,31,100,101]. In these cases, the "non-ligatable" DNA breaks must be repaired by DNA damage response and recombination pathways [3,5,6,17,31,100,101]. These actions can trigger unwanted chromosomal insertions, deletions, translocations, and cell death pathways [6,31,100,101]. Because DNA found ahead of tracking systems is usually overwound, cleavage complexes formed with (+)SC DNA are potentially the most lethal to cells [5,31,100–102].

9. The Effects of DNA Superhelicity on the Actions of Type II Topoisomerases

Type II topoisomerases play critical roles in a variety of essential nucleic acid processes. Because these enzymes work on negatively and positively supercoiled DNA substrates and removal of these supercoils results in the formation of relaxed DNA, it is important to understand how these enzymes distinguish their substrates and products and how supercoil handedness affects their actions.

9.1. Recognition of DNA Substrate versus Product

The earliest experiments on DNA topology recognition by type II topoisomerases focused on the abilities of these enzymes to distinguish their nucleic acid substrates from products. These studies found that type II enzymes interacted more tightly with their DNA substrates. Gyrase was found to bind relaxed DNA (substrate) ~10-fold more tightly than (−)SC DNA (product) [103,104]. For the canonical type II topoisomerases, topoisomerase IV also binds (−)SC (substrate) ~5-fold over relaxed DNA (product) [105]. Both yeast and *Drosophila* topoisomerase II can sense DNA supercoiling, preferentially binding with (−)SC over relaxed DNA [106,107]. *Drosophila* topoisomerase II also hydrolyzes ATP more rapidly in the presence of underwound DNA substrates [108]. Later, it was found that human topoisomerase IIα displayed higher affinities for supercoiled over relaxed DNA substrates [108,109]. Finally, human topoisomerase IIα maintains higher levels of cleavage complexes with (−)SC over relaxed DNA molecules, although the sites of cleavage remain the same [110].

It has been proposed that canonical type II topoisomerases distinguish supercoiled molecules from relaxed molecules by recognizing the presence or absence of DNA crossovers (i.e., writhe) during binding. Electron microscopy studies of *Drosophila* topoisomerase II-DNA complexes have shown that the enzymes strongly prefer to bind at sites of DNA juxtaposition independent of torsional stress [111]. A later study demonstrated that topoisomerase II simultaneously bound two double-stranded DNA segments and that the binding activity was independent of catalytic activity [112].

9.2. Recognition of Supercoil Handedness during DNA Strand Passage

9.2.1. DNA Relaxation and Supercoiling

Eventually, the field of type II topoisomerases transitioned away from distinguishing supercoiled DNA from relaxed DNA and toward the recognition of supercoil handedness. The global underwinding (negative supercoiling) of the genome puts energy into the DNA and enables the separation of the double helix, whereas DNA overwinding (positive supercoiling) ahead of tracking systems has the capacity to impede essential nucleic acid processes. Consequently, it is critical to understand how type II topoisomerases distinguish supercoil handedness during catalysis.

Early works on *Drosophila* and yeast topoisomerase II found that both enzymes were unable to distinguish (−)SC or (+)SC DNA during catalytic reactions and that they relaxed both substrates at comparable rates [107,109,113,114].

Similar to the type II topoisomerases from lower eukaryotes, human topoisomerase IIβ removes (−)SCs and (+)SCs at similar rates [109,115]. However, a major distinguishing characteristic between human topoisomerase IIα and topoisomerase IIβ is that the α isoform can discern supercoil handedness during strand passage and relaxes (+)SC 10–fold faster than it does (−)SC DNA [109,115]. Several lines of evidence indicate that this difference between topoisomerase IIα and topoisomerase IIβ results from elements in their respective C-terminal domains [115–117]. First, the C-terminal domain is the most varied region of type II topoisomerases. For example, the C-terminal domains of human topoisomerase IIα and topoisomerase IIβ display only ~31% sequence similarly, whereas the other regions of the enzymes display ~79% sequence similarity [30,37,40]. Second, the deletion of the C-terminal domain of topoisomerase IIα abrogates the ability of the enzyme to preferentially relax (+)SC substrates [109,115]. Even the deletion of a single "pinwheel blade" from the C-terminal domain of the enzyme decreases its ability to distinguish supercoil handedness [115]. Third, in experiments that switched the C-terminal domains of topoisomerase IIα and topoisomerase IIβ, a topoisomerase IIα enzyme that carried the C-terminal domain of the IIβ isoform lost the ability to distinguish supercoil handedness, whereas a chimeric topoisomerase IIβ enzyme that carried the C-terminal domain of the α isoform gained the ability to preferentially relax (+)SC DNA [109,115].

The ability of topoisomerase IIα to preferentially remove (+)SCs is due to a recognition of writhe, rather than twist, in its DNA substrate [118]. Presumably, the path that (+)SC DNA follows on the enzyme interacts with the C-terminal domain in a manner (not yet understood) that enhances the rate of strand passage.

Later studies examined the ability of the bacterial type II topoisomerases to distinguish supercoil geometry during strand passage. Gyrase removes (+)SC DNA at least 10-fold faster than it introduces (−)SCs into relaxed substrates [82,119,120]. This rapid removal of positive supercoils requires the GyrA box in the C-terminal domain of gyrase, which enables the DNA wrapping mechanism of the enzyme [82]. However, DNA wrapping cannot completely explain the recognition of supercoil handedness by gyrase, as mutant enzymes that lack this feature can still relax (+)SC DNA ~two-fold faster than (−)SC substrates [note that because wild-type gyrase normally underwinds DNA, it is not able to relax (−)SCs] [82,119–121].

Although topoisomerase IV primarily works behind replication forks as a decatenase, it also preferentially removes (+)SC versus (−)SC DNA [82,119]. Similar to findings with human topoisomerase IIα, the loss of the C-terminal domain impedes the ability of *E. coli* topoisomerase IV to distinguish DNA supercoil geometry during strand passage [64,66]. The ability of topoisomerase IV to distinguish supercoil handedness during DNA strand passage also appears to be based on writhe [118,122].

9.2.2. DNA Catenation/Decatenation

While DNA relaxation is performed via intramolecular strand passage of supercoiled substrates, intermolecular strand passage is necessary to resolve catenanes (i.e., tangles). Type II topoisomerases can also recognize topology during the catenation/decatenation reaction. Yeast topoisomerase II is capable of sensing supercoil handedness during the catenation/decatenation reaction [113]. The enzyme preferentially decatenates (−)SC over (+)SC DNA but favors catenating (+)SC over (−)SC DNA [113]. Similar to results with DNA relaxation, human topoisomerase IIα is able to distinguish between different supercoiled states of DNA during catenation, whereas topoisomerase IIβ cannot. Unexpectedly, the α isoform catenates underwound molecules faster than it does overwound substrates, which is antithetical to the preference during relaxation reactions (human topoisomerase IIα relaxes overwound substrates faster) [123]. In contrast to results with human topoisomerase IIα, topoisomerase IV preferentially catenates (+)SC DNA, which parallels results with relaxation experiments [123].

9.3. Recognition of Supercoil Handedness during DNA Cleavage

Type II topoisomerases can also recognize supercoil geometry during DNA cleavage [63,82,109,117,119,124,125]. However, this recognition differs from that which occurs during the DNA strand passage reaction. For example, even though topoisomerase IIα is the only human type II enzyme that can distinguish DNA topology during strand passage, both topoisomerase IIα and IIβ maintain two- to four-fold higher levels of cleavage complexes on (−)SC versus (+)SC DNA [109,115,117]. Furthermore, the ability to recognize supercoil handedness during DNA cleavage lies within the catalytic core of the enzymes as opposed to the C-terminal domain [110]. Consequently, type II topoisomerases appear to recognize supercoil handedness in a bimodal manner, using different mechanisms to distinguish DNA geometry during different catalytic events.

Similar to the human type II enzymes, gyrase also maintains two- to four-fold higher levels of cleavage complexes with (−)SC over (+)SC DNA. However, gyrase (at least the enzyme from *M. tuberculosis*) requires elements in the N-terminal domain to enable the recognition of supercoil handedness during DNA cleavage [119]. Because cleavage complexes formed on (+)SC DNA are the most dangerous, the fact that the human type II enzymes and gyrase generate lower levels of cleavage on overwound substrates make them safer for the cell.

In contrast to the above enzymes, topoisomerase IV shows no large difference in levels of cleavage generated with (−)SC versus (+)SC DNA [82,119,126,127]. However, because topoisomerase IV appears to work behind replication forks, this lack of discrimination during DNA cleavage may have less impact on the cell [82].

It is notable that the recognition of DNA topology during cleavage is not altered by the presence of anticancer or antibacterial drugs; human topoisomerase IIα, topoisomerase IIβ, and gyrase maintain higher levels of cleavage complexes with (−)SC over (+)SC substrates, whereas topoisomerase IV maintains similar levels [128].

The differential recognition of supercoil handedness by type II topoisomerases during DNA cleavage cannot be explained by rates of religation of the cleaved DNA, either in the absence or presence of drugs [128]. Topoisomerase IIα and gyrase form more stable cleavage complexes with (−)SC DNA in the presence of anticancer and antibacterial drugs, respectively [128]. However, in the absence of drugs, the lifetimes of cleavage complexes for all the human and bacterial type II topoisomerases are short and do not vary due to supercoil handedness [128]. Thus, while the stability of cleavage complexes may, under specific circumstances, contribute to the recognition of supercoil handedness, it cannot fully explain how the type II enzymes distinguish DNA geometry during cleavage [128]. Rather, the abilities of the human type II topoisomerases and bacterial gyrase to discern supercoil handedness during scission appear to reflect the forward rates of DNA cleavage. These enzymes all cleave (−)SC DNA faster than they do (+)SC substrates [128]. In contrast, topoisomerase IV, which does not discriminate supercoil handedness during DNA cleavage, cleaves underwound and overwound substrates at similar rates [128]. Again, the above relationships hold in the absence of anticancer or antibacterial drugs [128].

It is not known whether type II topoisomerases utilize twist or writhe to recognize supercoil handedness during DNA cleavage. However, because the portions of the enzymes involved in this recognition are so limited compared to those required during strand passage, it is not clear how DNA writhe could contribute to the recognition of supercoil geometry during cleavage. An intriguing possibility is that this recognition is dependent on DNA twist. To this point, the twist of underwound DNA aligns with the angle of gate opening during the double-stranded DNA passage reaction [129], whereas the twist in overwound DNA should oppose gate opening. Further studies are necessary to better understand the roles of twist and writhe in the recognition of supercoil geometry during DNA cleavage.

10. Conclusions

The globally underwound state of DNA in cells and the formation of overwound nucleic acid structures ahead of replication forks, transcription complexes, and other DNA tracking systems have important ramifications for proper biological function. To perform their critical cellular roles, type II topoisomerases transiently cut both strands of the DNA, open the double helix, and pass another nucleic acid segment through the DNA gate. Similar to other enzymes, the type II topoisomerases can distinguish their substrates from their products. In most cases, these enzymes can also distinguish between different DNA substrates. Type II topoisomerases that have been implicated in functioning on overwound DNA during replication or transcription often remove (+)SCs faster than (−)SCs and maintain lower levels of cleavage complexes with overwound substrates. These properties make these enzymes safer for the cell. Conversely, those that act primarily behind replication forks do not share these abilities. Thus, the ability of type II topoisomerases to recognize DNA supercoil geometry appears to have adapted to their unique cellular functions.

Author Contributions: Writing—review and editing, J.Y.J. and N.O.; funding acquisition, N.O. All authors have read and agreed to the published version of the manuscript.

Funding: This work was funded by the National Institute of Health (grant R01 GM126363 and R01 AI170546) and US Veterans Administration Merit Review (Award I01 Bx002198).

Institutional Review Board Statement: Not applicable.

Informed Consent Statement: Not applicable.

Data Availability Statement: Data sharing not applicable.

Acknowledgments: We thank Alexandria A. Oviatt, Jessica A. Collins, and Jillian F. Armenia for critical reviewing the manuscript.

Conflicts of Interest: The authors declare no conflict of interest.

References

1. Watson, J.D.; Crick, F.H. Molecular structure of nucleic acids; A structure for deoxyribose nucleic acid. *Nature* **1953**, *171*, 737–738. [CrossRef] [PubMed]
2. Bates, A.D.; Maxwell, A. *DNA Topology*; Oxford University Press: New York, NY, USA, 2005; p. 220.
3. Deweese, J.E.; Osheroff, M.A.; Osheroff, N. DNA topology and topoisomerases: Teaching a "knotty" subject. *Biochem. Mol. Biol. Educ.* **2008**, *37*, 2–10. [CrossRef] [PubMed]
4. Pommier, Y.; Sun, Y.; Huang, S.-Y.N.; Nitiss, J.L. Roles of eukaryotic topoisomerases in transcription, replication and genomic stability. *Nat. Rev. Mol. Cell Biol.* **2016**, *17*, 703–721. [CrossRef] [PubMed]
5. Ashley, R.E.; Osheroff, N. Regulation of DNA topology by topoisomerases: Mathematics at the molecular level. In *Knots, Low-dimensional Topology and Applications*; Adams, C.C., Gordon, C.M., Jones, V.F.R., Kauffman, L.H., Lambropoulou, S., Millett, K., Przytycki, J.H., Ricca, R., Sazdanovic, R., Eds.; Springer Proceedings in Mathematics & Statistics; Springer: New York, NY, USA, 2019; Volume 284, pp. 411–433.
6. Deweese, J.E.; Osheroff, N. The DNA cleavage reaction of topoisomerase II: Wolf in sheep's clothing. *Nucleic Acids Res.* **2009**, *37*, 738–748. [CrossRef]
7. Liu, Z.; Deibler, R.W.; Chan, H.S.; Zechiedrich, L. The why and how of DNA unlinking. *Nucleic Acids Res.* **2009**, *37*, 661–671. [CrossRef]
8. Finzi, L.; Olson, W.K. The emerging role of DNA supercoiling as a dynamic player in genomic structure and function. *Biophys. Rev.* **2016**, *8*, 1–3. [CrossRef]
9. Bauer, W.R.; Crick, F.H.; White, J.H. Supercoiled DNA. *Sci. Am.* **1980**, *243*, 100–113.
10. White, J.H.; Cozzarelli, N.R. A simple topological method for describing stereoisomers of DNA catenanes and knots. *Proc. Natl. Acad. Sci. USA* **1984**, *81*, 3322–3326. [CrossRef]
11. Shore, D.; Baldwin, R.L. Energetics of DNA twisting. II. Topoisomer analysis. *J. Mol. Biol.* **1983**, *170*, 983–1007. [CrossRef]
12. Espeli, O.; Marians, K.J. Untangling intracellular DNA topology. *Mol. Microbiol.* **2004**, *52*, 925–931. [CrossRef]
13. Falaschi, A.; Abdurashidova, G.; Sandoval, O.; Radulescu, S.; Biamonti, G.; Riva, S. Molecular and structural transactions at human DNA replication origins. *Cell Cycle* **2007**, *6*, 1705–1712. [CrossRef] [PubMed]
14. Travers, A.; Muskhelishvili, G. A common topology for bacterial and eukaryotic transcription initiation? *EMBO Rep.* **2007**, *8*, 147–151. [CrossRef]
15. Liu, L.F.; Wang, J.C. Supercoiling of the DNA template during transcription. *Proc. Natl. Acad. Sci. USA* **1987**, *84*, 7024–7027. [CrossRef]
16. Schvartzman, J.B.; Stasiak, A. A topological view of the replicon. *EMBO Rep.* **2004**, *5*, 256–261. [CrossRef]
17. Nitiss, J.L. Targeting DNA topoisomerase II in cancer chemotherapy. *Nat. Rev. Cancer* **2009**, *9*, 338–350. [CrossRef] [PubMed]
18. Pommier, Y.; Leo, E.; Zhang, H.; Marchand, C. DNA topoisomerases and their poisoning by anticancer and antibacterial drugs. *Chem. Biol.* **2010**, *17*, 421–433. [CrossRef]
19. Postow, L.; Crisona, N.J.; Peter, B.J.; Hardy, C.D.; Cozzarelli, N.R. Topological challenges to DNA replication: Conformations at the fork. *Proc. Natl. Acad. Sci. USA* **2001**, *98*, 8219–8226. [CrossRef]
20. Brill, S.J.; DiNardo, S.; Voelkel-Meiman, K.; Sternglanz, R. Need for DNA topoisomerase activity as a swivel for DNA replication for transcription of ribosomal RNA. *Nature* **1987**, *326*, 414–416. [CrossRef] [PubMed]
21. Kim, R.A.; Wang, J.C. Function of DNA topoisomerases as replication swivels in *Saccharomyces cerevisiae*. *J. Mol. Biol.* **1989**, *208*, 257–267. [CrossRef]
22. Peter, B.J.; Ullsperger, C.; Hiasa, H.; Marians, K.J.; Cozzarelli, N.R. The structure of supercoiled intermediates in DNA replication. *Cell* **1998**, *94*, 819–827. [CrossRef]
23. Holm, C.; Goto, T.; Wang, J.C.; Botstein, D. DNA topoisomerase II is required at the time of mitosis in yeast. *Cell* **1985**, *41*, 553–563. [CrossRef]
24. Baxter, J.; Diffley, J.F. Topoisomerase II inactivation prevents the completion of DNA replication in budding yeast. *Mol. Cell* **2008**, *30*, 790–802. [CrossRef]

25. Baxter, J.; Sen, N.; Martinez, V.L.; De Carandini, M.E.; Schvartzman, J.B.; Diffley, J.F.; Aragon, L. Positive supercoiling of mitotic DNA drives decatenation by topoisomerase II in eukaryotes. *Science* **2011**, *331*, 1328–1332. [CrossRef]
26. Sen, N.; Leonard, J.; Torres, R.; Garcia-Luis, J.; Palou-Marin, G.; Aragon, L. Physical Proximity of Sister Chromatids Promotes Top2-Dependent Intertwining. *Mol. Cell* **2016**, *64*, 134–147. [CrossRef] [PubMed]
27. Bauer, D.L.V.; Marie, R.; Rasmussen, K.H.; Kristensen, A.; Mir, K.U. DNA catenation maintains structure of human metaphase chromosomes. *Nucleic Acids Res.* **2012**, *40*, 11428–11434. [CrossRef]
28. Uemura, T.; Ohkura, H.; Adachi, Y.; Morino, K.; Shiozaki, K.; Yanagida, M. DNA topoisomerase II is required for condensation and separation of mitotic chromosomes in *S. pombe*. *Cell* **1987**, *50*, 917–925. [CrossRef]
29. Vos, S.M.; Tretter, E.M.; Schmidt, B.H.; Berger, J.M. All tangled up: How cells direct, manage and exploit topoisomerase function. *Nat. Rev. Mol. Cell. Biol.* **2011**, *12*, 827–841. [CrossRef]
30. Chen, S.H.; Chan, N.L.; Hsieh, T.S. New mechanistic and functional insights into DNA topoisomerases. *Annu. Rev. Biochem.* **2013**, *82*, 139–170. [CrossRef]
31. Vann, K.R.; Oviatt, A.A.; Osheroff, N. Topoisomerase II poisons: Converting essential enzymes into molecular scissors. *Biochemistry* **2021**, *60*, 1630–1641. [CrossRef] [PubMed]
32. Corbett, K.D.; Berger, J.M. Structure of the topoisomerase VI-B subunit: Implications for type II topoisomerase mechanism and evolution. *EMBO J.* **2003**, *22*, 151–163. [CrossRef] [PubMed]
33. Forterre, P.; Gadelle, D. Phylogenomics of DNA topoisomerases: Their origin and putative roles in the emergence of modern organisms. *Nucleic Acids Res.* **2009**, *37*, 679–692. [CrossRef]
34. Gellert, M.; Mizuuchi, K.; O'Dea, M.H.; Nash, H.A. DNA gyrase: An enzyme that introduces superhelical turns into DNA. *Proc. Natl. Acad. Sci. USA* **1976**, *73*, 3872–3876. [CrossRef]
35. Kato, J.; Nishimura, Y.; Imamura, R.; Niki, H.; Hiraga, S.; Suzuki, H. New topoisomerase essential for chromosome segregation in *E. coli*. *Cell* **1990**, *63*, 393–404. [CrossRef]
36. Sissi, C.; Palumbo, M. In front of and behind the replication fork: Bacterial type IIA topoisomerases. *Cell. Mol. Life. Sci.* **2010**, *67*, 2001–2024. [CrossRef]
37. Forterre, P.; Gribaldo, S.; Gadelle, D.; Serre, M.C. Origin and evolution of DNA topoisomerases. *Biochimie* **2007**, *89*, 427–446. [CrossRef]
38. Cole, S.T.; Brosch, R.; Parkhill, J.; Garnier, T.; Churcher, C.; Harris, D.; Gordon, S.V.; Eiglmeier, K.; Gas, S.; Barry, C.E., 3rd; et al. Deciphering the biology of *Mycobacterium tuberculosis* from the complete genome sequence. *Nature* **1998**, *393*, 537–544. [CrossRef]
39. Hsieh, T.; Brutlag, D.L. ATP-dependent DNA topoisomerase from *D. melanogaster* reversibly catenates duplex DNA rings. *Cell* **1980**, *21*, 115–125. [CrossRef]
40. Austin, C.A.; Lee, K.C.; Swan, R.L.; Khazeem, M.M.; Manville, C.M.; Cridland, P.; Treumann, A.; Porter, A.; Morris, N.J.; Cowell, I.G. TOP2B: The first thirty years. *Int. J. Mol. Sci.* **2018**, *19*, 2765. [CrossRef]
41. Tsai-Pflugfelder, M.; Liu, L.F.; Liu, A.A.; Tewey, K.M.; Whang-Peng, J.; Knutsen, T.; Huebner, K.; Croce, C.M.; Wang, J.C. Cloning and sequencing of cDNA encoding human DNA topoisomerase II and localization of the gene to chromosome region 17q21-22. *Proc. Natl. Acad. Sci. USA* **1988**, *85*, 7177–7181. [CrossRef]
42. Drake, F.H.; Hofmann, G.A.; Bartus, H.F.; Mattern, M.R.; Crooke, S.T.; Mirabelli, C.K. Biochemical and pharmacological properties of p170 and p180 forms of topoisomerase II. *Biochemistry* **1989**, *28*, 8154–8160. [CrossRef]
43. Chung, T.D.; Drake, F.H.; Tan, K.B.; Per, S.R.; Crooke, S.T.; Mirabelli, C.K. Characterization and immunological identification of cDNA clones encoding two human DNA topoisomerase II isozymes. *Proc. Natl. Acad. Sci. USA* **1989**, *86*, 9431–9435. [CrossRef]
44. Kato, J.; Suzuki, H.; Ikeda, H. Purification and characterization of DNA topoisomerase IV in *Escherichia coli*. *J. Biol. Chem.* **1992**, *267*, 25676–25684. [CrossRef]
45. Dalvie, E.D.; Osheroff, N. DNA topoisomerases: Type II. In *Encyclopedia of Biological Chemistry III*, 3rd ed.; Jez, J., Ed.; Elsevier: Oxford, UK, 2021; pp. 479–486.
46. Corbett, K.D.; Berger, J.M. Structure, molecular mechanisms, and evolutionary relationships in DNA topoisomerases. *Annu. Rev. Biophys. Biomol. Struct.* **2004**, *33*, 95–118. [CrossRef]
47. McKie, S.J.; Neuman, K.C.; Maxwell, A. DNA topoisomerases: Advances in understanding of cellular roles and multi-protein complexes via structure-function analysis. *Bioessays* **2021**, *43*, 2000286. [CrossRef]
48. Wendorff, T.J.; Schmidt, B.H.; Heslop, P.; Austin, C.A.; Berger, J.M. The structure of DNA-bound human topoisomerase IIα: Conformational mechanisms for coordinating inter-subunit interactions with DNA cleavage. *J. Mol. Biol.* **2012**, *424*, 109–124. [CrossRef]
49. Bjergbaek, L.; Kingma, P.; Nielsen, I.S.; Wang, Y.; Westergaard, O.; Osheroff, N.; Andersen, A.H. Communication between the ATPase and cleavage/religation domains of human topoisomerase IIα. *J. Biol. Chem.* **2000**, *275*, 13041–13048. [CrossRef]
50. Chang, C.C.; Wang, Y.R.; Chen, S.F.; Wu, C.C.; Chan, N.L. New insights into DNA-binding by type IIA topoisomerases. *Curr. Opin. Struct. Biol.* **2013**, *23*, 125–133. [CrossRef]
51. Schmidt, B.H.; Burgin, A.B.; Deweese, J.E.; Osheroff, N.; Berger, J.M. A novel and unified two-metal mechanism for DNA cleavage by type II and IA topoisomerases. *Nature* **2010**, *465*, 641–644. [CrossRef]
52. Aravind, L.; Leipe, D.D.; Koonin, E.V. Toprim—A conserved catalytic domain in type IA and II topoisomerases, DnaG-type primases, OLD family nucleases and RecR proteins. *Nucleic Acids Res.* **1998**, *26*, 4205–4213. [CrossRef]

53. Sissi, C.; Palumbo, M. Effects of magnesium and related divalent metal ions in topoisomerase structure and function. *Nucleic Acids Res.* **2009**, *37*, 702–711. [CrossRef]
54. Deweese, J.E.; Burgin, A.B.; Osheroff, N. Human topoisomerase IIα uses a two-metal-ion mechanism for DNA cleavage. *Nucleic Acids Res.* **2008**, *36*, 4883–4893. [CrossRef]
55. Pitts, S.L.; Liou, G.F.; Mitchenall, L.A.; Burgin, A.B.; Maxwell, A.; Neuman, K.C.; Osheroff, N. Use of divalent metal ions in the DNA cleavage reaction of topoisomerase IV. *Nucleic Acids Res.* **2011**, *39*, 4808–4817. [CrossRef]
56. Dong, K.C.; Berger, J.M. Structural basis for gate-DNA recognition and bending by type IIA topoisomerases. *Nature* **2007**, *450*, 1201–1205. [CrossRef]
57. Jang, Y.; Son, H.; Lee, S.W.; Hwang, W.; Jung, S.R.; Byl, J.A.W.; Osheroff, N.; Lee, S. Selection of DNA cleavage sites by topoisomerase II results from enzyme-induced flexibility of DNA. *Cell. Chem. Biol.* **2019**, *26*, 502–511.e3. [CrossRef]
58. Lee, S.; Jung, S.R.; Heo, K.; Byl, J.A.; Deweese, J.E.; Osheroff, N.; Hohng, S. DNA cleavage and opening reactions of human topoisomerase IIα are regulated via Mg^{2+}-mediated dynamic bending of gate-DNA. *Proc. Natl. Acad. Sci. USA* **2012**, *109*, 2925–2930. [CrossRef]
59. Lee, I.; Dong, K.C.; Berger, J.M. The role of DNA bending in type IIA topoisomerase function. *Nucleic Acids Res.* **2013**, *41*, 5444–5456. [CrossRef]
60. Corbett, K.D.; Shultzaberger, R.K.; Berger, J.M. The C-terminal domain of DNA gyrase A adopts a DNA-bending beta-pinwheel fold. *Proc. Natl. Acad. Sci. USA* **2004**, *101*, 7293–7298. [CrossRef]
61. Kramlinger, V.M.; Hiasa, H. The "GyrA-box" is required for the ability of DNA gyrase to wrap DNA and catalyze the supercoiling reaction. *J. Biol. Chem.* **2006**, *281*, 3738–3742. [CrossRef]
62. Lanz, M.A.; Klostermeier, D. The GyrA-box determines the geometry of DNA bound to gyrase and couples DNA binding to the nucleotide cycle. *Nucleic Acids Res.* **2012**, *40*, 10893–10903. [CrossRef]
63. Gibson, E.G.; Ashley, R.E.; Kerns, R.J.; Osheroff, N. Fluoroquinolone interactions with bacterial type II topoisomerases and target-mediated drug resistance. In *Antimicrobial Resistance and Implications for the 21st Century*; Drlica, K., Shlaes, D., Fong, I.W., Eds.; Springer: New York, NY, USA, 2018; pp. 507–529.
64. Vos, S.M.; Lee, I.; Berger, J.M. Distinct regions of the *Escherichia coli* ParC C-terminal domain are required for substrate discrimination by topoisomerase IV. *J. Mol. Biol.* **2013**, *425*, 3029–3045. [CrossRef]
65. Tretter, E.M.; Lerman, J.C.; Berger, J.M. A naturally chimeric type IIA topoisomerase in *Aquifex aeolicus* highlights an evolutionary path for the emergence of functional paralogs. *Proc. Natl. Acad. Sci. USA* **2010**, *107*, 22055–22059. [CrossRef]
66. Corbett, K.D.; Schoeffler, A.J.; Thomsen, N.D.; Berger, J.M. The structural basis for substrate specificity in DNA topoisomerase IV. *J. Mol. Biol.* **2005**, *351*, 545–561. [CrossRef]
67. Linka, R.M.; Porter, A.C.; Volkov, A.; Mielke, C.; Boege, F.; Christensen, M.O. C-terminal regions of topoisomerase IIα and IIβ determine isoform-specific functioning of the enzymes in vivo. *Nucleic Acids Res.* **2007**, *35*, 3810–3822. [CrossRef]
68. Broeck, A.V.; Lotz, C.; Drillien, R.; Haas, L.; Bedez, C.; Lamour, V. Structural basis for allosteric regulation of human topoisomerase IIα. *Nat. Commun.* **2021**, *12*, 2962. [CrossRef]
69. Antoniou-Kourounioti, M.; Mimmack, M.L.; Porter, A.C.G.; Farr, C.J. The impact of the C-terminal region on the interaction of topoisomerase IIα with mitotic chromatin. *Int. J. Mol. Sci.* **2019**, *20*, 1238. [CrossRef]
70. Deweese, J.E.; Osheroff, N. The use of divalent metal ions by type II topoisomerases. *Metallomics* **2010**, *2*, 450–459. [CrossRef]
71. Deweese, J.E.; Guengerich, F.P.; Burgin, A.B.; Osheroff, N. Metal ion interactions in the DNA cleavage/ligation active site of human topoisomerase IIα. *Biochemistry* **2009**, *48*, 8940–8947. [CrossRef]
72. Lindsley, J.E.; Wang, J.C. On the coupling between ATP usage and DNA transport by yeast DNA topoisomerase II. *J. Biol. Chem.* **1993**, *268*, 8096–8104. [CrossRef]
73. Marians, K.J. DNA gyrase-catalyzed decatenation of multiply linked DNA dimers. *J. Biol. Chem.* **1987**, *262*, 10362–10368. [CrossRef]
74. Mirkin, S.M.; Zaitsev, E.N.; Panyutin, I.G.; Lyamichev, V.I. Native supercoiling of DNA: The effects of DNA gyrase and ω protein in *E. coli*. *Mol. Gen. Genet.* **1984**, *196*, 508–512. [CrossRef]
75. Ullsperger, C.; Cozzarelli, N.R. Contrasting enzymatic activities of topoisomerase IV and DNA gyrase from *Escherichia coli*. *J. Biol. Chem.* **1996**, *271*, 31549–31555. [CrossRef]
76. Levine, C.; Hiasa, H.; Marians, K.J. DNA gyrase and topoisomerase IV: Biochemical activities, physiological roles during chromosome replication, and drug sensitivities. *Biochim. Biophys. Acta* **1998**, *1400*, 29–43. [CrossRef]
77. Khodursky, A.B.; Peter, B.J.; Schmidt, M.B.; DeRisi, J.; Botstein, D.; Brown, P.O.; Cozzarelli, N.R. Analysis of topoisomerase function in bacterial replication fork movement: Use of DNA microarrays. *Proc. Natl. Acad. Sci. USA* **2000**, *97*, 9419–9424. [CrossRef]
78. Hsu, Y.H.; Chung, M.W.; Li, T.K. Distribution of gyrase and topoisomerase IV on bacterial nucleoid: Implications for nucleoid organization. *Nucleic Acids Res.* **2006**, *34*, 3128–3138. [CrossRef]
79. Tadesse, S.; Graumann, P.L. Differential and dynamic localization of topoisomerases in *Bacillus subtilis*. *J. Bacteriol.* **2006**, *188*, 3002–3011. [CrossRef]
80. Zawadzki, P.; Stracy, M.; Ginda, K.; Zawadzka, K.; Lesterlin, C.; Kapanidis, A.N.; Sherratt, D.J. The localization and action of topoisomerase IV in *Escherichia coli* chromosome segregation Is coordinated by the SMC complex, MukBEF. *Cell Rep.* **2015**, *13*, 2587–2596. [CrossRef]

81. Zechiedrich, E.L.; Cozzarelli, N.R. Roles of topoisomerase IV and DNA gyrase in DNA unlinking during replication in *Escherichia coli*. *Genes Dev.* **1995**, *9*, 2859–2869. [CrossRef]
82. Ashley, R.E.; Dittmore, A.; McPherson, S.A.; Turnbough, C.L., Jr.; Neuman, K.C.; Osheroff, N. Activities of gyrase and topoisomerase IV on positively supercoiled DNA. *Nucleic Acids Res.* **2017**, *45*, 9611–9624. [CrossRef]
83. Heck, M.M.; Earnshaw, W.C. Topoisomerase II: A specific marker for cell proliferation. *J. Cell. Biol.* **1986**, *103*, 2569–2581. [CrossRef]
84. Heck, M.M.; Hittelman, W.N.; Earnshaw, W.C. Differential expression of DNA topoisomerases I and II during the eukaryotic cell cycle. *Proc. Natl. Acad. Sci. USA* **1988**, *85*, 1086–1090. [CrossRef]
85. Woessner, R.D.; Mattern, M.R.; Mirabelli, C.K.; Johnson, R.K.; Drake, F.H. Proliferation- and cell cycle-dependent differences in expression of the 170 kilodalton and 180 kilodalton forms of topoisomerase II in NIH-3T3 cells. *Cell Growth Differ.* **1991**, *2*, 209–214.
86. Kimura, K.; Saijo, M.; Ui, M.; Enomoto, T. Growth state- and cell cycle-dependent fluctuation in the expression of two forms of DNA topoisomerase II and possible specific modification of the higher molecular weight form in the M phase. *J. Biol. Chem.* **1994**, *269*, 1173–1176. [CrossRef] [PubMed]
87. Mirski, S.E.; Gerlach, J.H.; Cummings, H.J.; Zirngibl, R.; Greer, P.A.; Cole, S.P. Bipartite nuclear localization signals in the C terminus of human topoisomerase IIα. *Exp. Cell Res.* **1997**, *237*, 452–455. [CrossRef] [PubMed]
88. Grue, P.; Grasser, A.; Sehested, M.; Jensen, P.B.; Uhse, A.; Straub, T.; Ness, W.; Boege, F. Essential mitotic functions of DNA topoisomerase IIα are not adopted by topoisomerase IIβ in human H69 cells. *J. Biol. Chem.* **1998**, *273*, 33660–33666. [CrossRef]
89. Mirski, S.E.; Gerlach, J.H.; Cole, S.P. Sequence determinants of nuclear localization in the α and β isoforms of human topoisomerase II. *Exp. Cell Res.* **1999**, *251*, 329–339. [CrossRef]
90. Lee, J.H.; Berger, J.M. Cell cycle-dependent control and roles of DNA topoisomerase II. *Genes* **2019**, *10*, 859. [CrossRef]
91. Nitiss, J.L. DNA topoisomerase II and its growing repertoire of biological functions. *Nat. Rev. Cancer* **2009**, *9*, 327–337. [CrossRef]
92. Heintzman, D.R.; Campos, L.V.; Byl, J.A.W.; Osheroff, N.; Dewar, J.M. Topoisomerase II is crucial for fork convergence during vertebrate replication termination. *Cell Rep.* **2019**, *29*, 422–436.e5. [CrossRef]
93. Ray, S.; Panova, T.; Miller, G.; Volkov, A.; Porter, A.C.G.; Russell, J.; Panov, K.I.; Zomerdijk, J.C.B.M. Topoisomerase IIα promotes activation of RNA polymerase I transcription by facilitating pre-initiation complex formation. *Nat. Commun.* **2013**, *4*, 1598. [CrossRef]
94. Yu, X.; Davenport, J.W.; Urtishak, K.A.; Carillo, M.L.; Gosai, S.J.; Kolaris, C.P.; Byl, J.A.W.; Rappaport, E.F.; Osheroff, N.; Gregory, B.D.; et al. Genome-wide TOP2A DNA cleavage is biased toward translocated and highly transcribed loci. *Genome Res.* **2017**, *27*, 1238–1249. [CrossRef]
95. Bossaert, M.; Pipier, A.; Riou, J.; Noirot, C.; Nguyên, L.; Serre, R.; Bouchez, O.; Defranq, E.; Calsou, P.; Britton, S.; et al. Transcription-associated topoisomerase 2α (TOP2A) activity is a major effector of cytotoxicity induced by G-quadruplex ligands. *eLife* **2021**, *10*, e65184. [CrossRef]
96. Dereuddre, S.; Delaporte, C.; Jacquemin-Sablon, A. Role of topoisomerase IIβ in the resistance of 9-OH-ellipticine-resistant Chinese hamster fibroblasts to topoisomerase II inhibitors. *Cancer Res.* **1997**, *57*, 4301–4308.
97. Christensen, M.O.; Larsen, M.K.; Barthelmes, H.U.; Hock, R.; Andersen, C.L.; Kjeldsen, E.; Knudsen, B.R.; Westergaard, O.; Boege, F.; Mielke, C. Dynamics of human DNA topoisomerases IIα and IIβ in living cells. *J. Cell Biol.* **2002**, *157*, 31–44. [CrossRef]
98. Ju, B.G.; Lunyak, V.V.; Perissi, V.; Garcia-Bassets, I.; Rose, D.W.; Glass, C.K.; Rosenfeld, M.G. A topoisomerase IIβ-mediated dsDNA break required for regulated transcription. *Science* **2006**, *312*, 1798–1802. [CrossRef]
99. Yang, X.; Li, W.; Prescott, E.D.; Burden, S.J.; Wang, J.C. DNA topoisomerase IIβ and neural development. *Science* **2000**, *287*, 131–134. [CrossRef]
100. Pendleton, M.; Lindsey, R.H., Jr.; Felix, C.A.; Grimwade, D.; Osheroff, N. Topoisomerase II and leukemia. *Ann. N. Y. Acad. Sci.* **2014**, *1310*, 98–110. [CrossRef]
101. Aldred, K.J.; Kerns, R.J.; Osheroff, N. Mechanism of quinolone action and resistance. *Biochemistry* **2014**, *53*, 1565–1574. [CrossRef]
102. Ketron, A.C.; Osheroff, N. DNA topology and topoisomerases. In *Molecular Life Sciences: An Encyclopedic Reference*; Bell, E., Ed.; Springer: New York, NY, USA, 2014; pp. 1–19.
103. Sugino, A.; Higgins, N.P.; Cozzarelli, N.R. DNA gyrase subunit stoichiometry and the covalent attachment of subunit A to DNA during DNA cleavage. *Nucleic Acids Res.* **1980**, *8*, 3865–3874. [CrossRef]
104. Morrison, A.; Higgins, N.P.; Cozzarelli, N.R. Interaction between DNA gyrase and its cleavage site on DNA. *J. Biol. Chem.* **1980**, *255*, 2211–2219. [CrossRef]
105. Peng, H.; Marians, K.J. The interaction of topoisomerase IV with DNA. *J. Biol. Chem.* **1995**, *270*, 25286–25290. [CrossRef]
106. Brill, S.J.; Sternglanz, R. Transcription-dependent DNA supercoiling in yeast DNA topoisomerase mutants. *Cell* **1988**, *54*, 403–411. [CrossRef]
107. Osheroff, N.; Shelton, E.R.; Brutlag, D.L. DNA topoisomerase II from *Drosophila melanogaster*. Relaxation of supercoiled DNA. *J. Biol. Chem.* **1983**, *258*, 9536–9543. [CrossRef] [PubMed]
108. Osheroff, N. Eukaryotic topoisomerase II. Characterization of enzyme turnover. *J. Biol. Chem.* **1986**, *261*, 9944–9950. [CrossRef]
109. McClendon, A.K.; Rodriguez, A.C.; Osheroff, N. Human topoisomerase IIα rapidly relaxes positively supercoiled DNA: Implications for enzyme action ahead of replication forks. *J. Biol. Chem.* **2005**, *280*, 39337–39345. [CrossRef]

110. Lindsey, R.H., Jr.; Pendleton, M.; Ashley, R.E.; Mercer, S.L.; Deweese, J.E.; Osheroff, N. Catalytic core of human topoisomerase IIα: Insights into enzyme-DNA interactions and drug mechanism. *Biochemistry* **2014**, *53*, 6595–65602. [CrossRef]
111. Zechiedrich, L.; Osheroff, N. Eukaryotic topoisomerases recognize nucleic acid topology by preferentially interacting with DNA crossovers. *EMBO J.* **1990**, *9*, 4555–4562. [CrossRef]
112. Roca, J.; Berger, J.M.; Wang, J.C. On the simultaneous binding of eukaryotic DNA topoisomerase II to a pair of double-stranded DNA helices. *J. Biol. Chem.* **1993**, *268*, 14250–14255. [CrossRef]
113. Roca, J. Varying levels of positive and negative supercoiling differently affect the efficiency with which topoisomerase II catenates and decatenates DNA. *J. Mol. Biol.* **2001**, *305*, 441–450. [CrossRef]
114. Roca, J.; Berger, J.M.; Harrison, S.C.; Wang, J.C. DNA transport by a type II topoisomerase: Direct evidence for a two-gate mechanism. *Proc. Natl. Acad. Sci. USA* **1996**, *93*, 4057–4062. [CrossRef]
115. McClendon, A.K.; Gentry, A.C.; Dickey, J.S.; Brinch, M.; Bendsen, S.; Andersen, A.H.; Osheroff, N. Bimodal recognition of DNA geometry by human topoisomerase IIα: Preferential relaxation of positively supercoiled DNA requires elements in the C-terminal domain. *Biochemistry* **2008**, *47*, 13169–13178. [CrossRef]
116. McClendon, A.K.; Dickey, J.S.; Osheroff, N. Ability of viral topoisomerase II to discern the handedness of supercoiled DNA: Bimodal recognition of DNA geometry by type II enzymes. *Biochemistry* **2006**, *45*, 11674–11680. [CrossRef]
117. McClendon, A.K.; Osheroff, N. The geometry of DNA supercoils modulates topoisomerase-mediated DNA cleavage and enzyme response to anticancer drugs. *Biochemistry* **2006**, *45*, 3040–3050. [CrossRef] [PubMed]
118. Seol, Y.; Gentry, A.C.; Osheroff, N.; Neuman, K.C. Chiral discrimination and writhe-dependent relaxation mechanism of human topoisomerase IIα. *J. Biol. Chem.* **2013**, *288*, 13695–13703. [CrossRef] [PubMed]
119. Ashley, R.E.; Blower, T.R.; Berger, J.M.; Osheroff, N. Recognition of DNA supercoil geometry by *Mycobacterium tuberculosis* gyrase. *Biochemistry* **2017**, *56*, 5440–5448. [CrossRef]
120. Ashley, R.E.; Lindsey, R.H., Jr.; McPherson, S.A.; Turnbough, C.L., Jr.; Kerns, R.J.; Osheroff, N. Interactions between quinolones and *Bacillus anthracis* gyrase and the basis of drug resistance. *Biochemistry* **2017**, *56*, 4191–4200. [CrossRef] [PubMed]
121. Kampranis, S.C.; Maxwell, A. Conversion of DNA gyrase into a conventional type II topoisomerase. *Proc. Natl. Acad. Sci. USA* **1996**, *93*, 14416–14421. [CrossRef]
122. Neuman, K.C.; Charvin, G.; Bensimon, D.; Croquette, V. Mechanisms of chiral discrimination by topoisomerase IV. *Proc. Natl. Acad. Sci. USA* **2009**, *106*, 6986–6991. [CrossRef]
123. Dalvie, E.D.; Stacy, J.C.; Neuman, K.C.; Osheroff, N. Recognition of DNA supercoil handedness during catenation catalyzed by type II topoisomerases. *Biochemistry* **2022**, *61*, 2148–2158. [CrossRef]
124. Gibson, E.G.; Bax, B.; Chan, P.F.; Osheroff, N. Mechanistic and structural basis for the actions of the antibacterial gepotidacin against *Staphylococcus aureus* gyrase. *ACS Infect. Dis.* **2019**, *5*, 570–581. [CrossRef]
125. Gentry, A.C.; Pitts, S.L.; Jablonsky, M.J.; Bailly, C.; Graves, D.E.; Osheroff, N. Interactions between the etoposide derivative F14512 and human type II topoisomerases: Implications for the C4 spermine moiety in promoting enzyme-mediated DNA cleavage. *Biochemistry* **2011**, *50*, 3240–3249. [CrossRef]
126. Stone, M.D.; Bryant, Z.; Crisona, N.J.; Smith, S.B.; Vologodskii, A.; Bustamante, C.; Cozzarelli, N.R. Chirality sensing by *Escherichia coli* topoisomerase IV and the mechanism of type II topoisomerases. *Proc. Natl. Acad. Sci. USA* **2003**, *100*, 8654–8659. [CrossRef]
127. Crisona, N.J.; Strick, T.R.; Bensimon, D.; Croquette, V.; Cozzarelli, N.R. Preferential relaxation of positively supercoiled DNA by *E. coli* topoisomerase IV in single-molecule and ensemble measurements. *Genes Dev.* **2000**, *14*, 2881–2892. [CrossRef]
128. Jian, J.; McCarty, K.D.; Byl, J.A.; Guengerich, F.P.; Neuman, K.C.; Osheroff, N. Basis for the discrimination of supercoil handedness during DNA cleavage by human and bacterial type II topoisomerases. *Nucleic Acids Res.* **2023**, *51*, 3888–3902. [CrossRef]
129. Schmidt, B.H.; Osheroff, N.; Berger, J.M. Structure of a topoisomerase II-DNA-nucleotide complex reveals a new control mechanism for ATPase activity. *Nat. Struct. Mol. Biol.* **2012**, *19*, 1147–1154. [CrossRef]

International Journal of
Molecular Sciences

MDPI

Article

Extensive Bioinformatics Analyses Reveal a Phylogenetically Conserved Winged Helix (WH) Domain (Zτ) of Topoisomerase IIα, Elucidating Its Very High Affinity for Left-Handed Z-DNA and Suggesting Novel Putative Functions

Martin Bartas [1,*], Kristyna Slychko [1], Jiří Červeň [1], Petr Pečinka [1], Donna J. Arndt-Jovin [2] and Thomas M. Jovin [2,*]

1 Department of Biology and Ecology, University of Ostrava, 710 00 Ostrava, Czech Republic
2 Emeritus Laboratory of Cellular Dynamics, Max Planck Institute for Multidisciplinary Sciences, 37077 Göttingen, Germany
* Correspondence: martin.bartas@osu.cz (M.B.); tjovin@mpinat.mpg.de (T.M.J.)

Abstract: The dynamic processes operating on genomic DNA, such as gene expression and cellular division, lead inexorably to topological challenges in the form of entanglements, catenanes, knots, "bubbles", R-loops, and other outcomes of supercoiling and helical disruption. The resolution of toxic topological stress is the function attributed to DNA topoisomerases. A prominent example is the negative supercoiling (nsc) trailing processive enzymes such as DNA and RNA polymerases. The multiple equilibrium states that nscDNA can adopt by redistribution of helical twist and writhe include the left-handed double-helical conformation known as Z-DNA. Thirty years ago, one of our labs isolated a protein from *Drosophila* cells and embryos with a 100-fold greater affinity for Z-DNA than for B-DNA, and identified it as topoisomerase II (gene Top2, orthologous to the human UniProt proteins TOP2A and TOP2B). GTP increased the affinity and selectivity for Z-DNA even further and also led to inhibition of the isomerase enzymatic activity. An allosteric mechanism was proposed, in which topoII acts as a Z-DNA-binding protein (ZBP) to stabilize given states of topological (sub)domains and associated multiprotein complexes. We have now explored this possibility by comprehensive bioinformatic analyses of the available protein sequences of topoII representing organisms covering the whole tree of life. Multiple alignment of these sequences revealed an extremely high level of evolutionary conservation, including a winged-helix protein segment, here denoted as Zτ, constituting the putative structural homolog of Zα, the canonical Z-DNA/Z-RNA binding domain previously identified in the interferon-inducible RNA Adenosine-to-Inosine-editing deaminase, ADAR1p150. In contrast to Zα, which is separate from the protein segment responsible for catalysis, Zτ encompasses the active site tyrosine of topoII; a GTP-binding site and a GxxG sequence motif are in close proximity. Quantitative Zτ-Zα similarity comparisons and molecular docking with interaction scoring further supported the "B-Z-topoII hypothesis" and has led to an expanded mechanism for topoII function incorporating the recognition of Z-DNA segments ("Z-flipons") as an inherent and essential element. We further propose that the two Zτ domains of the topoII homodimer exhibit a single-turnover "conformase" activity on given G(ate) B-DNA segments ("Z-flipins"), inducing their transition to the left-handed Z-conformation. Inasmuch as the topoII-Z-DNA complexes are isomerase inactive, we infer that they fulfill important structural roles in key processes such as mitosis. Topoisomerases are preeminent targets of anti-cancer drug discovery, and we anticipate that detailed elucidation of their structural–functional interactions with Z-DNA and GTP will facilitate the design of novel, more potent and selective anti-cancer chemotherapeutic agents.

Keywords: Z-DNA; topoisomerase IIα; topoII; GTP; bioinformatics

Citation: Bartas, M.; Slychko, K.; Červeň, J.; Pečinka, P.; Arndt-Jovin, D.J.; Jovin, T.M. Extensive Bioinformatics Analyses Reveal a Phylogenetically Conserved Winged Helix (WH) Domain (Zτ) of Topoisomerase IIα, Elucidating Its Very High Affinity for Left-Handed Z-DNA and Suggesting Novel Putative Functions. *Int. J. Mol. Sci.* **2023**, *24*, 10740. https://doi.org/10.3390/ijms241310740

Academic Editors: Jesus Vicente De Julián Ortiz, Joseph E. Deweese and Neil Osheroff

Received: 19 April 2023
Revised: 13 June 2023
Accepted: 22 June 2023
Published: 27 June 2023

1. Introduction

This report represents the convergence of two molecular biological "currents". The first is the focus of this Special Issue of IJMS, the enzymatic activities discovered and denoted as "DNA topoisomerase(s)" by James Wang in 1979 [1] and extensively characterized structurally and functionally since then (reviewed in Refs. [2–4]). The second is the family of "non-B" DNA structures [5] deemed to intervene in cellular processes [6], one of which is the "Z-DNA" double helix with a left-handed helical sense, first proposed in 1970–1972 based on solution studies of alternating pur-pyr d[G-C] sequences [7,8] and confirmed by X-ray crystallography a decade later [9,10]. The structures and generation of Z-DNA and related Z-RNA have been reviewed recently [11–13], as have the biological roles attributed to them [14–19].

In the cell biological context, DNA and RNA can adopt and maintain the left-handed Z-conformation, but usually only when stabilized by one or more Z-DNA- or Z-RNA-binding domains (ZBDs). Excluding antibodies, the RNA A-to-I-editing adenosine deaminase, ADAR1, was the second (see below) protein reported to strongly bind Z-DNA (not a substrate but a presumed targeting moiety [20]) and Z~RNA (the substrate) [21]). The editing function is key in the mediation of innate immunity directed against pathogens, e.g., RNA viruses, and endogenous retroviral elements. ADAR1 is also a major mediator of resistance to immunotherapies based on Immune Checkpoint Blockade (ICB) [22]. Zα, the winged helix-turn-helix (wHTH, WHD) DNA recognition domain of the interferon-induced ADAR1p150 isoform (residues 121–197) was identified in 1997 [21]. The crystal structure of its complex with a hexameric Z-DNA revealed key interactions with the characteristic zig-zag backbone and the pur-pyr alternation of glycosyl linkages and sugar puckering of Z-DNA [23] (Figure 1). Initial sequence searches based on Zα led to the identification of other host and pathogen proteins ("ZBPs") exhibiting distinct, affine, functional interactions with Z-DNA and/or Z-RNA and likewise involved in host-pathogen response but also in stress response, cancer, autoimmunity, and germ cell DNA remodeling: ZBP1, PKZ, E3L, ORF112, RBP7910, ZBTB43 ([24] and references therein; [25]). In a recent study, we extended these findings by searching for new ZBPs with homologous Zα domains in the complete PDB structure database and in AlphaFold2 protein models [26]. A structure-based similarity search identified putative Zα ZBDs in 14 proteins with assigned structures and in 185 proteins modeled with AlphaFold2. STRING interaction networks revealed numerous functional clusters, one of which included HOP2 (with the highest Q score in the Zα domain search), a protein involved in stimulating strand exchange underlying homologous chromosome pairing in meiosis [27]. A HOP2-Z-DNA docking exercise led to the interaction image depicted in Figure 2A, in which the interaction is provided by α-helix 3 (α3) and supported by α-helix 1 (α1). Key amino acid interaction residues in this model are three glutamines (Q), three lysines (K), two glutamic acids (E), one arginine (R), one alanine (A), and one serine (S). A tyrosine residue (Y) believed to be crucial for Zα-Z-DNA interaction is located in β-sheet β1 but seems not to be directly involved in binding with our particular docking model. The Zα-ZBP family may be even more extensive; according to the current SMART non-redundant database (nrdb), there are 934 Zα domains in 478 Zα-protein homologs in various organisms [28]. The challenge is to establish their relevance, or the lack thereof, as functional ZBPs.

Following the advent of the Z-DNA crystallographic structures [9,10], extensive attempts were initiated (and persist) to define biological function(s) and identify specific protein interaction partners of left-handed dsDNA. In 1993, we reported the biochemical isolation, based on Z-DNA binding, of a ~165 kDa protein from *Drosophila melanogaster* cells and embryos with a 100-fold greater affinity for Z-DNA than for B-DNA [29]. It was unexpectedly yet unambiguously identified as the known cellular topoisomerase II (Top2; we will use the term topoII as generic for Type II, and particularly IIα, topoisomerases), a key member of the family of enzymes which resolve the inherent topological problems that arise with cellular nucleic acids [3,30]. TopoII was the first-reported non-antibody ZBP and exhibited a number of intriguing properties involving left-handed DNA, particularly the very high affinity and the pronounced allosteric influence of GTP. The latter was manifested by a significant enhancement of Z-DNA binding and the time-dependent emergence of enzyme inhibition and salt-resistant complexes (Table 1). Many properties were shared by the isoform topoisomerase IIβ and the topoII from other species, confirming that topoII primarily recognizes DNA secondary and/or tertiary structure rather than the primary sequence of DNA. Thus, the previously demonstrated higher affinity of the enzyme for bent or kinked DNA [31] was now extended to the non-B conformation Z-DNA, suggesting attractive models for accurately targeting the key topoII topology simplification activity to defined cellular loci. Unfortunately, the information of Table 1 languished until being revisited in a current retrospective account of left-handed DNA [8], incorporating an updated model for topoII function (Figure 2B). It was suggested that the techniques of bioinformatics and structural biology should be employed for elucidating the role(s) of topoII as a ZBP in the creation—as well as disassembly—of the DNA–protein complexes underlying topological (sub)domains. Such "topoclamps" could also serve as recognition targets and spatial delimiters of linear diffusion or "hopping" routes [32] for DNA-bound proteins, including topoII itself.

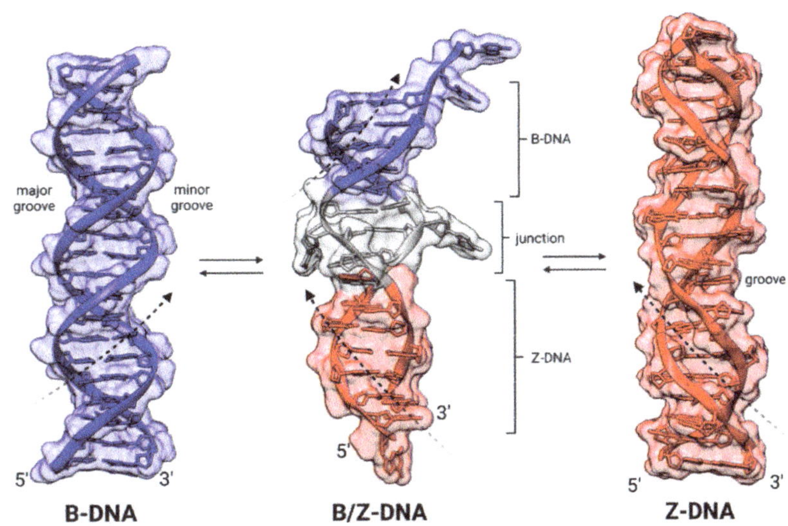

Figure 1. Comparative structures of alternative conformations of dsDNA: **left**, B-DNA; **right**, Z-DNA; **center**, a B-Z composite DNA with an intervening junction. Black dashed arrows indicate handedness (B-DNA, **right**; Z-DNA, **left**). Horizontal arrows indicate transitions between depicted dsDNA conformations. The figure was constructed with UCSF Chimera: B-DNA and Z-DNA are modeled structures, and B/Z-DNA is a crystal structure (PDB: 5zup). The helical pitch of B-DNA is 33 nm (10.5 bp/turn), and for Z-DNA, it is 46 nm (12 bp/turn).

These results and the extensive in silico search for new Z-DNA/Z-RNA binding proteins mentioned above [26] led to the convergence of interests of the presenting labs with an initial focus on the verification of topoII as a ZBP candidate based on comprehensive Zα homology screening of the topoII family. This effort has revealed a novel, highly conserved "active zone" encompassing a winged-helix Zτ domain flanked by a GTP-binding site and a pervasive GxxG motif.

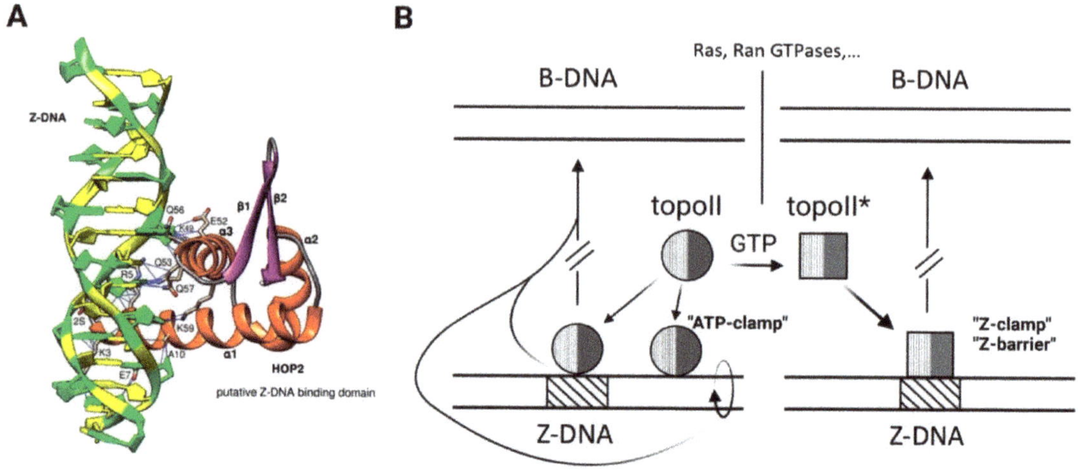

Figure 2. Prior bioinformatic and biochemical modeling instigating the search for a Z-DNA recognition domain of topoII. (**A**) Representative molecular docking of human HOP2 Zα region (aa 1–74) to Z-DNA [26]. Potential key amino acid interactions are depicted by thin blue lines. (**B**) Example of topoII as an allosteric ZBP (left and right). It is subject to competition by molecules (middle) exerting isomerase activity on nsc B-DNA segments. The relaxation process (ellipse with arrow, long curved line) abrogates the Z-conformation (short curved line) in the designated topologically linked segments. The affinity of topoII for Z-DNA is much greater than for B-DNA and increases further in the presence of GTP (topoII*), which also inhibits isomerase function (property 12, Table 1). These binding sites are deemed to constitute potential clamps, barriers, and crosslinkers, for example, in chromatin remodeling and mitosis/meiosis. Adapted from Figure 9 of Ref. [8].

Table 1. Paralogous topoisomerases IIα (topoII) and IIβ are Z-DNA-binding proteins (ZBPs).

No.	Properties (1991–1994) [a]	Ref.
	Drosophila (D) topoisomerase II and human (H) topoisomerase IIα (topoII)	
1	Two orders of magnitude higher binding affinity for Z-DNA than for B-DNA (D);	[29]
2	Complexes with Z-DNA salt resistant after 5 min;	[29]
3	Inhibition by linear Z-DNA of relaxation of co-incubated nsc minicircles (D);	[33]
4	Preferential affinity for and enhanced relaxation of ns (D) minicircles with Z-DNA forming insert (D, H);	[33]
5	Distinct DNA loci of binding and scission (cleavage/resealing);	[33]
6	VM-26 inhibitor-induced covalent DNA–protein complexes with minicircles ± Z-DNA forming insert (D);	[33]
7	Much greater affinity for intrinsically curved compared to linear B-DNA (D, H);	[34]
8	Hierarchy of DNA affinity: linear Z-DNA ≈ curved B-DNA ≥ nscDNA >> linear B-DNA (D, H);	[34]
9	No binding of ssDNA (D);	[29]
10	Increased formation of aggregates of nsc minicircles with Z-DNA forming inserts (D).	[33]
	Effects of GTP or non-hydrolyzable GTPγS (much more effective)	
11	Persistent, time-dependent, temperature-dependent inactivation of enzyme activity (D); incubation ± nscDNA;	[29]
12	Inhibition of DNA relaxation activity (D, H, calf thymus) via a proposed allosteric mechanism;	[29]
13	A 5–10 increase in affinity for Z-DNA and decreased affinity for B-DNA (D);	[29]
14	Inhibition of ATPase activity (D);	[34]
15	Relaxation inhibited by >4 mM ATP and >0.5 mM ITP but not by UTP or CTP;	[29]
16	Limited DNA compaction (knotting, catenation) by stoichiometric *Bombyx* and human topoII; GTPase activity [b].	[35]
	human isoform topoisomerase IIβ	
17	Hierarchy of DNA affinity: linear Z-DNA > nscDNA ≥ curved DNA >> poly[d(A-T)] > poly[d(G-C)].	[34]

[a] Mg^{2+} always present. [b] a study not incorporating Z-DNA. nsc, negatively supercoiled.

2. Results and Discussion

2.1. TopoII Contains a Putative Z-DNA-Binding Domain (Zτ)

The two human paralogs TOP2A (170 kDa) and TOP2B (183 kDa) share ~70% sequence identity. TOP2A is expressed predominantly in proliferating cells, while TOP2B is present in all cells, including those in quiescent or differentiated states. Thus, TOP2A mediates DNA replication, chromosome condensation and decondensation, and sister chromatid segregation, whereas TOP2B is key in transcription and differentiation, particularly during neuronal development [36]. The fundamental linkage between DNA topological states and topoisomerase function [37–39] is reflected in the architecture of these enzymes (see Figure 5A below). The enzymatic core of both isoforms comprises three functional regions: the N-terminal N-gate/ATPase, the DNA-gate, and the C-gate. The isoforms also contain differing C-terminal domains (CTDs) that are largely unstructured and deemed to fulfil regulatory and targeting roles [40–42]. We briefly recapitulate the currently accepted [2–4,30,43–48] 3-gate (N, DNA, C) *isomerase* catalytic mechanism of topoII in the schematic representation of Figure 3. A molecular depiction of topoII is given in Figure 4B.

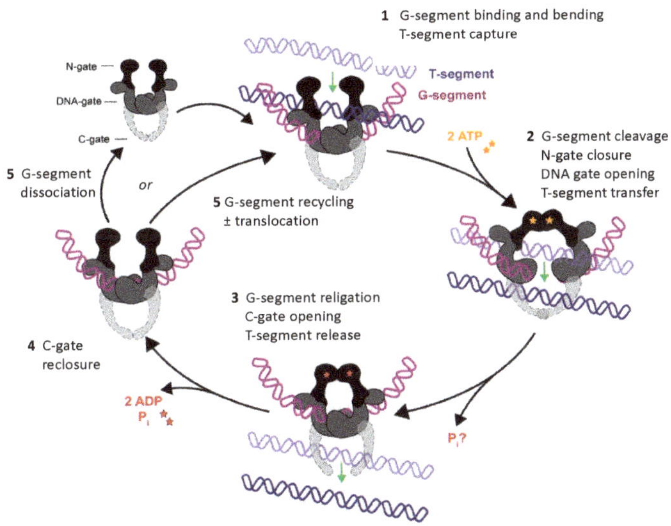

Figure 3. Canonical triple-gate isomerase mechanism for topoII. A double-helical DNA segment (G) binds to the topoII homodimer (upper left), is bent in the process, and is then cleaved, resulting in covalent protein (tyrosine)-DNA intermediates demarcating a double-strand break (DSB). A second "captured" DNA segment (T) traverses the DSB and is then released, while the G segment is religated, thereby restoring the integrity of the double helix. The intricate, sequential, concerted process is under allosteric control [48] mediated by ATP binding and turnover ([49], colored asterisks), divalent cations [50,51], and protein domains subject to post-translational modification, notably of the CTD [48]. The open and closed clamps of stages 1 and 3, respectively, are well depicted in a model of the tobacco enzyme (Figure 6 of Ref. [52]). Each cycle comprises a dual strand passage and thus changes the topological linking number *Lk* by ±2. The juxtaposition (a more appropriate term might be apposition) of the G and T segments at the crossover locus is dictated by the 3D structure of the local DNA domain, leading to numerous alternative topological outcomes [53]: resolution/simplification (relaxation, disentanglement) of plectonemic and toroidal supercoiled (+,−) substructures and reversal/formation of knots and catenanes arising during the processes of DNA transcription, replication, repair, recombination, higher-order chromosomal restructuring during mitosis and meiosis, and processing of closed circular DNA. Interference with DSB formation and resealing is highly genotoxic, and thus, steps 2 and 3 are key targets of antimicrobial and anticancer drugs [54–58]. Adapted from Figure 4 of Ref. [30].

The vast majority of known protein sequences of Class IIA topoisomerases (>10^5) are present in the domain Bacteria (88.5%), followed by the domain Eukaryota (7.2%), Archaea (2.0%), and Viruses (1.0%) (see Figure 7A below). The remainder of the protein sequences are still unclassified (1.3%). Interestingly, except for Viruses, the total number of protein sequences in each Domain of life is more than double the number of species, indicating that each species has more than two different types of Class II DNA topoisomerases on the average. This phenomenon most probably reflects the association of two separate protein subunits, as in bacterial gyrases (GyrA and GyrB) and the coexistence of topoisomerase IV (another member of the type IIA class) [42].

Based on our previous experimental findings and working hypotheses outlined in the Introduction, we searched for and identified a winged-helix domain, Zτ (Figure 5A,B), in human topoII (TOP2A) that is structurally similar, albeit distinct, to Zα of ADAR1p150 (Figure 4). The WHD already identified in topoII (Figure 5A) overlaps Zτ but is not strictly defined in the literature, being assigned to aa721—820 [59], to aa731—906 [45], or to an arbitrary region encompassing the active site catalytic tyrosine Y805. For the purposes of superposition, we selected particular regions of Zα and Zτ and modeled them with AlphaFold to reconcile differences in the dozens of experimental structures available. The superposition of these (Figure 4C) revealed significant structural similarity with a *p*-value of $1.1 \cdot 10^{-4}$ and root-mean-square deviation (RMSD) of atomic positions of 2.4 Å, based on 64 equivalent positions, and employing FATCAT flexible structural alignment [60] (Figure 4D). The slight discrepancies can be explained by the different lengths of the compared protein regions (represented by vertical lines in the figure): 67 aa for Zα from human ADAR1p150 and the significantly longer 91 aa for the Zτ segment of human topoII. The long wing between the β1 and β2 sequences of Zτ may confer greater affinity and specificity in its interaction with Z-DNA.

Table 2 quantifies the structural similarity of Zτ (from human topoII) to Zα of known Z-DNA/Z-RNA binding proteins (ADAR1, ZBP1, PKZ, E3, and ORF112) according to a number of parameters. In order to eliminate variations in different crystallographic approaches/quality of structures deposited in the RCSB database, AlphaFold models were always used since particular protein regions corresponding to Zτ or Zα/Zβ were predicted to have high, or even very high confidence. The highest number of aligned residues and the best RMSD and *p*-values scores are found in a pairwise comparison between Zτ (TOP2A) and Zα of human ADAR1p150. In contrast, sequence identity is very low (<10%) in all pairwise comparisons, suggesting that there is no detectable sequence homology. The *p*-value denotes the statistical significance of structural similarity. Human proteins Ubiquitin Fold Modifier 1 (UFM1, P61960) and histone H4 (P62805) served as negative controls.

Table 2. Structural parameters of similarity between Zτ of human topoII (P11388) and Zα/Zβ of various proteins: human ADAR1p150 (P55265), human ZBP1 (Q9H171), PKZ from *Danio rerio* (Q5NE12), E3 from *Vaccinia virus* (P21605), and ORF112 from *Cyprinid herpesvirus 3* (A4FTK7), together with negative controls (n.c.) UFM1 (P61960) and H4 (P62805) from *Homo sapiens*. *p*-values lower than the 0.05 threshold are in italics.

Feature	Parameter	Units				Domain					
			Zα	Zβ	Zα1	Zα2	Zα	Zα	Zα	n.c.	n.c.
protein			ADAR1p150		ZBP1		PKZ	E3	ORF112	UFM1	H4
RMSD	structure	Å	2.4	2.7	2.7	3.0	2.6	2.6	2.7	3.2	2.8
p-value	structure	10^{-3}	*0.11*	*1.3*	*0.39*	*2.8*	*0.24*	*0.34*	*0.70*	200	55
equivalent positions	structure	no.	64	62	60	61	60	61	60	39	50
gaps	sequence	%	30	32	34	33	34	32	34	39	7
identity	sequence	%	7.7	8.8	3.3	5.5	2.2	3.3	6.6	9.4	7.4
similarity	sequence	%	26	24	17	18	18	19	22	19	22

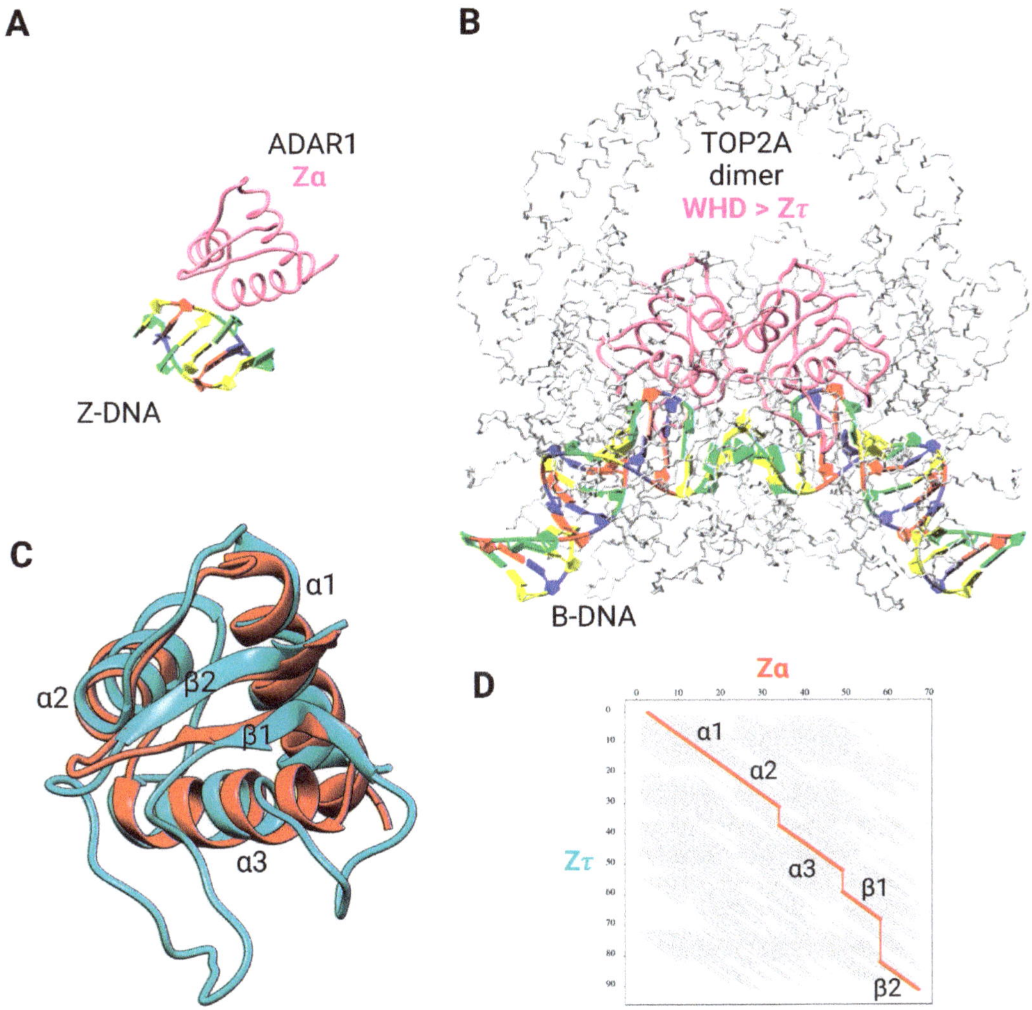

Figure 4. Structural comparison of Zα domain from human ADAR1p150 and putative Zτ domain from human topoII. (**A**) Crystal structure of Zα in complex with d(CACGTG) (PDB: 3f21). (**B**) Cryo-EM structure of the human TOP2A DNA-binding/cleavage domain in State 1 (PDB: 6zy5). Zα or Zτ domains are colored in hot pink, and DNA is colored according to NDB standards (A in red, T in blue, C in yellow, and G in green). (**C**) Superposition of Zα (ADAR1p150, red) and Zτ (TOP2A 722–812, cyan) domains, canonical designation of helices and β-sheets is indicated. (**D**) Graph of FATCAT [60] chaining result intuitively showing structural similarity across all 91 aa-long alignment (thick red diagonals); three gaps are depicted using thin vertical red lines, and non-significant structural similarity is depicted by gray diagonals.

All results, including the *p*-value, are much better for DNA-binding H4 than for UFM1, a membrane protein. Similar results were also obtained for Zτ from the human isoform TOP2B (Supplementary Material File S4), as expected in view of the high sequence similarity outside of the C-terminal domain (CTD). Only eight substitutions are present in the region corresponding to Zτ: L722F, S756A, M762Q, S763A, I769V, L781I, S800A, and S812T.

2.2. TopoII Contains a Major GTP-Binding Site

We next searched for significant GTP-binding sites in the human topoII, prompted by the results summarized in Table 1. For this purpose, we used the NSitePred tool [61], which was developed to accurately predict binding residues for ATP, ADP, AMP, GTP, and GDP via a sequence-based approach [60]. At first, we verified that the NSitePred tool is able to predict previously known and experimentally verified ATP-binding sites. We then directed our attention to GTP-binding sites. A very significant GTP-binding site (having the maximum score of 0.68 out of 1) was found in the TOP2A sequence at position I864, but the nearby sequence (852GAxGIxTGWxxKIPNF867) also showed GTP-binding potential. Worthy of note is that this region (particularly I856) is responsible for DNA bending, a key feature of the functional TOP2A dimer. Two isoleucines (one on each protomer) intercalate into the minor groove of DNA, bending the duplex by 130° [44]. A greatly reduced potential for GDP, ATP, and ADP binding at the same locus was predicted (Supplementary Material File S2). The results for topoII from diverse species including human are depicted in Figure 5C.

A second large cluster (149SNxDxxxxxVxxGRNGYGxKxCxxxxT175) of 13 GTP-binding sites was found in the region 149–175, spanning the known and experimentally validated ATP-binding region [62,63]. GTP was predicted to also bind strongly to this region (maximum binding score of 0.97), possibly even better than ATP (maximum binding score of 0.78; Supplementary Material File S2). These results suggest an even greater potential for strong allosteric control by GTP: promoting Z-DNA interactions (at the new GTP site) while concurrently inhibiting ATPase (property 14, Table 1) and thus isomerase (at the ATP sites). As a control of our calculations, the de novo predicted ATP-binding sites were in excellent congruence with the previously known sites. No GTP-binding site was found in human topoisomerase I (TOP1).

To put our data into further perspective, we extracted deleterious SNPs from the ENSEMBL Variant Database and filtered the most significant missense mutations using a strict threshold. There is a significant enrichment of these deleterious SNPs in the predicted GTP-binding locus and at the origin of the known ATP-binding domain, indicating a high functional relevance of these protein sites (red Ds in Figure 5A). There are also two such SNPs within the newly identified Zτ domain. Complete information about all 33 highly deleterious SNPs within human TOP2A is supplied in Supplementary Material File S1.

2.3. Both Zτ and GTP-Binding Site Are Phylogenetically Conserved across the Tree of Life

To depict the phylogenetical conservation of identified features (Zτ and putative GTP-binding site) in human TOP2A, we made a multiple sequence alignment of five representative eukaryotic species (Human, Zebrafish, Drosophila, Yeast, and Arabidopsis) (Figure 6). We then constructed a comprehensive multiple sequence alignment of all known metazoan topoisomerases of type IIA, particularly focused on their DNA-binding region (~400 aa). Nearly 1500 sequences were inspected and aligned to the Hidden Markov model logo (HMM) and about 350 sequences containing artifacts or truncated N or C ends were manually removed. In the rest of the sequences (1131), the putative Z-DNA-binding domain Zτ together with the newly identified GTP-binding region were most conserved.

DNA topoisomerases of type IIA (TOP2A, InterPro Domain ID: IPR001241) are highly conserved in Metazoa [64] but can be found across the whole tree of life, as depicted in Figure 7A. In Bacteria, the related gyrase and TopoIV are also found (with some exceptions, such as the order *Corynebacteriales*). They have quite distinct properties and cellular functions [65]. In eukaryotic organisms, it is quite often that particular species contain several duplicated copies of TOP2A. Probably the most important event (from the anthropocentric point of view) occurred early in vertebrate evolution: the duplication leading to the paralog TOP2A and TOP2B genes [64,66]. A viral origin of eukaryotic topoisomerases was recently proposed [67].

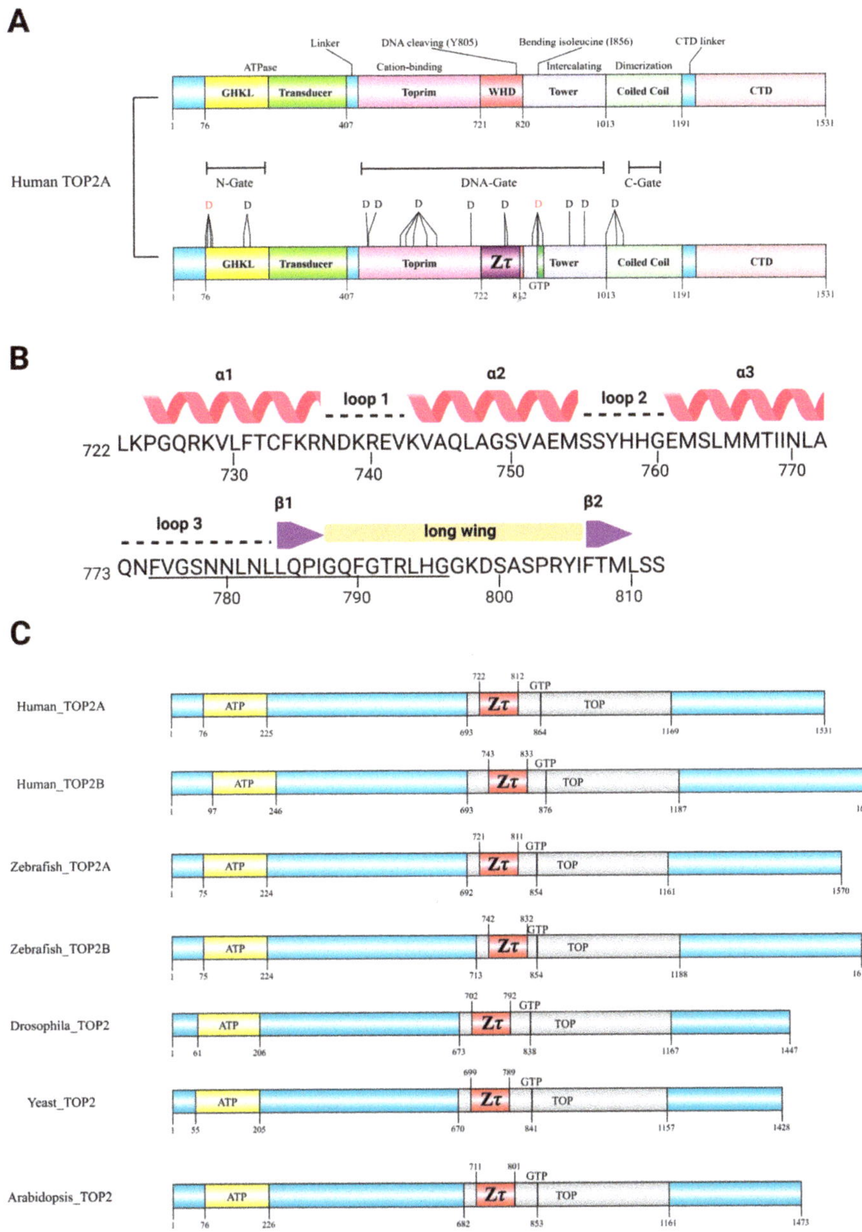

Figure 5. Domain composition of topoII together with putative Z-DNA-binding domains, Zτ (red), and predicted GTP-binding sites. ATP-binding sites (HATPase_c) and TOP (TOP4C) were annotated using a SMART web server [68]. (**A**) Known and newly identified features (Zτ, GTP-binding site, and deleterious SNPs, here depicted as Ds) in human TOP2A. (**B**) Zτ sequence (722–812), secondary structure, including the proposed "Z-discrimination region" (775–796, underlined residues; Table 3 below). (**C**) Evolutionary conservation of Zτ and GTP-binding sites in diverse eukaryotic species.

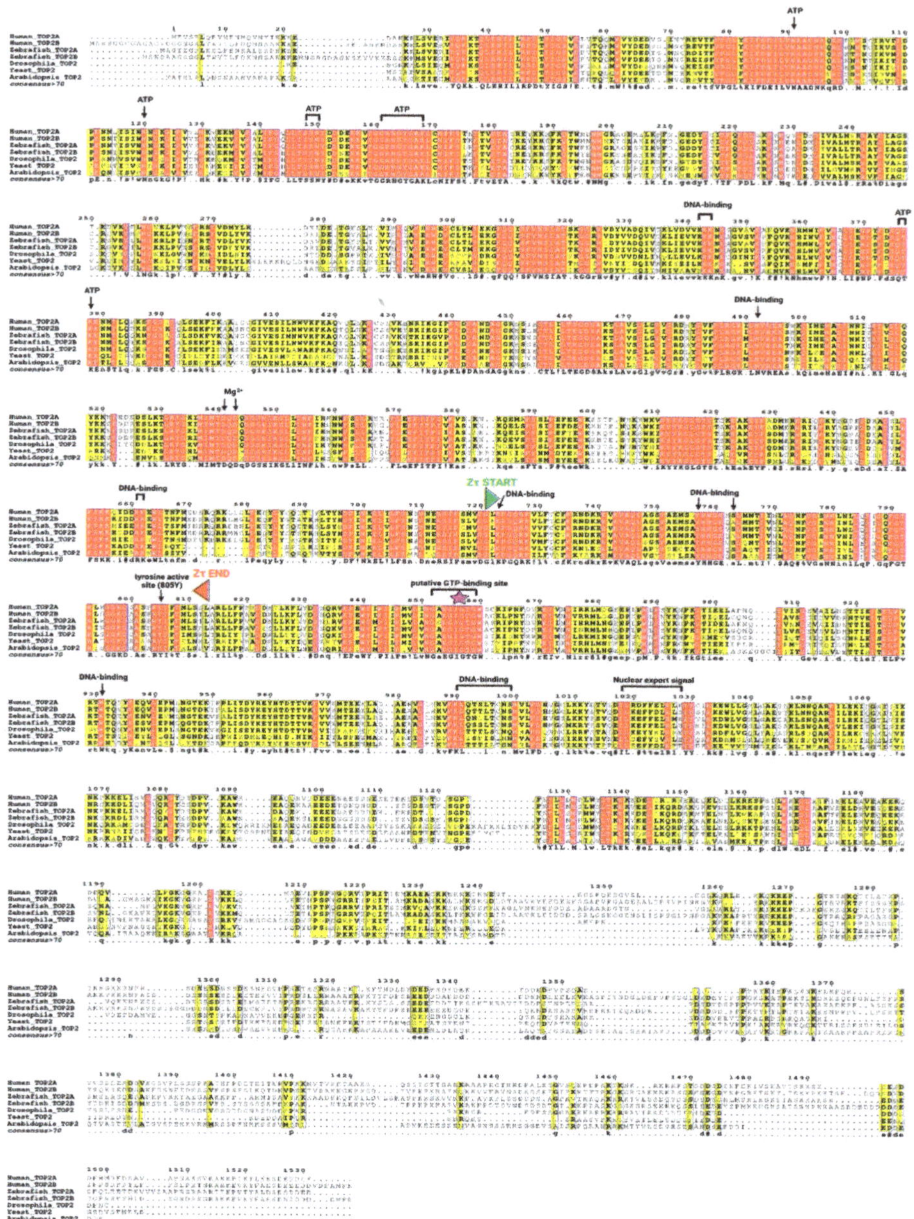

Figure 6. Multiple sequence alignment of TOP2 protein sequence in representative species showing very high conservation of the region $Z\tau$ constituting the $Z\alpha$ structural homolog (bounded by green and red triangle marks) and putative GTP-binding site. The purple asterisk mark highlights the locus critical for DNA bending. Experimentally validated ATP, Mg^{2+}, and DNA-binding sites, together with the critical tyrosine 805 active site, are depicted as well. Columns highlighted in red and yellow show evolutionarily conserved regions/amino acid positions (primary sequence); the resulting consensus sequence is displayed in the bottom row, using criteria from MultAlin [69]: uppercase is identity, lowercase is consensus level > 0.5, ! is any-one of IV, $ is anyone of LM, % is anyone of FY, # is anyone of NDQE.

2.4. Molecular Docking of Various DNA Types to Zτ

We further explored the possible GTP-binding potential of the human TOP2A protein by carrying out a computational docking procedure using a representative crystal structure of human TOP2A with bound DNA (PDB: 4fm9) [44] as a receptor, and GTP as a small ligand. One should note that the DNA in this case was in the B form, inasmuch as an experimental structure with bound Z-DNA is not yet available. Figure 7B indicates GTP docked very close to the predicted GTP-binding region, which also contains a highly conserved GxxG motif (Figure 7C), a key feature of K Homology (KH) domains and one which can provide local stereochemical flexibility [70].

We docked various nucleic acid structures (B-DNA, Z-DNA, Z-RNA, and B/Z-DNA) to the isolated Zτ domain of TOP2A (AlphaFold structure) (Figure 8). It appears that the Zτ domain of human TOP2A may interact with Z-DNA mainly through its α-helix α3 and β1-β2 loop, without the involvement of α-helices α1 and α2. Such results are illustrative yet inconclusive, inasmuch as the full extent of the protein–protein and protein–nucleic acid interactions of the homodimer is not represented. Fortunately, a control parallel exercise with the Zα domain (Table 3) reproduced the majority of residues and contacts denoted as idiosyncratic of the Zα family, based on extensive biophysical characterization [14–19]. The differences in the orientations of the two Z-domains relative to the DNAs are remarkable and await elucidation by high-resolution structure determinations.

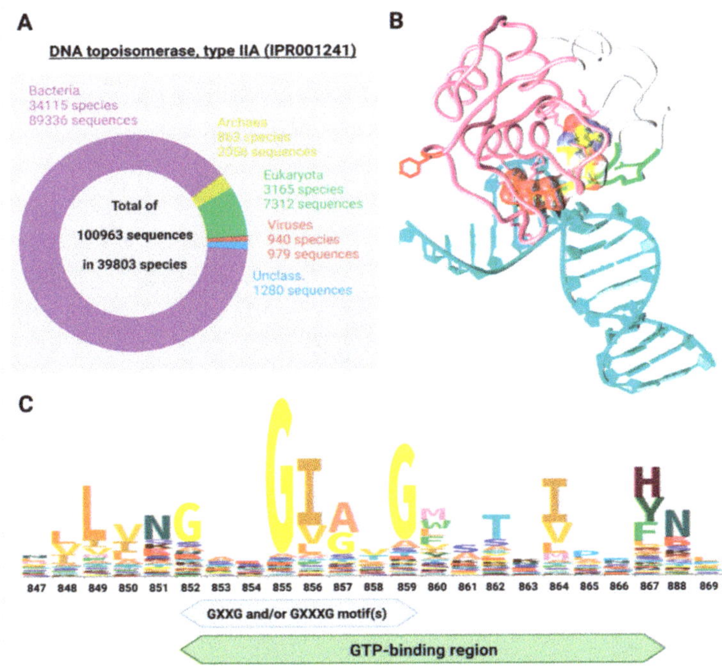

Figure 7. DNA Topoisomerase type IIA diversity (**A**) and GTP-binding (**B**,**C**). (**A**) Known diversity of Topoisomerase type IIA (IPR001241). According to InterPro Database, there are more than 10^5 protein sequences in more than 50,000 different species across the whole tree of life. (**B**) GTP docked to the crystal structure of human topoII (PDB: 4fm9). Only the immediate surrounding of the docked GTP molecule is shown, but it comprises both the predicted GTP-binding region (in green) and Zτ. The tyrosine active site of TOP2A is in red. (**C**) Sequence logo of the region corresponding to the putative GTP-binding site, and the GxxG/GxxxG motif(s) based on seed alignment of the DNA topoisomerase IV (PF00521) domain. The logo was produced using the Skylign tool [71].

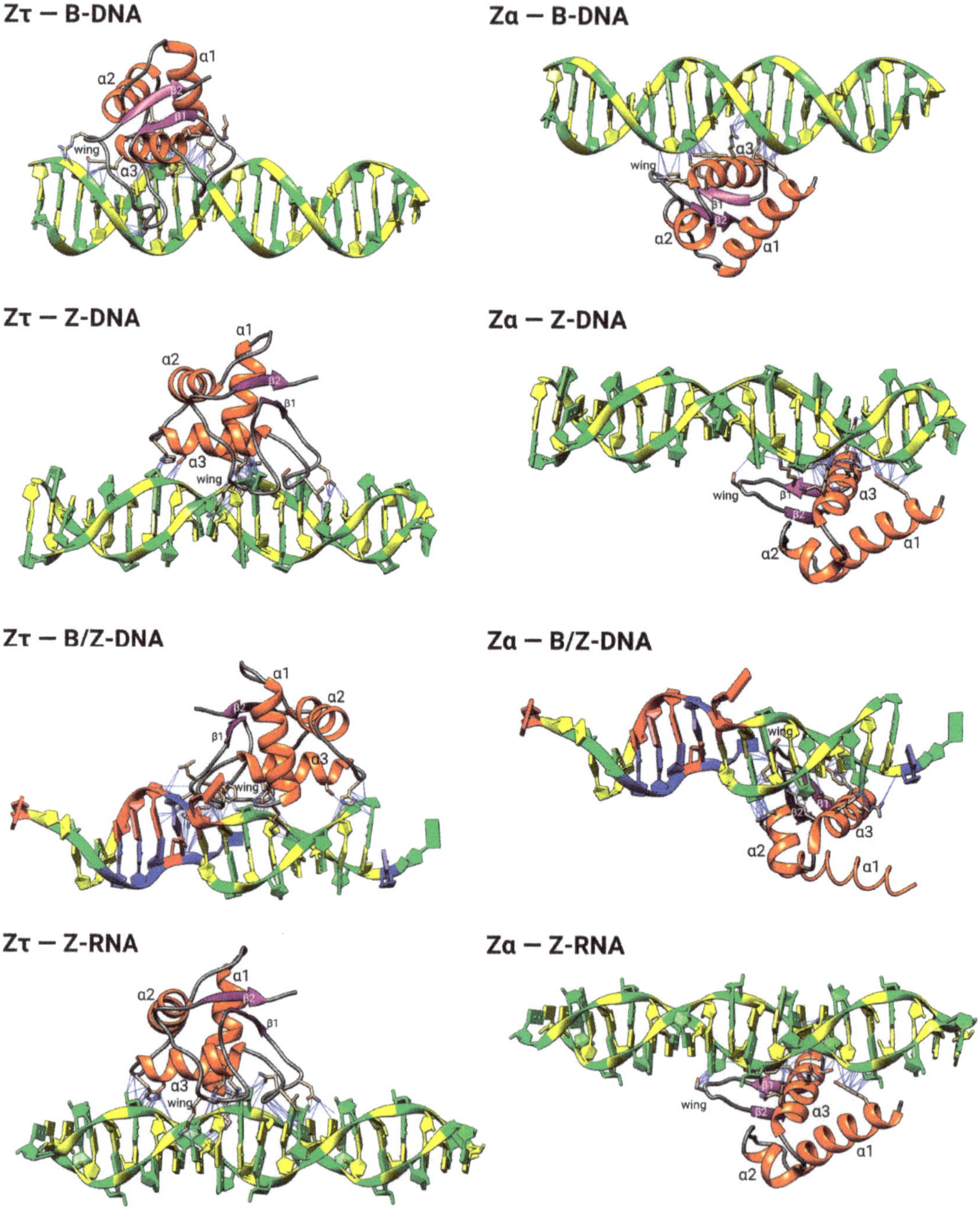

Figure 8. Molecular docking for various interacting protein–DNA pairs. The left column is for Zτ from the human protein TOP2A and the right column for Zα from the human protein ADAR1p150, both docked to the indicated nucleic acid structures (B-DNA, Z-DNA, B/Z-DNA, and Z-RNA).

Table 3. Parameters of protein–nucleic acid complexes obtained by molecular docking of Zτ from human TOP2A and Zα from human ADAR1p150 to left-handed Z-DNA and Z-RNA. Docking procedure: HDOCK. The docking score is calculated by a knowledge-based iterative scoring function; greater negativity usually implies a more feasible binding model. The confidence score empirically indicates the binding likeliness of two molecules (in the range of 0–1). Interacting aa residues in crystal structures 4fm9 (TOP2A in complex with B-DNA) and 3f21 (Zα in complex with Z-DNA) are shown as well. Shaded area: "Z-discrimination region", common for all 3 docking interactions of Zτ with left-handed species (Figure 5B).

	Zτ Docking Model					Zα Docking Model				Crystal	
Parameter	**B-DNA**	**Z-DNA**	**B/Z-DNA**	**Z-RNA**		**B-DNA**	**Z-DNA**	**B/Z-DNA**	**Z-RNA**	**4fm9**	**3f21**
docking score [1]	−172	−175	−213	−208		−136	−179	−165	−193	-	-
confidence score [2]	0.61	0.62	0.78	0.76		0.43	0.64	0.57	0.70	-	-
no. contacts	105	69	91	74		51	63	67	65	39	53
aa residue [3]											
L722 *n*	1				Y136 *a*	1	2		1	K723	K169
K723 *b*	1, 2	1, 2	1	1, 2	H159 *b*		2			Y757	K170
Q726 *p*	2	1	2	2	K169 *b*	2		1, 2		H759	N173
Y757 *a*	2				K170 *b*		1	1	2	S763	R174
H759 *b*	2	2	1	1	N173 *p*		1	1	2	N770	Y177
G760 *n*			1	1	R174 *b*	1	1, 2	1	2	K798	T191
M762 *n*	1				Y177 *a*	2	1, 2	1	1, 2		P192
S763 *p*	2	2	1	1	S178 *p*	1	2		1		P193
T767 *p*	2				K181 *b*	1	2		1		
N770 *p*	2	1	1	2	K187 *b*	2	1		2		
L771 *n*	2				G190 *n*				2		
F775 *n*			1		T191 *p*		1	1 2			
V776 *n*		2	1	1	P192 *n*		2				
G777 *n*		1	1, 2	1, 2	P193 *n*			1			
S778 *p*		2	2								
N779 *p*			2	2							
N780 *p*		1, 2	1, 2	1, 2							
L781 *n*			2								
R793 *b*			1								
G796 *n*			1								
K798 *b*	1	1		2							
S802 *p*	1										
R804 *b*	1										
strand 1/(1 + 2)	6/14	6/12	10/17	7/14		4/7	6/13	7/9	4/10	1	1

[1], more negative, better; [2], higher, better; [3], interaction with strands of nucleic acid double helix: 1, strand 1; 2, strand 2.; crystal structures, strand 1. Amino acid type: *a*, aromatic; *b*, basic; *n*, nonpolar; *p*, polar.

Interesting trends were observed for Zτ (Table 3). The best docking and confidence scores were obtained for Zτ with B/Z-DNA, followed by Z-RNA and Z-DNA. Zτ–B-DNA, a model of the canonical interaction of topoII with DNA, scored worse, a result compatible with the relative affinities for B- and Z-DNAs established for topoII (Table 1). Surprisingly, the docking model for B-DNA had 2.7× the number of interacting amino acid residues in the corresponding crystal structure 4fm9. The relatively high scores obtained with Zτ–Z-RNA raise the question as to whether class II topoisomerases (TOP2A, TOP2B) can bind productively to Z-RNA. In fact, topoII has been implicated in the regulation of viral replication [72], and most identified Z-RNA binding proteins to date have a role in (anti)viral mechanisms. GTP-binding proteins often engage in guanylate-mediated dimerization that endows them with antiviral properties [73].

In addition, Zα in the known ZBPs (ADAR1, DAI, ORF112, E3, PKZ) and the 14 best ZBPs we have predicted previously [26] are invariably located near the N- or C- terminus and are thus spatially exposed. Other regions of these proteins are not presumed to play a key role in the interaction with Z-DNA/Z-RNA.

In contrast, the Zτ of TOP2A occupies a central part of the protein, as does the DNA (G segment), such that other amino acid residues in the TOPRIM and TOWER domains are in close contact with the DNA [74]. For example, two tryptophan residues (W860 and W931) are involved in the crystal structure with bent B-DNA (4fm9). Tryptophan is a critical and well-described Z-DNA-binding residue in the Zα domain [18,20,23], and in 4fm9, W860 is in direct contact with the DNA backbone. In addition, it is at the center of the putative GTP-binding region predicted in this report. Interestingly, a conserved tryptophan in the core domain of rat transglutaminase (TGM2) is essential for catalytic activity [75]. TGM2 is a GTP-binding and hydrolyzing protein as well, interacting with topoII to promote DNA damage repair of DSBs in lung cancer cells [76].

2.5. Expanded "B-Z TopoII" Reaction Mechanism

The results and interpretations of the bioinformatics search featured above coupled with the prior biochemical data summarized in Table 1 constitute compelling evidence for the assertion that topoII possesses an inherent and pronounced affinity for left-handed Z-DNA. We also invoke below a putative capacity of topoII for catalyzing the right-to-left reversal in the helical sense of an attached DNA segment. If such an activity exists, topoII would represent a separate class of ZBPs, distinct from the family of proteins featuring the Zα domain and identified to date [14]. The implications regarding the functional repertoire of both partners (protein, nucleic acid) in the biological cell are profound.

The new mechanism ("B-Z TopoII" scheme) proposed for topoII is depicted in Figure 9 and summarized in Table 4, significantly expands its known repertoire as a topo*isomerase* (topo function 1, **tf1**) by incorporating three new features (**tf2, tf3, tf4**) into the standard topoII model of Figure 3. Acting in concert, these functions are deemed to fulfill essential requirements for maintenance of genomic DNA integrity and function: topological resolution, structural demarcation, and 3D (de)condensation and segregation. One should note that **tf1** targets 2 DNA segments, whereas **tf3** is considered to act on only one. In other words, the two activities "target" writhe ("writhase" or "crossover invertase", [3]) and helical twist ("twistase"), respectively. ATP hydrolysis is essential for the catalytic function of **tf1** [77], but it is unclear whether it would also be required in **tf3**.

Table 4. Expanded functionality of topoII ("B-Z topoII"). The question marks indicate potential RNA targets that have yet to be investigated.

Function	Activity	Target(s)
tf1	*isomerase*: double-helix passage (ΔLk)	B-DNA
tf2	high affinity recognition and stabilization of left-handed double helix; no covalent protein–DNA intermediate	Z-DNA, Z-RNA?
tf3	*conformase*: induction of the right-to-left transition in double-helical sense	B-DNA, A-RNA?
tf4	pronounced positive heterotropic allosteric role of GTP in **tf2** and **tf3**	topoII

Panel A of Figure 9 provides an overview of the scheme, and panel B depicts certain features in greater detail. The apparent "B-Z" symmetry is more apparent than real, because the outcomes of the alternative pathways are quite distinct. In the "B-mode" of action, the topoII homodimer ($\mathbf{T_o}$) is shown to bind and process a B-DNA G-segment by adopting a quaternary configuration, $\mathbf{T_B}$, under allosteric control by ATP. The interaction with DNA leads to a complex, $\mathbf{T_B B}$, with two feasible fates, the first of which is to proceed through the isomerase cycle (Figure 3). The second fate arises if the proposed topo*conformase* activity (**tf3**) is manifested, such that $\mathbf{T_B B}$ undergoes the transformation to $\mathbf{T_Z Z}$. In the alternative "Z-mode" of action, $\mathbf{T_o}$ binds to a *preexistent* Z-DNA G-segment (see below), and the DNA gate (Figure 3) adopts an alternative configuration, $\mathbf{T_Z}$, in the stable complex $\mathbf{T_Z Z}$ and does not proceed beyond stage 1 of Figure 3. $\mathbf{T_B}$ incorporates the conformational mechanisms coordinating the inter-subunit interactions required for DNA cleavage [44]; $\mathbf{T_Z}$ extends this notion to the quaternary structure favored for Z-DNA recognition.

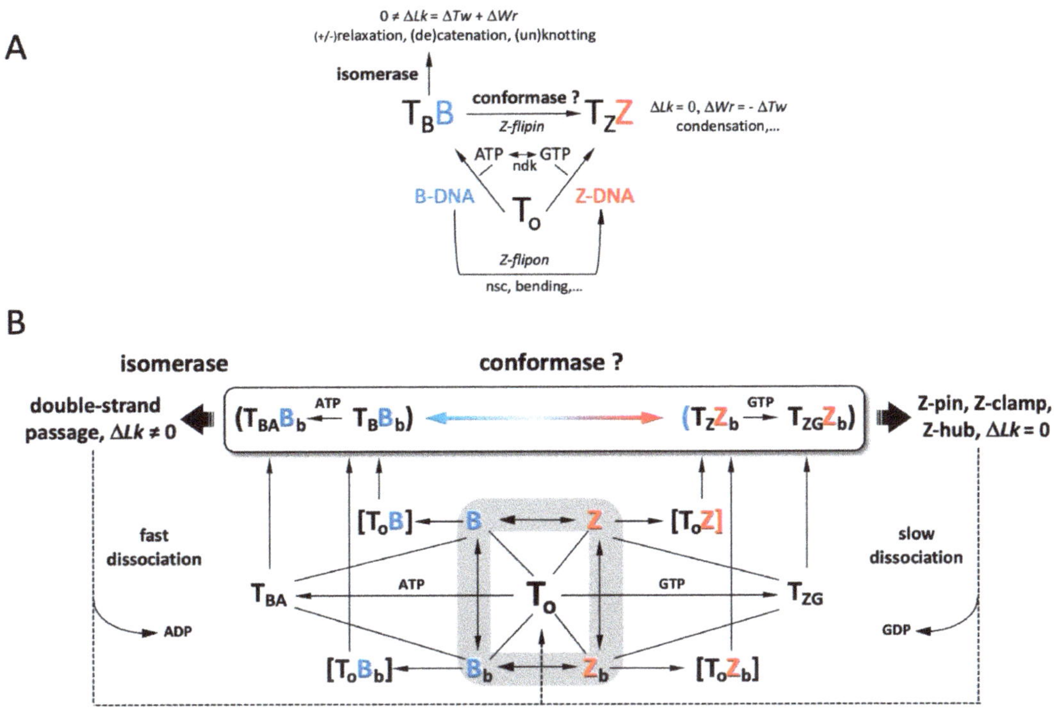

Figure 9. "B-Z-TopoII": expanded reaction mechanism of topoII incorporating Z-DNA binding and a postulated conformase activity in addition to its canonical isomerase function. (**A**) Functional scheme, explained in the text. (**B**) Details of intermediate states (in square brackets) and outcomes of the isomerase and conformase pathways. The gray shaded area comprises the "DNA manifold" with interconversions between linear and bent conformations that depend on sequence, topological state, solution conditions, and external factors. Straight lines denote interactions between binding partners, leading to reactions (lines with arrows; in reversible reactions, a larger arrowhead indicates preferential state). Configurations of the topoII homodimer: $\mathbf{T_o}$, free; $\mathbf{T_B}$; bound bent ($\mathbf{B_b}$) B-DNA (G-segment); $\mathbf{T_Z}$; configuration bound to bent ($\mathbf{Z_b}$) (G-segment); $\mathbf{T_{BA}}$, $\mathbf{T_B}$ with bound ATP; $\mathbf{T_{ZG}}$, $\mathbf{T_Z}$ with bound GTP. Adapted from Figure 1 in Ref. [78].

GTP exerts a positive heterotropic allosteric influence on $\mathbf{T_Z Z}$, increasing its thermodynamic stability even more (property 13, Table 1). At the same time, it profoundly inhibits isomerase activity (properties 11, 12, 14, Table 1). $\mathbf{T_Z Z_b}$ $\pm$ GTP constitutes a highly stable topoII-Z-DNA end-state complex, with three possible consequences, neither of which leads to a change in the global ΔLk inasmuch as strand/helix passage is not involved. One eventuality is to clamp the distribution of supercoiled states within the topological domain encompassing the site of action. A second possible consequence is to act as a temporary, local storage site, maintaining the temporarily inactivated enzyme in nearby proximity for a subsequent required function. In this connection, it is relevant that the active site tyrosine lies *within* the Zτ segment of topoII, in contrast to the sequence-separated DNA-binding and catalytic elements of ADAR1p150. In the case of topoII, Z-DNA binding leads to enzymatic *inactivation*, whereas with ADAR1p150, Z-RNA recognition leads to *activation*.

TopoII possesses both ATPase [77] and GTPase (Figure 7 of Ref. [34]) activities. Considering the much greater efficacy of non-hydrolyzable GTP (Table 1), the scheme of Panel A presumes that the disruption of the $\mathbf{T_Z Z}$ complexes occurs (albeit slowly, Figure 9B) upon hydrolysis of GTP. After the release of GDP, the protein reverts to its initial $\mathbf{T_o}$ state and the DNA to its basal conformation dictated by the microenvironment. Finally, the $\mathbf{T_Z Z}$

complex is deemed to serve as a recognition and/or structural element for establishing higher order single- or multicomponent complexes. There are two additional features in the scheme of panel A to note. One of them is the potential equilibration between local B- and Z-conformations in the *absence* of topoII, i.e., depending on factors such as DNA sequence, state of deformation from torsion (supercoiling), tension (stretching), bending, and solution parameters (temperature, salt composition, small effector molecules such as polyamines, ionic strength, dielectric constant). Such sequence domains that can fluctuate relatively easily between the left- and right-handed helical conformations constitute the class of ***Z-flipons*** introduced by Allan Herbert for DNA (and RNA) sequences that "flip" into the Z-conformation, are recognized by ZBPs [79] and thereby exert a number of cell biological functions [17,18,20]. For sequences that *require* the putative conformase function of topoII to adopt and stabilize a left-handed conformation (**fp3**, Table 4), we propose the new term ***Z-flipins***. The other noteworthy feature is the exchange reaction mediated by ubiquitous nucleoside diphosphate kinases (ndks), permitting the facile interconversion of ATP and GTP via their respective diphosphates.

Panel B of Figure 9 provides a more detailed view of "B-Z-TopoII". It stresses the key role of a particular, essential feature of topoII isomerase function (Figure 3), the *bending* of the G-segment [80–82]. Such a "bind-then-bend" mechanism [83] is shared with many nucleic-acid-binding proteins [84–86], and bending has been newly demonstrated to constitute a physical means for promoting *by itself* the B-to-Z transition under physiological conditions [87]. The latter finding provided a major impetus for proposing a *conformase* capability of topoII. Pre-bent DNA is a preferred binding target for topoII ([31]; property 7, Table 1) and enzyme-induced flexibility is invoked as the means for selecting cleavage sites [88], but the claim of a potential conformase activity presumes an *obligatory* intervention by the protein in order for the deformation (bending)-facilitated B-Z transition to occur.

In the absence of direct structural data, we can only speculate as to whether the entire length of the bound G-segment would adopt the Z-conformation in a concerted reaction [7] or whether a sequential transformation, such as BjBjB → BjZjB → ZjZjZ → Z (j, unpaired junction) would be more likely, perhaps even involving the Z(WC)-DNA alternative left-handed double helix proposed to overcome the inherent "chain-sense paradox" of crystallographic Z-DNA [89,90]. It is also conceivable that the isomerase prescission intermediate exhibiting the A-DNA conformation within the G-segment DNA [80] may participate in a BjBjB → BjAjB → BjZjB → ZjZjZ → Z conformase reaction sequence. A- and Z-DNA share certain features: dependencies on hydration state and counter-cations, water bridging of free phosphate oxygens, and purine sugar pucker, and the B, A, Z interconversion landscape is exceedingly complex (90], Figure 2 of Ref. [91]). An intriguing question is whether the existence of an obligatory A-conformation intermediate in the isomerase cycle implies a B → A "conformase" property for this core function (**tf1**) of topoII as well as B → Z in **tf3**.

The other currently known class of Zα based ZBPs share many structural and functional properties [14–19], but these do not include DNA bending. Z-DNA "inducibility" is also ascribed to some of these ZBPs, yet in our estimation, unambiguous experimental demonstration is lacking for a *catalytic* activity (with turnover), instead of, or in addition to, the selective binding preference for the left-handed conformation. This issue arose early in Z-DNA research in relation to anti-Z-DNA antibodies, but can and has been resolved kinetically (Figure 10 and Figure 11 of Ref. [92]).

In the isomerase pathway, the "cleavage-competent" bending of the G-segment DNA (step 2, Figure 8) [51] is accomplished by Mg^{2+} coordination to the TOPRIM domain [93] without involvement of direct amino acid side chain-base contacts [80]. Instead, a topoII-invariant isoleucine intercalates into and widens the minor groove, thereby increasing DNA rise and roll while decreasing twist and, thus, the charge density of the helix [94], effects which by themselves would also favor the B-to-Z transition and thereby enable the alternative conformase pathway. This "local conformational micropolymorphism" [95] provides the flexibility required to achieve a bend of ~120–150° [80,96]. Furthermore, in vitro studies have demonstrated that the application of moderate tension (stretching) greatly reduces the requirement for torsional stress (untwisting) in the B-Z transition induced by supercoiling [97]. Structurally, the finding that GTP binding dramatically increases the affinity for Z-DNA while decreasing the affinity for B-DNA (Table 1) implies that the quaternary conformations $\mathbf{T_Z}$ and $\mathbf{T_{ZG}}$, as well as $\mathbf{T_B}$ and $\mathbf{T_{BA}}$, differ in significant ways, and DNA stretching (longitudinal tension) may well be involved. More generally, it appears that topoII is representative of proteins that engage nucleic acids in a manner that exploits the capacity of *both* macromolecules to undergo mutual conformational adaptations that provide thermodynamic stability and specific recognition via enthalpic-entropic compensation [94,98,99]. In the case of the DNA, the nucleotide sequence is a (the) major factor. Yet, in a real way, the conformase activity that we are invoking for topoII encompasses *both* the protein and DNA, which is to say that "the enzyme and the substrate are one" (the biochemical equivalent of "it takes two to tango"). CRISPR-Cas9 is a prominent albeit complex example of such concerted conformational adaptation. Protein recognition and catalytic activation ensue upon an open-to-closed domain rearrangement in concert with DNA twisting, bending, and base flipping, all pursuant to initial successful pairing of the guide RNA and R-loop formation [100]. An intriguing Z-to-B-DNA remodeling protein (ZBTB43) has also been reported [25].

Another notable feature of the Z-mode function depicted in Figure 9 is its temporal behavior. Simulations of the system originating from B-DNA and the $\mathbf{T_0}$ state can exhibit damped oscillatory responses culminating in steady-state levels of free and protein-associated Z-DNA and topoII-Z-DNA complexes. The system may thus comprise an inherent memory" property, distinct yet related to the rationalizations of the capacity of topoII to generate topological distributions "beyond thermodynamic equilibrium" [101]. In the latter case, the inherent supercoil-dictated directionality of the DNA-gating mechanisms (Figure 8, [2,45]) and the perturbed counterion distributions of juxtaposed helices [94] undoubtedly contribute. However, in the Z-mode of topoII, the longevity of exposed (B)-Z-(B) segments would be a primary factor, and they may account for the hysteretic behavior reported for complexes of anti-Z-DNA antibody with supercoiled ccDNA carrying Z-forming inserts [102].

2.6. A Case Study of the "B-Z TopoII" Mechanism: Mitosis

We now test the applicability of the "B-Z TopoII" mechanism to rationalize aspects of cellular mitosis, the autopoietic [103] process underlying cell division in which the expression of topoII peaks (at G2-M, there are ~10^6 molecules/cell [104]) as it executes the essential functions of chromosomal DNA condensation and then segregation [46,105,106], distinct from its contributions to genome stability and organization in interphase [3]. The sequential progressive stages of the mitotic cell cycle (G2-interphase → prophase → prometaphase → metaphase → anaphase + telophase → cytokinesis) are precisely choreographed [105–112] and are accessible to high-resolution microscopy [105]. In prophase, the topologically associated domains (TADs) of interphase are disrupted, and the 6.3 Gbps (human diploid) DNA is organized by condensin II in a process of loop extrusion (LE) into ~$4{\cdot}10^4$ loops of ~450 kb. These are fixated at their base by dynamic [113] ring-shaped protein complexes (SMC, structural maintenance of chromosomes) aligned so as to form the axes of the sister chromatids. In prometaphase (and again in anaphase), the primary condensin II loops are further partitioned, 5–10-fold, by condensin I association into ~$2{\cdot}10^5$ nested ~90 kb off-axis subloops (the ~1 topoII/condensin-1-loop stoichiometry is intriguing), resulting in pronounced DNA overwinding and progressive compaction. The latter continues and achieves a maximum, ~10^4-fold, *after* sister chromatid separation in anaphase [105,114].

In metaphase, TopoII is the most abundant protein component of the chromosome scaffold, followed by condensins I and II and chromokinesin KIF4 [106,107]. TopoII is bound to the chromosome axes and centromeres and is a key and indispensable participant in the processes outlined in the preceding paragraph [46,115,116]. Particular emphasis has been placed on the interplay between the respective roles of topoII and condensin II/KIF4 [112]. This focus arises because the fundamental activities of the two components (LE/compaction vs. topological simplification) are seemingly antagonistic in the coordination of DNA condensation with the equally requisite and concurrent elimination of spurious knots, tangles and sister chromatid interlinks [115]. Lateral chromosomal compaction has been attributed to condensin and KIF4 and axial compression to topoII [117]. Adding to the complexity of the system are: a dual driver–damper role of two condensin ATPases [108]; the intervention of other topoisomerases, chromokinesins, cohesin, and cyclins; and extensive temporally synchronized protein modifications [118], notably (de)phosphorylation and sumoylation, such as of the DNA-gate and the C-terminal domain (CTD) of topoII [41,112]. The mitosis literature abounds with conundrums, assertions, and still open questions, including the following:

- What are the ultrastable topoII-DNA complexes that play a structural role in chromosome architecture? [43]
- Do centromeres *drive* chromosome compaction?
- How do non-B-DNA centromere sequences participate to (de)condensation?
- How does topoII contribute to axial shortening of the chromosomes [117]?
- How is cohesin release coordinated spatiotemporally with the actions of condensin and topoII in sister chromatid resolution [119]?
- What are the kinetic pathways of topology simplification in metaphase chromosomes [47,120]?
- How is *large-scale* compaction and spatial arrangement achieved [109]?

A detailed treatment of the above is beyond the scope of this publication. Yet, we can invoke features of the "B-Z-TopoII" mechanism of Figure 9 to address some of the issues. We start by noting that evidence exists for localized protein interactions with genomic DNA, including: the recruitment of topoII to SAR/MAR (nuclear scaffold/matrix attachment) sites [43], which can assume a variety of non-B-DNA conformations [121]; the interactions of flipons and nucleosomes [122]; and 40 years of chromosome immunochemistry with anti-Z antibodies, revealing localized binding to heterochromatin. At this juncture, we propose the following scenarios incorporating B-Z-TopoII in mitosis as worthy targets for experimental verification. The aim is to specify a robust mechanism, one applicable to all chromosomes and organism expressing a topoII, and mindful of Ref. [106]: "Our data point to a role for TOP2A as a structural chromosome maintenance enzyme locking in condensation states once adequate compaction is achieved".

GTP fulfills multiple functions in the cell: nucleic acid precursor, energy source, and messenger/allosteric regulator of protein synthesis, cytoskeleton dynamics, intracellular transport, signaling, and organelle function [123]. In contrast, ATP is utilized as a (the) general cellular energy carrier and phosphoryl donor. The mean cellular GTP concentration is ~1/10th that of ATP (<1 mM, >1 mM, respectively [124]); both are under tight metabolic regulation [103]. However, the synthesis of GTP is compartmentalized, leading to the notion that its production—by nucleotide salvage, de novo biosynthesis, and nucleoside diphosphate kinase activity—and consumption may generate gradients that affect cellular phenotypes in accordance to the immediate spatiotemporal demands of the cell [123]. The metaphase–anaphase stages of mitosis are such a case because chromosome segregation requires sister kinetochores at the centromere to attach microtubules emanating from opposite spindle poles. The small GTPase, Ran-GTP, promotes spindle assembly around chromosomes [125,126] by locally delivering cargoes (importin-bound spindle assembly factors, SAFs) that regulate microtubule dynamics and organization [127]. Because RCC1, the RanGEF (Ran guanine nucleotide exchange factor), is chromatin associated, a strong negative gradient of activated Ran-GTP is established, radiating from the kinetochores to the spindle poles [126,127]. Growing microtubules, associated motor proteins, and Ran-GTP require GTP hydrolysis for function. The local levels of GTP must be accordingly high.

We recall from Table 1 (properties 7, 8) that topoII exhibits a graded affinity for non-canonical DNAs and now postulate that high prometaphase levels of GTP at the centromere convert a substantial fraction of the resident topoII molecules to the $\mathbf{T_{ZG}}$ species via function **tf4** (Table 4). These lead to chromosomal compaction at the centromeres, where topoII and preexistent and/or potential Z-form segments of α-satellite DNA are concentrated [128–131]. This process occurs via (a) function **tf3** (binding to flipons, e.g., at the base of condensin-1,2 loops); and (b) function **tf2** at suitable loci (e.g., flipins at loop apices), given appropriate conditions of DNA helical bending, tension, torsion, and sequence. Isomerase function **tf1** is inoperative except at positions of high topoII occupancy [132], where limited GTP-driven rounds of catenation and knotting, both contributing to compaction, can occur (property 16, Table 1). The dimerization capacity of certain GTP proteins [73] alluded to earlier implies that topo-topo crosslinks, as well as Z*-DNA, a self-associated form of Z-DNA [8], may contribute to the axial compression evident during and after metaphase. Topological "redistribution" is also a key factor in normal compaction (from the Abstract to Ref. [133]: "The results suggest that the local deformation caused by protein binding can yield a global configurational change, dominated by slithering, which brings two (originally) remote sites to close proximity, and that the nature of such effect is related to the sequence architecture."). The great affinity of topoII for Z-DNA would confer a temporal stability in the metaphase stage, and it is perhaps indicated by the fractional non-recoverable population in FRAP determinations performed on mitotic chromosomes [40]. As in the case of the inhibitor, etoposide [53], loop trapping at Z-clamps may block sliding of topoII on the DNA, thereby increasing its action as a roadblock.

Upon exit from metaphase, the Ran-GTP gradient and high GTP concentration dissipate, and cohesin is released from the centromeres, unlocking the sister chromatids [116,134,135]. Flipins revert to the B-conformation as topoII is released, regaining the **To** conformation with isomerase (**tf1**) functionality. It can thus proceed to decatenate and unknot residual inter-chromosomal links, insuring error-free segregation. The chirality dependence of human topoII dynamics (+ over − supercoiled DNA) may also be a factor [53].

Is there any evidence for the mechanisms proposed above? Mutants with phenotypes indicative of selective inactivation of the individual functions of Table 4 would be relevant. The literature is indeed replete with mutations of topoII, particularly in reference to topo inhibitors/"poisons", and their distinctive influence on isomerase function and processivity [136]. However, it is difficult to conceive of unambiguous selection strategies for isomerase+/Z-binding- mutants in view of the functional overlaps envisioned in the proposed "B-Z topoII" scheme. Nonetheless, such a phenotype may apply in the case of a reported allele, *top-2(it7ts)*, of TOP-2, the single topoisomerase II homolog in *C. elegans* [137]. An arginine → cysteine (R → C) missense mutation at residue 828 (corresponding to residue 793 of huTOPIIα) leads to failure of segregation during anaphase I of meiosis, resulting in anucleate sperm. The segregation defects are not due to residual entanglements incurred during meiotic DNA replication, implying a possible **tf1+/tf2(tf3,tf4)-** (Table 4) status of topoII. The authors write: "We propose that TOP-2 localization during late pachytene positions the protein to function in chromosome condensation/karyosome formation prior to the meiotic divisions. When TOP-2 localization is disrupted in the *top-2(it7ts)* mutant, either abnormal or insufficient chromatin remodeling occurs during late prophase resulting in aberrant chromosome segregation". A second publication [138] deals with the sensitivity of wild strains of *C. elegans* to etoposide poisons depending upon whether they carry a methionine → glutamine (M → Q) substitution in TOP-2, residue 797. The non-polar methionine increases hydrophobic interactions between the protein and the etoposide, resulting in increased genomic instability. Residue 797 is conserved from yeast to humans but exhibits one of the few differences between the two human topoII isoforms (huTOPIIα M762, huTOPIIβ Q778). HuTOPIIα M762 and R793, featured in the two cited publications are identified as DNA interactors in our Zτ docking model (Table 3). R793 is located in what we have designated as the "Z-discrimination region" such that the charge altering R → C substitution would be very significant.

We conclude that while the above discussion of mitosis does not incorporate the complex interplay of myriad other proteins, including histones, and their programmed modifications, it provides a plausible cellular context for the B-Z-TopoII scheme of Figure 9. The potential for extending the concepts to detailed treatments of meiosis and interphase chromatin is obvious.

2.7. Perspectives and Biomedical Outlook

This publication offers new paradigms for the biological relevance of left-handed double-stranded DNA (RNA?) and for the functions of Type II (and possibly other) topoisomerases. Confirmation, elaboration, and extension will require substantial efforts in the fields of molecular, cellular, and structural biology, including ultrahigh-resolution imaging [139] but also in medical pharmacology. The Z-DNA related properties of topoII described in this publication, potentially shared with some of its interactome partners [140], offer the prospect of new antiproliferative compounds, pharmacologically complementary to the numerous existing anticancer drugs targeting the isomerase mechanism of the protein. Current strategies for topoII-based drug discovery [56–58,141,142] lend themselves to this goal. GTP-binding site-specific compounds based on non-hydrolyzable nucleotides, isomerase inhibitory purine scaffolds [143] or suitably adapted ATP-competitors [144] would introduce a new dimension of target selectivity. Small molecule Z-DNA interactors or inducers [22] are of potential interest as well. Combination therapy also lends itself to physical means for precision targeting minimizing off-target toxicity, for example, by exploiting superparamagnetic nanocarriers, click chemistry, and magnetic focusing.

3. Materials and Methods

3.1. Structural Similarity Analysis of Human Zτ (TOP2A) and Various Proteins Containing Zα Domains

As a representative example of Zτ (from human TOP2A) and Zα (from human ADAR1) structural similarity (Figure 3), the following structures were used: Cryo-EM structure of human topoisomerase IIα DNA-binding/cleavage domain in State 1 (PDB: 6zy5) [48], and crystal structure of Zα in complex with d(CACGTG) (PDB: 3f21) [145]. The structural similarity for statistical comparison (Table 2) was computed using the FATCAT approach [60] (accessed from https://fatcat.godziklab.org/fatcat/fatcat_pair.html, 25 February 2023) with the flexible alignment mode. AlphaFold-modelled [146] PDBs obtained from the UniProt database [147], i.e., human ADAR1p150 (P55265), human ZBP1 (Q9H171), PKZ from *Danio rerio* (Q5NE12), E3 from *Vaccinia virus* (P21605), and ORF112 from *Cyprinid herpesvirus 3* (A4FTK7), and negative controls UFM1 (P61960) and H4 (P62805) from *Homo sapiens*. Only regions corresponding to particular Zτ and Zα domains were always used as input (Supplementary Material File S3). To visually show the structural similarity of particular regions, UCSF Chimera molecular modeling system [148] and toolkit [149] were used.

3.2. Searching for Putative GTP-Binding Sites within Topoisomerases

Putative GTP-binding sites (together with ATP-, ADP-, AMP-, and GDP-binding sites) within protein sequences of interest were predicted using a Nsitepred web server [61] (accessed from http://biomine.cs.vcu.edu/servers/NsitePred/, 2 January 2023). This tool computes the so-called binding probability for both GTP/GDP/ATP/ADP/AMP (on a scale of 0–1) for each amino acid residue. Default parameters were used, and raw results were obtained in tabular format and further processed/filtered/described in Microsoft Excel (these processed data are available as Supplementary Material File S2).

3.3. Searching for Deleterious SNPs within Human TOP2A

Deleterious SNPs with missense consequence in the human *TOP2A* gene were extracted from the ENSEMBL Variation resources [150] accessed from https://www.ensembl.org/Homo_sapiens/Gene/Variation_Gene/Table?db=core;g=ENSG00000131747;r=17:40388525-40417896, 9 March 2023. Strict filtering criteria were used: SIFT [151] score ≤ 0.05;

PolyPhen [152] $\geq$0.95; REVEL [153] $\geq$0.65; and Mutation Assessor [154] score $\geq$ 0.9. Only SNPs meeting all criteria at the same time were chosen as highly deleterious ones.

3.4. Multiple Sequence Alignment of Full-Length TOP2 Protein Sequences

Seven representative and phylogenetically diverse protein sequences of Topo II, including Human TOP2A (UniProt Protein ID: P11388), Human TOP2B (UniProt Protein ID: Q02880), Zebrafish TOP2A (UniProt Protein ID: Q5PQY4), Zebrafish TOP2B (UniProt Protein ID: Q1LUT2), Drosophila Top2 (UniProt Protein ID: P15348), Yeast TOP2 (UniProt Protein ID: P06786), and Arabidopsis TOP2 (UniProt Protein ID: P30182), were used to construct multiple sequence alignment within UGENE standalone tool [155]. The following parameters were used: mode MUSCLE default, max iterations = 3, -stable (do not rearrange sequences). ESPript 3.0 tool [156] (accessed from https://espript.ibcp.fr/ESPript/ESPript/, 1 January 2023) was used for alignment rendering and producing figures for publication. The following sequence similarities depiction parameters were used: Similarity coloring scheme—%Equivalent (a percentage of equivalent residues was calculated considering physico-chemical properties); Global score = 0.7; Display consensus seq: -yes. Alignment color scheme: -flashy; the number of columns: 140. Additional functional features (active sites, etc.) were added manually using Biorender, according to UniProt [147] features (accessed from https://www.uniprot.org/uniprotkb/P11388/entry, 2 January 2023).

3.5. Molecular Docking

Molecular docking was performed using the HDOCK tool [157] (accessed from http://hdock.phys.hust.edu.cn/, 27 January 2023). The following structures were used as an input: AlphaFold structure of Zα domain of human protein ADAR1 (AF-P55265-F1, region corresponding to amino acid residues 133–199); AlphaFold structure of putative Zτ domain of Topo II (AF-P11388-F1v2, region corresponding to amino acid residues 722–812); CG dodecamer in B-DNA form modeled on x3DNA-DSSR webserver (accessed from https://x3dna.org/, 5 January 2023) [158]; CG dodecamer in Z-DNA or Z-RNA forms modeled using 3D-NuS webserver (accessed from https://iith.ac.in/3dnus/, 3 January 2023) [159]; and crystal structure of B-Z junction obtained from RCSB PDB database (accessed from https://www.rcsb.org/structure/5zup, accessed on 7 January 2023) [160]. As a "Receptor molecule", protein structures were used, and the structures of nucleic acids were always designated as "Ligand molecule". Default parameters for docking procedures were used, except for our choice to use template-free docking only. Obtained models were then visualized in UCSF Chimera [148] and are enclosed in pdb formats in Supplementary Material File S5. GTP molecule (obtained from https://pubchem.ncbi.nlm.nih.gov/compound/guanosine-triphosphate, 2 February 2023) was docked to the structure of human TOP2A (PDB: 4fm9) using the PATCHDOCK web server for rigid docking with default parameters [161], and the resulting model is also enclosed in pdb format as Supplementary Material File S6.

Supplementary Materials: The following supporting information can be downloaded at: https://www.mdpi.com/article/10.3390/ijms241310740/s1.

Author Contributions: Conceptualization, T.M.J. and M.B.; methodology, M.B. and T.M.J.; validation, D.J.A.-J., J.Č., P.P. and T.M.J.; formal analysis, M.B., K.S. and T.M.J.; investigation, M.B., K.S., J.Č., P.P., D.J.A.-J. and T.M.J.; resources, T.M.J. and M.B.; data curation, M.B. and K.S.; writing—original draft preparation, M.B. and T.M.J.; writing—review and editing, T.M.J., M.B., J.Č. and D.J.A.-J.; visualization, T.M.J., M.B. and K.S.; supervision, T.M.J. and P.P.; project administration, T.M.J. and P.P.; funding acquisition, T.M.J., D.J.A.-J. and P.P. All authors have read and agreed to the published version of the manuscript.

Funding: This research was funded by the University of Ostrava (SGS10/PřF/2022 to K.S. and P.P., and SGS05/PřF/2023 to K.S.), and by a Project Grant of the Manfred Eigen Foundation to T.M.J. and D.J.A.-J. Open Access funding provided by the Max Planck Society.

Institutional Review Board Statement: Not applicable.

Informed Consent Statement: Not applicable.

Data Availability Statement: All data are available in the manuscript and within Supplementary Materials.

Conflicts of Interest: The authors declare no conflict of interest. The funders had no role in the design of the study; in the collection, analyses, or interpretation of data; in the writing of the manuscript; or in the decision to publish the results.

References

1. Wang, J.C. *Untangling the Double Helix: DNA Entanglement and the Action of the DNA Topoisomerases*; Cold Spring Harbor Laboratory Press: New York, NY, USA, 2009; ISBN 978-0-87969-863-8.
2. Hanke, A.; Ziraldo, R.; Levene, S.D. DNA-Topology Simplification by Topoisomerases. *Molecules* **2021**, *26*, 3375. [CrossRef]
3. Pommier, Y.; Nussenzweig, A.; Takeda, S.; Austin, C. Human Topoisomerases and Their Roles in Genome Stability and Organization. *Nat. Rev. Mol. Cell Biol.* **2022**, *23*, 407–427. [CrossRef] [PubMed]
4. Vidmar, V.; Vayssières, M.; Lamour, V. What's on the Other Side of the Gate: A Structural Perspective on DNA Gate Opening of Type IA and IIA DNA Topoisomerases. *Int. J. Mol. Sci.* **2023**, *24*, 3986. [CrossRef] [PubMed]
5. Neidle, S.; Sanderson, M. *Principles of Nucleic Acid Structure*, 2nd ed.; Academic Press: Cambridge, MA, USA, 2021; ISBN 978-0-12-819678-6.
6. Du, Y.; Zhou, X. Targeting Non-B-Form DNA in Living Cells. *Chem. Rec.* **2013**, *13*, 371–384. [CrossRef]
7. Pohl, F.M.; Jovin, T.M. Salt-Induced Co-Operative Conformational Change of a Synthetic DNA: Equilibrium and Kinetic Studies with Poly(dG-dC). *J. Mol. Biol.* **1972**, *67*, 375–396. [CrossRef] [PubMed]
8. Jovin, T.M. The Origin of Left-Handed Poly[d(G-C)]. In *Z-DNA: Methods and Protocols*; KIm, K.K., Subramani, V.K., Eds.; Springer: New York, NY, USA, 2023; pp. 1–32. [CrossRef]
9. Wang, A.H.-J.; Quigley, G.J.; Kolpak, F.J.; Crawford, J.L.; van Boom, J.H.; van der Marel, G.; Rich, A. Molecular Structure of a Left-Handed Double Helical DNA Fragment at Atomic Resolution. *Nature* **1979**, *282*, 680–686. [CrossRef] [PubMed]
10. Drew, H.; Takano, T.; Tanaka, S.; Itakura, K.; Dickerson, R.E. High-Salt d(CpGpCpG), a Left-Handed Z′ DNA Double Helix. *Nature* **1980**, *286*, 567–573. [CrossRef]
11. Kim, D.; Subramani, V.K.; Park, S.; Lee, J.-H.; Kim, K.K. Z-DNA. In *Handbook of Chemical Biology of Nucleic Acids*; Sugimoto, N., Ed.; Springer Nature: Singapore, 2022; pp. 1–29. ISBN 9789811613135.
12. Krall, J.B.; Nichols, P.J.; Henen, M.A.; Vicens, Q.; Vögeli, B. Structure and Formation of Z-DNA and Z-RNA. *Molecules* **2023**, *28*, 843. [CrossRef]
13. KIm, K.K.; Subramani, V.K. (Eds.) *Z-DNA: Methods and Protocols*; Springer: New York, NY, USA, 2023. [CrossRef]
14. Herbert, A. Z-DNA and Z-RNA in Human Disease. *Commun. Biol.* **2019**, *2*, 7. [CrossRef]
15. Ravichandran, S.; Subramani, V.K.; Kim, K.K. Z-DNA in the Genome: From Structure to Disease. *Biophys. Rev.* **2019**, *11*, 383–387. [CrossRef]
16. Kim, C. How Z-DNA/RNA Binding Proteins Shape Homeostasis, Inflammation, and Immunity. *BMB Rep.* **2020**, *53*, 453–457. [CrossRef]
17. Herbert, A. The Simple Biology of Flipons and Condensates Enhances the Evolution of Complexity. *Molecules* **2021**, *26*, 4881. [CrossRef]
18. Herbert, A. Z-DNA and Z-RNA: Methods—Past and Future. In *Z-DNA: Methods and Protocols*; Springer: New York, NY, USA, 2023; pp. 295–329. [CrossRef]
19. Nichols, P.J.; Krall, J.B.; Henen, M.A.; Vögeli, B.; Vicens, Q. Z-RNA Biology: A Central Role in the Innate Immune Response? *RNA* **2023**, *29*, 273–281. [CrossRef]
20. Herbert, A.; Alfken, J.; Kim, Y.-G.; Mian, I.S.; Nishikura, K.; Rich, A. A Z-DNA Binding Domain Present in the Human Editing Enzyme, Double-Stranded RNA Adenosine Deaminase. *Proc. Natl. Acad. Sci. USA* **1997**, *94*, 8421–8426. [CrossRef] [PubMed]
21. Herbert, A.; Schade, M.; Lowenhaupt, K.; Alfken, J.; Schwartz, T.; Shlyakhtenko, L.S.; Lyubchenko, Y.L.; Rich, A. The Zα Domain from Human ADAR1 Binds to the Z-DNA Conformer of Many Different Sequences. *Nucleic Acids Res.* **1998**, *26*, 3486–3493. [CrossRef] [PubMed]
22. Zhang, T.; Yin, C.; Fedorov, A.; Qiao, L.; Bao, H.; Beknazarov, N.; Wang, S.; Gautam, A.; Williams, R.M.; Crawford, J.C. ADAR1 Masks the Cancer Immunotherapeutic Promise of ZBP1-Driven Necroptosis. *Nature* **2022**, *606*, 594–602. [CrossRef] [PubMed]
23. Schwartz, T.; Rould, M.A.; Lowenhaupt, K.; Herbert, A.; Rich, A. Crystal Structure of the Zα Domain of the Human Editing Enzyme ADAR1 Bound to Left-Handed Z-DNA. *Science* **1999**, *284*, 1841–1845. [CrossRef] [PubMed]
24. Chiang, D.C.; Li, Y.; Ng, S.K. The Role of the Z-DNA Binding Domain in Innate Immunity and Stress Granules. *Front. Immunol.* **2021**, *11*, 3779. [CrossRef]
25. Meng, Y.; Wang, G.; He, H.; Lau, K.H.; Hurt, A.; Bixler, B.J.; Parham, A.; Jin, S.-G.; Xu, X.; Vasquez, K.M. Z-DNA Is Remodelled by ZBTB43 in Prospermatogonia to Safeguard the Germline Genome and Epigenome. *Nat. Cell Biol.* **2022**, *24*, 1141–1153. [CrossRef]
26. Bartas, M.; Slychko, K.; Brázda, V.; Červeň, J.; Beaudoin, C.A.; Blundell, T.L.; Pečinka, P. Searching for New Z-DNA/Z-RNA Binding Proteins Based on Structural Similarity to Experimentally Validated Zα Domain. *Int. J. Mol. Sci.* **2022**, *23*, 768. [CrossRef] [PubMed]

27. Enomoto, R.; Kinebuchi, T.; Sato, M.; Yagi, H.; Kurumizaka, H.; Yokoyama, S. Stimulation of DNA Strand Exchange by the Human TBPIP/Hop2-Mnd1 Complex. *J. Biol. Chem.* **2006**, *281*, 5575–5581. [CrossRef] [PubMed]
28. EMBL SMART Z-Alpha. Available online: http://smart.embl-heidelberg.de/smart/selective.cgi?domains=Zalpha&terms=&taxon_text=&input=Architecture+query (accessed on 10 February 2023).
29. Arndt-Jovin, D.J.; Udvardy, A.; Garner, M.M.; Ritter, S.; Jovin, T.M. Z-DNA Binding and Inhibition by GTP of Drosophila Topoisomerase II. *Biochemistry* **1993**, *32*, 4862–4872. [CrossRef]
30. McKie, S.J.; Neuman, K.C.; Maxwell, A. DNA Topoisomerases: Advances in Understanding of Cellular Roles and Multi-Protein Complexes via Structure-Function Analysis. *BioEssays* **2021**, *43*, 2000286. [CrossRef] [PubMed]
31. Howard, M.T.; Lee, M.P.; Hsieh, T.; Griffith, J.D. Drosophila Topoisomerase II-DNA Interactions Are Affected by DNA Structure. *J. Mol. Biol.* **1991**, *217*, 53–62. [CrossRef] [PubMed]
32. Bigman, L.S.; Greenblatt, H.M.; Levy, Y. What Are the Molecular Requirements for Protein Sliding along DNA? *J. Phys. Chem. B* **2021**, *125*, 3119–3131. [CrossRef] [PubMed]
33. Glikin, G.C.; Jovin, T.M.; Arndt-Jovin, D.J. Interactions of Drosophilla DNA Topoisomerase II with Left-Handed Z-DNA in Supercoiled Minicircles. *Nucleic Acids Res.* **1991**, *19*, 7139–7144. [CrossRef]
34. Bechert, T.; Diekmann, S.; Arndt-Jovin, D.J. Human 170 KDa and 180 KDa Topoisomerases II Bind Preferentially to Curved and Left-Handed Linear DNA. *J. Biomol. Struct.* **1994**, *12*, 605–623. [CrossRef]
35. Hirose, S.; Tabuchi, H.; Yoshinaga, K. GTP Induces Knotting, Catenation, and Relaxation of DNA by Stoichiometric Amounts of DNA Topoisomerase II from Bombyx Mori and HeLa Cells. *J. Biol. Chem.* **1988**, *263*, 3805–3810. [CrossRef]
36. Bollimpelli, V.S.; Dholaniya, P.S.; Kondapi, A.K. Topoisomerase IIβ and Its Role in Different Biological Contexts. *Arch. Biochem. Biophys.* **2017**, *633*, 78–84. [CrossRef]
37. Deweese, J.E.; Osheroff, M.A.; Osheroff, N. DNA Topology and Topoisomerases: Teaching a "Knotty" Subject. *Biochem. Mol. Biol. Educ.* **2009**, *37*, 2–10. [CrossRef]
38. Seol, Y.; Neuman, K.C. The Dynamic Interplay between DNA Topoisomerases and DNA Topology. *Biophys. Rev.* **2016**, *8*, 101–111. [CrossRef]
39. Michieletto, D.; Fosado, Y.A.G.; Melas, E.; Baiesi, M.; Tubiana, L.; Orlandini, E. Dynamic and Facilitated Binding of Topoisomerase Accelerates Topological Relaxation. *Nucleic Acids Res.* **2022**, *50*, 4659–4668. [CrossRef]
40. Antoniou-Kourounioti, M.; Mimmack, M.L.; Porter, A.C.; Farr, C.J. The Impact of the C-Terminal Region on the Interaction of Topoisomerase II Alpha with Mitotic Chromatin. *Int. J. Mol. Sci.* **2019**, *20*, 1238. [CrossRef]
41. Hoang, K.G.; Menzie, R.A.; Rhoades, J.H.; Fief, C.A.; Deweese, J.E. Reviewing the Modification, Interactions, and Regulation of the C-Terminal Domain of Topoisomerase IIα as a Prospect for Future Therapeutic Targeting. *EC Pharmacol. Toxicol.* **2020**, *8*, 27–43.
42. Hirsch, J.; Klostermeier, D. What Makes a Type IIA Topoisomerase a Gyrase or a Topo IV? *Nucleic Acids Res.* **2021**, *49*, 6027–6042. [CrossRef] [PubMed]
43. Roca, J. Topoisomerase II: A Fitted Mechanism for the Chromatin Landscape. *Nucleic Acids Res.* **2009**, *37*, 721–730. [CrossRef] [PubMed]
44. Wendorff, T.J.; Schmidt, B.H.; Heslop, P.; Austin, C.A.; Berger, J.M. The Structure of DNA-Bound Human Topoisomerase II Alpha: Conformational Mechanisms for Coordinating Inter-Subunit Interactions with DNA Cleavage. *J. Mol. Biol.* **2012**, *424*, 109–124. [CrossRef]
45. Chen, S.-F.; Huang, N.-L.; Lin, J.-H.; Wu, C.-C.; Wang, Y.-R.; Yu, Y.-J.; Gilson, M.K.; Chan, N.-L. Structural Insights into the Gating of DNA Passage by the Topoisomerase II DNA-Gate. *Nat. Commun.* **2018**, *9*, 3085. [CrossRef] [PubMed]
46. Lee, J.H.; Berger, J.M. Cell Cycle-Dependent Control and Roles of DNA Topoisomerase II. *Genes* **2019**, *10*, 859. [CrossRef]
47. Ziraldo, R.; Hanke, A.; Levene, S.D. Kinetic Pathways of Topology Simplification by Type-II Topoisomerases in Knotted Supercoiled DNA. *Nucleic Acids Res.* **2019**, *47*, 69–84. [CrossRef]
48. Vanden Broeck, A.; Lotz, C.; Drillien, R.; Haas, L.; Bedez, C.; Lamour, V. Structural Basis for Allosteric Regulation of Human Topoisomerase IIα. *Nat. Commun.* **2021**, *12*, 2962. [CrossRef] [PubMed]
49. Schmidt, B.H.; Osheroff, N.; Berger, J.M. Structure of a Topoisomerase II–DNA–Nucleotide Complex Reveals a New Control Mechanism for ATPase Activity. *Nat. Struct. Mol. Biol.* **2012**, *19*, 1147–1154. [CrossRef] [PubMed]
50. Deweese, J.E.; Osheroff, N. The DNA Cleavage Reaction of Topoisomerase II: Wolf in Sheep's Clothing. *Nucleic Acids Res.* **2009**, *37*, 738–748. [CrossRef] [PubMed]
51. Lee, S.; Jung, S.-R.; Heo, K.; Byl, J.A.W.; Deweese, J.E.; Osheroff, N.; Hohng, S. DNA Cleavage and Opening Reactions of Human Topoisomerase IIα Are Regulated via Mg^{2+}-Mediated Dynamic Bending of Gate-DNA. *Proc. Natl. Acad. Sci. USA* **2012**, *109*, 2925–2930. [CrossRef]
52. Singh, B.N.; Achary, V.M.M.; Venkatapuram, A.K.; Parmar, H.; Karippadakam, S.; Sopory, S.K.; Reddy, M.K. Expression and Functional Analysis of Various Structural Domains of Tobacco Topoisomerase II: To Understand the Mechanistic Insights of Plant Type II Topoisomerases. *Plant Physiol. Biochem.* **2023**, *194*, 302–314. [CrossRef]
53. Le, T.T.; Wu, M.; Lee, J.H.; Bhatt, N.; Inman, J.T.; Berger, J.M.; Wang, M.D. Etoposide Promotes DNA Loop Trapping and Barrier Formation by Topoisomerase II. *Nat. Chem. Biol.* **2023**, *19*, 641–650. [CrossRef] [PubMed]
54. Pommier, Y.; Leo, E.; Zhang, H.; Marchand, C. DNA Topoisomerases and Their Poisoning by Anticancer and Antibacterial Drugs. *Chem. Biol.* **2010**, *17*, 421–433. [CrossRef]

55. Bax, B.D.; Murshudov, G.; Maxwell, A.; Germe, T. DNA Topoisomerase Inhibitors: Trapping a DNA-Cleaving Machine in Motion. *J. Mol. Biol.* **2019**, *431*, 3427–3449. [CrossRef]
56. Vann, K.R.; Oviatt, A.A.; Osheroff, N. Topoisomerase II Poisons: Converting Essential Enzymes into Molecular Scissors. *Biochemistry* **2021**, *60*, 1630–1641. [CrossRef]
57. Matias-Barrios, V.M.; Dong, X. The Implication of Topoisomerase II Inhibitors in Synthetic Lethality for Cancer Therapy. *Pharmaceuticals* **2023**, *16*, 94. [CrossRef]
58. Okoro, C.O.; Fatoki, T.H. A Mini Review of Novel Topoisomerase II Inhibitors as Future Anticancer Agents. *Int. J. Mol. Sci.* **2023**, *24*, 2532. [CrossRef] [PubMed]
59. Sun, Y.; Nitiss, J.L.; Pommier, Y. SUMO: A Swiss Army Knife for Eukaryotic Topoisomerases. *Front. Mol. Biosci.* **2022**, *9*, 871161. [CrossRef] [PubMed]
60. Li, Z.; Jaroszewski, L.; Iyer, M.; Sedova, M.; Godzik, A. FATCAT 2.0: Towards a Better Understanding of the Structural Diversity of Proteins. *Nucleic Acids Res.* **2020**, *48*, W60–W64. [CrossRef] [PubMed]
61. Chen, K.; Mizianty, M.J.; Kurgan, L. Prediction and Analysis of Nucleotide-Binding Residues Using Sequence and Sequence-Derived Structural Descriptors. *Bioinformatics* **2012**, *28*, 331–341. [CrossRef]
62. Wei, H.; Ruthenburg, A.J.; Bechis, S.K.; Verdine, G.L. Nucleotide-Dependent Domain Movement in the ATPase Domain of a Human Type IIA DNA Topoisomerase. *J. Biol. Chem.* **2005**, *280*, 37041–37047. [CrossRef]
63. Stanger, F.V.; Dehio, C.; Schirmer, T. Structure of the N-Terminal Gyrase B Fragment in Complex with ADP·Pi Reveals Rigid-Body Motion Induced by ATP Hydrolysis. *PLoS ONE* **2014**, *9*, e107289. [CrossRef]
64. Moreira, F.; Arenas, M.; Videira, A.; Pereira, F. Evolutionary History of TOPIIA Topoisomerases in Animals. *J. Mol. Evol.* **2022**, *90*, 149–165. [CrossRef]
65. Valdés, A.; Coronel, L.; Martínez-García, B.; Segura, J.; Dyson, S.; Díaz-Ingelmo, O.; Micheletti, C.; Roca, J. Transcriptional Supercoiling Boosts Topoisomerase II-Mediated Knotting of Intracellular DNA. *Nucleic Acids Res.* **2019**, *47*, 6946–6955. [CrossRef]
66. Lang, A.J.; Mirski, S.E.; Cummings, H.J.; Yu, Q.; Gerlach, J.H.; Cole, S.P. Structural Organization of the Human TOP2A and TOP2B Genes. *Gene* **1998**, *221*, 255–266. [CrossRef]
67. Guglielmini, J.; Gaia, M.; Da Cunha, V.; Criscuolo, A.; Krupovic, M.; Forterre, P. Viral Origin of Eukaryotic Type IIA DNA Topoisomerases. *Virus Evol.* **2022**, *8*, veac097. [CrossRef]
68. Letunic, I.; Bork, P. 20 Years of the SMART Protein Domain Annotation Resource. *Nucleic Acids Res.* **2018**, *46*, D493–D496. [CrossRef]
69. Corpet, F. Multiple Sequence Alignment with Hierarchical Clustering. *Nucleic Acids Res* **1988**, *16*, 10881–10890. [CrossRef]
70. Zhao, R.; Shin, D.S.; Fiser, A.; Goldman, I.D. Identification of a Functionally Critical GXXG Motif and Its Relationship to the Folate Binding Site of the Proton-Coupled Folate Transporter (PCFT-SLC46A1). *Am. J. Physiol. Cell Physiol.* **2012**, *303*, C673–C681. [CrossRef]
71. Wheeler, T.J.; Clements, J.; Finn, R.D. Skylign: A Tool for Creating Informative, Interactive Logos Representing Sequence Alignments and Profile Hidden Markov Models. *BMC Bioinform.* **2014**, *15*, 7. [CrossRef]
72. Afowowe, T.O.; Sakurai, Y.; Urata, S.; Zadeh, V.R.; Yasuda, J. Topoisomerase II as a Novel Antiviral Target against Panarenaviral Diseases. *Viruses* **2022**, *15*, 105. [CrossRef]
73. Cui, W.; Braun, E.; Wang, W.; Tang, J.; Zheng, Y.; Slater, B.; Li, N.; Chen, C.; Liu, Q.; Wang, B.; et al. Structural Basis for GTP-Induced Dimerization and Antiviral Function of Guanylate-Binding Proteins. *Proc. Natl. Acad. Sci. USA* **2021**, *118*, e2022269118. [CrossRef]
74. Chang, C.-C.; Wang, Y.-R.; Chen, S.-F.; Wu, C.-C.; Chan, N.-L. New Insights into DNA-Binding by Type IIA Topoisomerases. *Curr. Opin. Struct. Biol.* **2013**, *23*, 125–133. [CrossRef] [PubMed]
75. Murthy, S.N.P.; Iismaa, S.; Begg, G.; Freymann, D.M.; Graham, R.M.; Lorand, L. Conserved Tryptophan in the Core Domain of Transglutaminase Is Essential for Catalytic Activity. *Proc. Natl. Acad. Sci. USA* **2002**, *99*, 2738–2742. [CrossRef] [PubMed]
76. Lei, X.; Cao, K.; Chen, Y.; Shen, H.; Liu, Z.; Qin, H.; Cai, J.; Gao, F.; Yang, Y. Nuclear Transglutaminase 2 Interacts with Topoisomerase IIα to Promote DNA Damage Repair in Lung Cancer Cells. *J. Exp. Clin. Cancer Res.* **2021**, *40*, 224. [CrossRef]
77. Ogrizek, M.; Janežič, M.; Valjavec, K.; Perdih, A. Catalytic Mechanism of ATP Hydrolysis in the ATPase Domain of Human DNA Topoisomerase IIα. *J. Chem. Inf. Model.* **2022**, *62*, 3896–3909. [CrossRef] [PubMed]
78. Jovin, T.M. Recognition Mechanisms of DNA-Specific Enzymes. *Annu. Rev. Biochem.* **1976**, *45*, 889–920. [CrossRef]
79. Bae, S.; Kim, D.; Kim, K.K.; Kim, Y.-G.; Hohng, S. Intrinsic Z-DNA Is Stabilized by the Conformational Selection Mechanism of Z-DNA-Binding Proteins. *J. Am. Chem. Soc.* **2011**, *133*, 668–671. [CrossRef]
80. Dong, K.C.; Berger, J.M. Structural Basis for Gate-DNA Recognition and Bending by Type IIA Topoisomerases. *Nature* **2007**, *450*, 1201–1205. [CrossRef]
81. Lee, I.; Dong, K.C.; Berger, J.M. The Role of DNA Bending in Type IIA Topoisomerase Function. *Nucleic Acids Res.* **2013**, *41*, 5444–5456. [CrossRef] [PubMed]
82. Thomson, N.H.; Santos, S.; Mitchenall, L.A.; Stuchinskaya, T.; Taylor, J.A.; Maxwell, A. DNA G-Segment Bending Is Not the Sole Determinant of Topology Simplification by Type II DNA Topoisomerases. *Sci. Rep.* **2014**, *4*, 6158. [CrossRef] [PubMed]
83. Sarangi, M.K.; Zvoda, V.; Holte, M.N.; Becker, N.A.; Peters, J.P.; Maher, L.J., III; Ansari, A. Evidence for a Bind-Then-Bend Mechanism for Architectural DNA Binding Protein YNhp6A. *Nucleic Acids Res.* **2019**, *47*, 2871–2883. [CrossRef]

84. Dickerson, R.E. DNA Bending: The Prevalence of Kinkiness and the Virtues of Normality. *Nucleic Acids Res.* **1998**, *26*, 1906–1926. [CrossRef]
85. Raskó, T.; Finta, C.; Kiss, A. DNA Bending Induced by DNA (Cytosine-5) Methyltransferases. *Nucleic Acids Res.* **2000**, *28*, 3083–3091. [CrossRef] [PubMed]
86. Harteis, S.; Schneider, S. Making the Bend: DNA Tertiary Structure and Protein-DNA Interactions. *Int. J. Mol. Sci.* **2014**, *15*, 12335–12363. [CrossRef] [PubMed]
87. Yi, J.; Yeou, S.; Lee, N.K. DNA Bending Force Facilitates Z-DNA Formation under Physiological Salt Conditions. *J. Am. Chem. Soc.* **2022**, *144*, 13137–13145. [CrossRef]
88. Jang, Y.; Son, H.; Lee, S.-W.; Hwang, W.; Jung, S.-R.; Byl, J.A.W.; Osheroff, N.; Lee, S. Selection of DNA Cleavage Sites by Topoisomerase II Results from Enzyme-Induced Flexibility of DNA. *Cell Chem. Biol.* **2019**, *26*, 502–511.e3. [CrossRef]
89. Ansevin, A.T.; Wang, A.H. Evidence for a New Z-Type Left-Handed DNA Helix: Properties of Z(WC)-DNA. *Nucleic Acids Res.* **1990**, *18*, 6119–6126. [CrossRef]
90. Fuertes, M.A.; Cepeda, V.; Alonso, C.; Pérez, J.M. Molecular Mechanisms for the B−Z Transition in the Example of Poly[d(G−C)·d(G−C)] Polymers. *A Critical Review. Chem. Rev.* **2006**, *106*, 2045–2064. [CrossRef]
91. Chakraborty, D.; Wales, D.J. Probing Helical Transitions in a DNA Duplex. *Phys. Chem. Chem. Phys.* **2017**, *19*, 878–892. [CrossRef] [PubMed]
92. Jovin, T.M.; McIntosh, L.P.; Arndt-Jovin, D.J.; Zarling, D.A.; Robert-Nicoud, M.; van de Sande, J.H.; Jorgenson, K.F.; Eckstein, F. Left-Handed DNA: From Synthetic Polymers to Chromosomes. *J. Biomol. Struct. Dyn.* **1983**, *1*, 21–57. [CrossRef] [PubMed]
93. Deweese, J.E.; Osheroff, N. The Use of Divalent Metal Ions by Type II Topoisomerases. *Metallomics* **2010**, *2*, 450–459. [CrossRef]
94. Fogg, J.M.; Randall, G.L.; Pettitt, B.M.; Sumners, D.W.L.; Harris, S.A.; Zechiedrich, L. Bullied No More: When and How DNA Shoves Proteins Around. *Q. Rev. Biophys.* **2012**, *45*, 257–299. [CrossRef]
95. Vlahovicek, K.; Munteanu, M.G.; Pongor, S. Sequence-Dependent Modelling of Local DNA Bending Phenomena: Curvature Prediction and Vibrational Analysis. *Genetica* **1999**, *106*, 63–73. [CrossRef] [PubMed]
96. Hardin, A.H.; Sarkar, S.K.; Seol, Y.; Liou, G.F.; Osheroff, N.; Neuman, K.C. Direct Measurement of DNA Bending by Type IIA Topoisomerases: Implications for Non-Equilibrium Topology Simplification. *Nucleic Acids Res.* **2011**, *39*, 5729–5743. [CrossRef]
97. Lee, M.; Kim, S.H.; Hong, S.-C. Minute Negative Superhelicity Is Sufficient to Induce the B-Z Transition in the Presence of Low Tension. *Proc. Natl. Sci. Acad. USA* **2010**, *107*, 4985–4990. [CrossRef]
98. Fogg, J.M.; Catanese, D.J.; Randall, G.L.; Swick, M.C.; Zechiedrich, L. Differences Between Positively and Negatively Supercoiled DNA That Topoisomerases May Distinguish. In *Mathematics of DNA Structure, Function and Interactions*; Benham, C.J., Harvey, S., Olson, W.K., Sumners, D.W., Swigon, D., Eds.; The IMA Volumes in Mathematics and its Applications; Springer: New York, NY, USA, 2009; Volume 150, pp. 73–121. ISBN 978-1-4419-0669-4.
99. Vologodskii, A. Theoretical Models of DNA Topology Simplification by Type IIA DNA Topoisomerases. *Nucleic Acids Res.* **2009**, *37*, 3125–3133. [CrossRef]
100. Cofsky, J.C.; Soczek, K.M.; Knott, G.J.; Nogales, E.; Doudna, J.A. CRISPR-Cas9 Bends and Twists DNA to Read Its Sequence. *Nat. Struct. Mol. Biol.* **2022**, *29*, 395–402. [CrossRef]
101. Stuchinskaya, T.; Mitchenall, L.A.; Schoeffler, A.J.; Corbett, K.D.; Berger, J.M.; Bates, A.D.; Maxwell, A. How Do Type II Topoisomerases Use ATP Hydrolysis to Simplify DNA Topology beyond Equilibrium? Investigating the Relaxation Reaction of Nonsupercoiling Type II Topoisomerases. *J. Mol. Biol.* **2009**, *385*, 1397–1408. [CrossRef] [PubMed]
102. Pohl, F.M. Hysteretic Behaviour of a Z-DNA-Antibody Complex. *Biophys. Chem.* **1987**, *26*, 385–390. [CrossRef]
103. Igamberdiev, A.U.; Kleczkowski, L.A. Toward Understanding the Emergence of Life: A Dual Function of the System of Nucleotides in the Metabolically Closed Autopoietic Organization. *Biosystems* **2023**, *224*, 104837. [CrossRef]
104. Padget, K.; Pearson, A.D.; Austin, C.A. Quantitation of DNA Topoisomerase IIalpha and Beta in Human Leukaemia Cells by Immunoblotting. *Leukemia* **2000**, *14*, 1997–2005. [CrossRef]
105. Walther, N.; Hossain, M.J.; Politi, A.Z.; Koch, B.; Kueblbeck, M.; Ødegård-Fougner, Ø.; Lampe, M.; Ellenberg, J. A Quantitative Map of Human Condensins Provides New Insights into Mitotic Chromosome Architecture. *J. Cell Biol.* **2018**, *217*, 2309–2328. [CrossRef] [PubMed]
106. Nielsen, C.F.; Zhang, T.; Barisic, M.; Kalitsis, P.; Hudson, D.F. Topoisomerase IIα Is Essential for Maintenance of Mitotic Chromosome Structure. *Proc. Natl. Acad. Sci. USA* **2020**, *117*, 12131–12142. [CrossRef]
107. Gibcus, J.H.; Samejima, K.; Goloborodko, A.; Samejima, I.; Naumova, N.; Nuebler, J.; Kanemaki, M.T.; Xie, L.; Paulson, J.R.; Earnshaw, W.C.; et al. A Pathway for Mitotic Chromosome Formation. *Science* **2018**, *359*, eaao6135. [CrossRef]
108. Elbatsh, A.M.O.; Kim, E.; Eeftens, J.M.; Raaijmakers, J.A.; van der Weide, R.H.; García-Nieto, A.; Bravo, S.; Ganji, M.; Uit de Bos, J.; Teunissen, H.; et al. Distinct Roles for Condensin's Two ATPase Sites in Chromosome Condensation. *Mol. Cell* **2019**, *76*, 724–737.e5. [CrossRef]
109. Zhou, C.Y.; Heald, R. Emergent Properties of Mitotic Chromosomes. *Curr. Opin. Cell Biol.* **2020**, *64*, 43–49. [CrossRef] [PubMed]
110. Man, T.; Witt, H.; Peterman, E.J.G.; Wuite, G.J.L. The Mechanics of Mitotic Chromosomes. *Q. Rev. Biophys.* **2021**, *54*, e10. [CrossRef] [PubMed]
111. Paulson, J.R.; Hudson, D.F.; Cisneros-Soberanis, F.; Earnshaw, W.C. Mitotic Chromosomes. *Semin. Cell Dev. Biol.* **2021**, *117*, 7–29. [CrossRef]

112. Dekker, B.; Dekker, J. Regulation of the Mitotic Chromosome Folding Machines. *Biochemistry* **2022**, *479*, 2153–2173. [CrossRef] [PubMed]
113. Stray, J.E.; Lindsley, J.E. Biochemical Analysis of the Yeast Condensin Smc2/4 Complex: An ATPase That Promo.tes Knotting of Circular DNA. *J. Biol. Chem.* **2003**, *278*, 26238–26248. [CrossRef]
114. Mora-Bermúdez, F.; Gerlich, D.; Ellenberg, J. Maximal Chromosome Compaction Occurs by Axial Shortening in Anaphase and Depends on Aurora Kinase. *Nat. Cell Biol.* **2007**, *9*, 822–831. [CrossRef]
115. Baxter, J.; Aragón, L. A Model for Chromosome Condensation Based on the Interplay between Condensin and Topoisomerase II. *Trends Genet.* **2012**, *28*, 110–117. [CrossRef] [PubMed]
116. Roca, J.; Dyson, S.; Segura, J.; Valdés, A.; Martínez-García, B. Keeping Intracellular DNA Untangled: A New Role for Condensin? *BioEssays* **2022**, *44*, e2100187. [CrossRef] [PubMed]
117. Samejima, K.; Samejima, I.; Vagnarelli, P.; Ogawa, H.; Vargiu, G.; Kelly, D.A.; de Lima Alves, F.; Kerr, A.; Green, L.C.; Hudson, D.F.; et al. Mitotic Chromosomes Are Compacted Laterally by KIF4 and Condensin and Axially by Topoisomerase IIα. *J. Cell Biol.* **2012**, *199*, 755–770. [CrossRef]
118. Riccio, A.A.; Schellenberg, M.J.; Williams, R.S. Molecular Mechanisms of Topoisomerase 2 DNA–Protein Crosslink Resolution. *Cell. Mol. Life Sci.* **2020**, *77*, 81–91. [CrossRef]
119. Shintomi, K.; Hirano, T. Sister Chromatid Resolution: A Cohesin Releasing Network and Beyond. *Chromosoma* **2010**, *119*, 459–467. [CrossRef]
120. Le, T.T.; Wang, M.D. Topoisomerase II and Etoposide—A Tangled Tale. *Nat. Chem. Biol.* **2023**, *19*, 546–547. [CrossRef]
121. Boulikas, T. Nature of DNA Sequences at the Attachment Regions of Genes to the Nuclear Matrix. *J. Cell. Biochem.* **1993**, *52*, 14–22. [CrossRef] [PubMed]
122. Herbert, A. Nucleosomes and Flipons Exchange Energy to Alter Chromatin Conformation, the Readout of Genomic Information, and Cell Fate. *Bioessays* **2022**, *44*, e2200166. [CrossRef] [PubMed]
123. Wolff, D.W.; Bianchi-Smiraglia, A.; Nikiforov, M.A. Compartmentalization and Regulation of GTP in Control of Cellular Phenotypes. *Trends Mol. Med.* **2022**, *28*, 758–769. [CrossRef]
124. Traut, T.W. Physiological Concentrations of Purines and Pyrimidines. *Mol. Cell. Biochem.* **1994**, *140*, 1–22. [CrossRef]
125. Kaláb, P.; Pralle, A.; Isacoff, E.Y.; Heald, R.; Weis, K. Analysis of a RanGTP-Regulated Gradient in Mitotic Somatic Cells. *Nature* **2006**, *440*, 697–701. [CrossRef]
126. Kapoor, T.M. Metaphase Spindle Assembly. *Biology* **2017**, *6*, 8. [CrossRef]
127. Ozugergin, I.; Piekny, A. Complementary Functions for the Ran Gradient during Division. *Small GTPases* **2021**, *12*, 177–187. [CrossRef]
128. Spence, J.M.; Critcher, R.; Ebersole, T.A.; Valdivia, M.M.; Earnshaw, W.C.; Fukagawa, T.; Farr, C.J. Co-Localization of Centromere Activity, Proteins and Topoisomerase II within a Subdomain of the Major Human X Alpha-Satellite Array. *EMBO J.* **2002**, *21*, 5269–5280. [CrossRef]
129. Kasinathan, S.; Henikoff, S. Non-B-Form DNA Is Enriched at Centromeres. *Mol. Biol. Evol.* **2018**, *35*, 949–962. [CrossRef] [PubMed]
130. Mellone, B.G.; Fachinetti, D. Diverse Mechanisms of Centromere Specification. *Curr. Biol.* **2021**, *31*, R1491–R1504. [CrossRef] [PubMed]
131. Liu, Q.; Yi, C.; Zhang, Z.; Su, H.; Liu, C.; Huang, Y.; Li, W.; Hu, X.; Liu, C.; Birchler, J.A.; et al. Non-B-Form DNA Tends to Form in Centromeric Regions and Has Undergone Changes in Polyploid Oat Subgenomes. *Proc. Natl. Acad. Sci. USA* **2023**, *120*, e2211683120. [CrossRef]
132. Mills, W.E.; Spence, J.M.; Fukagawa, T.; Farr, C.J. Site-Specific Cleavage by Topoisomerase 2: A Mark of the Core Centromere. *Int. J. Mol. Sci.* **2018**, *19*, 534. [CrossRef] [PubMed]
133. Biton, Y.Y. Effects of Protein-Induced Local Bending and Sequence Dependence on the Configurations of Supercoiled DNA Minicircles. *J. Chem. Theory Comput.* **2018**, *14*, 2063–2075. [CrossRef]
134. Wang, L.H.-C.; Mayer, B.; Stemmann, O.; Nigg, E.A. Centromere DNA Decatenation Depends on Cohesin Removal and Is Required for Mammalian Cell Division. *J. Cell Sci.* **2010**, *123*, 806–813. [CrossRef]
135. Chu, L.; Zhang, Z.; Mukhina, M.; Zickler, D.; Kleckner, N. Sister Chromatids Separate during Anaphase in a Three-Stage Program as Directed by Interaxis Bridges. *Proc. Natl. Acad. Sci. USA* **2022**, *119*, e2123363119. [CrossRef]
136. Gibson, E.G.; Deweese, J.E. Structural and Biochemical Basis of Etoposide-Resistant Mutations in Topoisomerase IIα. *Symmetry* **2022**, *14*, 1309. [CrossRef]
137. Jaramillo-Lambert, A.; Fabritius, A.S.; Hansen, T.J.; Smith, H.E.; Golden, A. The Identification of a Novel Mutant Allele of Topoisomerase II in Caenorhabditis Elegans Reveals a Unique Role in Chromosome Segregation during Spermatogenesis. *Genetics* **2016**, *204*, 1407–1422. [CrossRef]
138. Zdraljevic, S.; Strand, C.; Seidel, H.S.; Cook, D.E.; Doench, J.G.; Andersen, E.C. Natural Variation in a Single Amino Acid Substitution Underlies Physiological Responses to Topoisomerase II Poisons. *PLoS Genet.* **2017**, *13*, e1006891. [CrossRef] [PubMed]
139. Masullo, L.A.; Lopez, L.F.; Stefani, F.D. A Common Framework for Single-Molecule Localization Using Sequential Structured Illumination. *Biophys. Rep.* **2022**, *2*, 100036. [CrossRef]

140. Austin, C.A.; Cowell, I.G.; Khazeem, M.M.; Lok, D.; Ng, H.T. TOP2B's Contributions to Transcription. *Biochem. Soc. Trans.* **2021**, *49*, 2483–2493. [CrossRef] [PubMed]
141. Blower, T.R.; Bandak, A.; Lee, A.S.Y.; Austin, C.A.; Nitiss, J.L.; Berger, J.M. A Complex Suite of Loci and Elements in Eukaryotic Type II Topoisomerases Determine Selective Sensitivity to Distinct Poisoning Agents. *Nucleic Acids Res.* **2019**, *47*, 8163–8179. [CrossRef]
142. Van Ravenstein, S.X.; Mehta, K.P.; Kavlashvili, T.; Byl, J.A.W.; Zhao, R.; Osheroff, N.; Cortez, D.; Dewar, J.M. Topoisomerase II Poisons Inhibit Vertebrate DNA Replication through Distinct Mechanisms. *EMBO J.* **2022**, *41*, e110632. [CrossRef] [PubMed]
143. Furet, P.; Schoepfer, J.; Radimerski, T.; Chène, P. Discovery of a New Class of Catalytic Topoisomerase II Inhibitors Targeting the ATP-Binding Site by Structure Based Design. *Part I. Bioorg. Med. Chem. Lett.* **2009**, *19*, 4014–4017. [CrossRef]
144. Park, S.; Hwang, S.-Y.; Shin, J.; Jo, H.; Na, Y.; Kwon, Y. A Chromenone Analog as an ATP-Competitive, DNA Non-Intercalative Topoisomerase II Catalytic Inhibitor with Preferences toward the Alpha Isoform. *ChemComm* **2019**, *55*, 12857–12860. [CrossRef] [PubMed]
145. Ha, S.C.; Choi, J.; Hwang, H.-Y.; Rich, A.; Kim, Y.-G.; Kim, K.K. The Structures of Non-CG-Repeat Z-DNAs Co-Crystallized with the Z-DNA-Binding Domain, HZα ADAR1. *Nucleic Acids Res.* **2009**, *37*, 629–637. [CrossRef]
146. Jumper, J.; Evans, R.; Pritzel, A.; Green, T.; Figurnov, M.; Ronneberger, O.; Tunyasuvunakool, K.; Bates, R.; Žídek, A.; Potapenko, A.; et al. Highly Accurate Protein Structure Prediction with AlphaFold. *Nature* **2021**, *596*, 583–589. [CrossRef]
147. UniProt: The Universal Protein Knowledgebase in 2021. *Nucleic Acids Res.* **2021**, *49*, D480–D489. [CrossRef]
148. Pettersen, E.F.; Goddard, T.D.; Huang, C.C.; Couch, G.S.; Greenblatt, D.M.; Meng, E.C.; Ferrin, T.E. UCSF Chimera—A Visualization System for Exploratory Research and Analysis. *J. Comput. Chem.* **2004**, *25*, 1605–1612. [CrossRef]
149. Meng, E.C.; Pettersen, E.F.; Couch, G.S.; Huang, C.C.; Ferrin, T.E. Tools for Integrated Sequence-Structure Analysis with UCSF Chimera. *BMC Bioinform.* **2006**, *7*, 339. [CrossRef]
150. Hunt, S.E.; McLaren, W.; Gil, L.; Thormann, A.; Schuilenburg, H.; Sheppard, D.; Parton, A.; Armean, I.M.; Trevanion, S.J.; Flicek, P.; et al. Ensembl Variation Resources. *Database* **2018**, *2018*, bay119. [CrossRef] [PubMed]
151. Ng, P.C.; Henikoff, S. SIFT: Predicting Amino Acid Changes That Affect Protein Function. *Nucleic Acids Res.* **2003**, *31*, 3812–3814. [CrossRef]
152. Adzhubei, I.; Jordan, D.M.; Sunyaev, S.R. Predicting Functional Effect of Human Missense Mutations Using PolyPhen-2. *Curr. Protoc. Hum.* **2013**, *76*, 7–20. [CrossRef] [PubMed]
153. Ioannidis, N.M.; Rothstein, J.H.; Pejaver, V.; Middha, S.; McDonnell, S.K.; Baheti, S.; Musolf, A.; Li, Q.; Holzinger, E.; Karyadi, D. REVEL: An Ensemble Method for Predicting the Pathogenicity of Rare Missense Variants. *Am. J. Hum. Genet.* **2016**, *99*, 877–885. [CrossRef]
154. Reva, B.; Antipin, Y.; Sander, C. Predicting the Functional Impact of Protein Mutations: Application to Cancer Genomics. *Nucleic Acids Res.* **2011**, *39*, e118. [CrossRef]
155. Okonechnikov, K.; Golosova, O.; Fursov, M.; Team, U. Unipro UGENE: A Unified Bioinformatics Toolkit. *Bioinformatics* **2012**, *28*, 1166–1167. [CrossRef] [PubMed]
156. Gouet, P.; Robert, X.; Courcelle, E. ESPript/ENDscript: Extracting and Rendering Sequence and 3D Information from Atomic Structures of Proteins. *Nucleic Acids Res.* **2003**, *31*, 3320–3323. [CrossRef]
157. Yan, Y.; Zhang, D.; Zhou, P.; Li, B.; Huang, S.-Y. HDOCK: A Web Server for Protein–Protein and Protein–DNA/RNA Docking Based on a Hybrid Strategy. *Nucleic Acids Res.* **2017**, *45*, W365–W373. [CrossRef]
158. Zheng, G.; Lu, X.-J.; Olson, W.K. Web 3DNA—A Web Server for the Analysis, Reconstruction, and Visualization of Three-Dimensional Nucleic-Acid Structures. *Nucleic Acids Res.* **2009**, *37*, W240–W246. [CrossRef]
159. Patro, L.P.P.; Kumar, A.; Kolimi, N.; Rathinavelan, T. 3D-NuS: A Web Server for Automated Modeling and Visualization of Non-Canonical 3-Dimensional Nucleic Acid Structures. *J. Mol. Biol.* **2017**, *429*, 2438–2448. [CrossRef]
160. Kim, D.; Hur, J.; Han, J.H.; Ha, S.C.; Shin, D.; Lee, S.; Park, S.; Sugiyama, H.; Kim, K.K. Sequence Preference and Structural Heterogeneity of BZ Junctions. *Nucleic Acids Res.* **2018**, *46*, 10504–10513. [CrossRef] [PubMed]
161. Schneidman-Duhovny, D.; Inbar, Y.; Nussinov, R.; Wolfson, H.J. PatchDock and SymmDock: Servers for Rigid and Symmetric Docking. *Nucleic Acids Res.* **2005**, *33*, W363–W367. [CrossRef] [PubMed]

International Journal of *Molecular Sciences*

MDPI

Article

StaR Is a Positive Regulator of Topoisomerase I Activity Involved in Supercoiling Maintenance in *Streptococcus pneumoniae*

Antonio A. de Vasconcelos Junior [1], Jose M. Tirado-Vélez [1], Antonio J. Martín-Galiano [2], Diego Megias [3], María-José Ferrándiz [1], Pablo Hernández [4], Mónica Amblar [1,*] and Adela G. de la Campa [1,5,*]

[1] Centro Nacional de Microbiología, Instituto de Salud Carlos III, Majadahonda, 28220 Madrid, Spain
[2] Unidades Centrales Científico-Técnicas, Instituto de Salud Carlos III, Majadahonda, 28220 Madrid, Spain
[3] Unidad de Microscopía Confocal, Instituto de Salud Carlos III, Majadahonda, 28220 Madrid, Spain
[4] Centro de Investigaciones Biológicas Margarita Salas, Consejo Superior de Investigaciones Científicas, 28040 Madrid, Spain
[5] Presidencia, Consejo Superior de Investigaciones Científicas, 28006 Madrid, Spain
* Correspondence: mamblar@isciii.es (M.A.); agcampa@isciii.es (A.G.d.l.C.); Tel.: +34-91-448-283 (M.A.); +34-91-448-944 (A.G.d.l.C.)

Abstract: The DNA topoisomerases gyrase and topoisomerase I as well as the nucleoid-associated protein HU maintain supercoiling levels in *Streptococcus pneumoniae*, a main human pathogen. Here, we characterized, for the first time, a topoisomerase I regulator protein (StaR). In the presence of sub-inhibitory novobiocin concentrations, which inhibit gyrase activity, higher doubling times were observed in a strain lacking *staR*, and in two strains in which StaR was over-expressed either under the control of the $ZnSO_4$-inducible P_{Zn} promoter (strain Δ*staR*P_{Zn}*staR*) or of the maltose-inducible P_{Mal} promoter (strain Δ*staR*pLS1ROM*staR*). These results suggest that StaR has a direct role in novobiocin susceptibility and that the StaR level needs to be maintained within a narrow range. Treatment of Δ*staR*P_{Zn}*staR* with inhibitory novobiocin concentrations resulted in a change of the negative DNA supercoiling density (σ) in vivo, which was higher in the absence of StaR ($\sigma = -0.049$) than when StaR was overproduced ($\sigma = -0.045$). We have located this protein in the nucleoid by using super-resolution confocal microscopy. Through in vitro activity assays, we demonstrated that StaR stimulates TopoI relaxation activity, while it has no effect on gyrase activity. Interaction between TopoI and StaR was detected both in vitro and in vivo by co-immunoprecipitation. No alteration of the transcriptome was associated with StaR amount variation. The results suggest that StaR is a new streptococcal nucleoid-associated protein that activates topoisomerase I activity by direct protein-protein interaction.

Keywords: DNA supercoiling; DNA topoisomerase I; DNA gyrase; supercoiling regulation; supercoiling homeostasis; nucleoid-associated proteins; topsoisomerase regulator; StaR

Citation: de Vasconcelos Junior, A.A.; Tirado-Vélez, J.M.; Martín-Galiano, A.J.; Megias, D.; Ferrándiz, M.-J.; Hernández, P.; Amblar, M.; de la Campa, A.G. StaR Is a Positive Regulator of Topoisomerase I Activity Involved in Supercoiling Maintenance in *Streptococcus pneumoniae*. *Int. J. Mol. Sci.* **2023**, *24*, 5973. https://doi.org/10.3390/ijms24065973

Academic Editors: Joseph E. Deweese and Neil Osheroff

Received: 20 February 2023
Revised: 16 March 2023
Accepted: 18 March 2023
Published: 22 March 2023

1. Introduction

The *Streptococcus pneumoniae* chromosome is confined within the nucleoid, in which it is compacted by up to 1000-fold [1], allowing 1.3 Mb to be accommodated within cells just 1–2 µm long. Chromosome topology depends on DNA supercoiling (Sc), which is controlled by DNA topoisomerases [2]. *S. pneumoniae* possesses three of these enzymes: two type II topoisomerases (topoisomerase IV and gyrase), which cleave both DNA strands, and a single type I enzyme (topoisomerase I, TopoI), which cleaves only on one strand. Gyrase and TopoI actively regulate the degree of Sc solving topological problems associated with dynamic DNA remodeling in *Escherichia coli* [3–6] as well as in *S. pneumoniae* [7–9]. In addition, in other bacteria, a number of nucleoid-associated proteins (NAPs) [10] also control Sc, forming a functional network that maintains DNA topology via bending, wrapping, bridging and constraining supercoils [11]. Transcription also regulates Sc, given that domains of negative and positive Sc are generated, behind and ahead of the moving RNA polymerase (RNAP), respectively [12]. In fact, physical interaction between TopoI and RNAP has been detected in vitro in both *E. coli* [13] and *S. pneumoniae* [14]. In addition, ChIP-Seq experiments have revealed in vivo co-localization of RNAP and TopoI in *Mycobacterium tuberculosis* [15] and *S. pneumoniae* [14], as well as co-localization of RNAP and gyrase in *M. tuberculosis*. Likewise, Sc also regulates transcription, structuring the bacterial chromosome into domains with intrinsic topological behavior [7,16–19]. Genes within the Sc domains of the pneumococcal chromosome have a coordinated transcription [7,16] and similar functions [20]. Four types of domains, defined by their transcriptional response to DNA relaxation and their AT content, have been identified: up-regulated, down-regulated, non-regulated and AT-rich. The genes encoding the topoisomerases are themselves located in regulated Sc domains: *topA* in a down-regulated domain [7], and *gyrB* in an up-regulated domain [21]. The control of Sc in *S. pneumoniae* occurs mainly through the regulation of transcription of the topoisomerase genes. Relaxation triggers up-regulation of gyrase and down-regulation of topoisomerases I and IV, while hyper-negative Sc down-regulates the expression of TopoI. Therefore, TopoI plays a fundamental role in the regulation of transcript levels by Sc, since transcription levels of *topA* in homeostasis correlate with the induced variation in the density of Sc [16].

S. pneumoniae is a devastating infectious human pathogen, causing community-acquired pneumonia, meningitis, bacteremia and otitis media and producing the death of one million children worldwide annually [22]. Although resistance in this bacterium to beta-lactams and macrolides antibiotics, which are directed against cell-wall and protein synthesis, respectively, has spread globally [23], no worrisome levels of resistance to antibiotics directed to DNA topology maintenance (fluoroquinolones) [24], which target the type II DNA topoisomerases, have been detected. However, an increase in resistance may occur in tandem with the increased use of fluoroquinolones either due to alterations in their target topoisomerases genes or to the action of active efflux ([25–30]. Therefore, knowledge of the molecular basis of Sc control in this important pathogen is essential for finding new antibiotic targets and adequate antibiotic therapies. As mentioned above, in addition to topoisomerases, Sc is also modulated by NAPs. Although several NAPs have been characterized in the Gram-negative bacterium, *E. coli*, very few have been detected in Gram-positive species, including *S. pneumoniae* [31]. HU [32] and SMC [33] are the only NAPs identified in *S. pneumoniae* so far. HU is essential for growth and for the preservation of Sc in this bacterium [32].

The role of many proteins, including NAPs, in bacterial pathogens remains to be described. A previous study identified 44 hypothetical proteins conserved in *S. pneumoniae* and other Gram-positive pathogens, with potential physiological and biomedical value [34]. One of these proteins, a 324-residue-long polypeptide encoded by the *spr0929* locus, was predicted to interact with three protein partners, and was preliminarily classified as a DNA binding protein [34]. Interestingly, the sequence of this protein is highly conserved and present in virtually all *S. pneumoniae* isolates. Notably, genes coding for Spr0929 homologs are found in other Gram-positive pathogens, including *Clostridium*, *Enterococcus* and *Listeria* genera. In this study, we investigated the role of this hypothetical protein in *S. pneumoniae*. We demonstrated that the product of *spr0929* localizes to the nucleoid and activates TopoI relaxation activity affecting Sc homeostasis when DNA relaxation increases. This topoisomerase regulator of *S. pneumoniae* can be considered a new NAP and will be herein renamed as StaR (streptococcal topoisomerase activity regulator).

2. Results

2.1. The Absence of StaR Affects Novobiocin Susceptibility in S. pneumoniae

To investigate the role of the hypothetical protein encoded by *spr0929*, herein named *staR*, we performed an InterPro analysis to search for bacterial homologues. The best hit found was the NA37 (Nucleotide-Associated proteins of 37 kDa) family. The YejK protein of *E. coli*, the representative member of this family and the only one with experimental information, had 19.4% identity and a 46.6% similarity with pneumococcal StaR (Figure 1A). YejK has been described as a NAP [35]. To decipher the function of StaR, the growth of R6 (wild type) and the previously constructed Δ*staR* deletion mutant [34] was analyzed with or without NOV. NOV inhibits the ATPase activity of gyrase, which is necessary for its Sc activity. However, gyrase has relaxing activity in the absence of ATP hydrolysis, which could also contribute to the final effect in Sc relaxation induced by NOV [36,37]. While in the absence of NOV, both strains showed equivalent growth rates (Figure 1B); in the presence of sub-inhibitory NOV concentrations (0.5 × MIC), Δ*staR* grew slower than R6: doubling time 112 ± 6 min versus 89 ± 6 min ($p = 0.01$), suggesting a role of StaR in Sc maintenance under challenging conditions. In order to purify and characterize StaR, a chromosomal fragment containing its coding gene (*staR*) was cloned into the *E. coli* pET28a BL21-CodonPlus (DE3) expression system, and overproduction of histidine-tagged Star (H_6-StaR) was achieved by induction with IPTG (Figure 1C). A protein with an apparent molecular weight of 37.1 kDa, corresponding to the predicted molecular weight of the H_6-StaR protein (39.3 kDa: 37 kDa + 6 His), was detected and further purified by affinity chromatography. Protein from fraction 10 in Figure 1C was collected and used in subsequent in vitro experiments.

2.2. The Expression Level of StaR Affects Novobiocin Susceptibility

To elucidate the role of StaR in NOV susceptibility, the Δ*staR*P_{Zn}*staR* strain was constructed by introducing an ectopic copy of *staR* under the control of a promoter inducible by zinc (P_{Zn}) in the Δ*staR* strain. The growth of this strain was analyzed in the presence of 150 μM $ZnSO_4$ as the inductor of *staR* transcription and compared to R6 as control. In addition, the levels of StaR and topoisomerases involved in Sc homeostasis (TopoI and gyrase) were determined by Western blot along the growth curve. Quantification was performed using RpoB as an internal control since no change in its transcription was observed under NOV treatments [7]. While, in the absence of NOV, these strains showed equivalent growth rates both in the absence and in the presence of $ZnSO_4$, in the presence of 0.5 × MIC of NOV, Δ*staR*P_{Zn}*staR* grew slower than R6 (Figure 2A). The doubling times in the absence of $ZnSO_4$ were 215 min ± 32 and 94 min ± 10 for Δ*staR*P_{Zn}*staR* and R6 (average ± SD, $n = 3$), respectively. In the presence of $ZnSO_4$, these differences were similar: 154 min ± 19 for Δ*staR*P_{Zn}*staR* and 89 min ± 10 for R6. Growth in the presence of 150 μM of $ZnSO_4$ yielded StaR increases of about 280-fold at all times considered, either in the presence or absence of NOV. However, StaR levels in strain Δ*staR*P_{Zn}*staR* with $ZnSO_4$ were only about 2-fold higher than those in R6 (Figure 2B,C). Thus, a 2-fold increase in the amount of StaR as well as its absence produced a higher susceptibility to NOV.

Regarding expression levels of topoisomerases, treatment with NOV decreased the level of TopoI about 4-fold at time points 90 and 150 min, and 6-fold at 210 min, in R6 compared to non-treated cells. This was expected considering previous results [7]. Similar decreases were observed in the strain Δ*staR*P_{Zn}*staR*, which showed a reduction of about 7-fold in TopoI level in NOV-treated versus non-treated cells at all times tested (90, 150, and 210 min; Figure 2B,C). Therefore, the differences observed in NOV susceptibility between both strains were not associated with a topoisomerase unbalance, but they were due to the distinct levels of StaR as a consequence of its ectopic location. Therefore, appropriate levels of StaR are important for the Sc homeostatic response against the topological stress imposed by sub-inhibitory NOV concentrations [7].

To test whether higher levels of StaR affect cell viability, we used a plasmid cloning approach. We cloned *staR* into pLS1ROM, rendering pLS1ROMstaR, in which *staR* is under the control of the maltose-inducible P_M promoter. Accordingly, StaR was induced in the presence of maltose independently on NOV treatment (Figure 3B). The growth rate and protein levels of Δ*staR* strain containing pLS1ROM*staR* were analyzed upon induction with maltose in the presence or in the absence of NOV. Growth in the presence of 0.4% sucrose + 0.4% maltose (SM) yielded a StaR increase of about 10-fold at 150 and 210 min with respect to sucrose (S)-grown cultures, although no effect on NOV susceptibility was observed (Figure 3A). However, induction with 0.8% maltose (M) increased NOV susceptibility, a condition yielding the increase of StaR to about 30-fold (150 min and 210 min) with respect to S-grown cultures (Figure 3B,C). StaR protein levels after induction with SM were similar to those observed in R6, consistent with the absence of any effect on growth (Figure 3B). In contrast, treatment with M induced a 3.5-fold change at all times tested. Altogether, our data indicate that NOV susceptibility increased both in the absence of StaR (Figures 1 and 2A) and when its levels were increased by 2- to 3.5-fold (Figures 2 and 3).

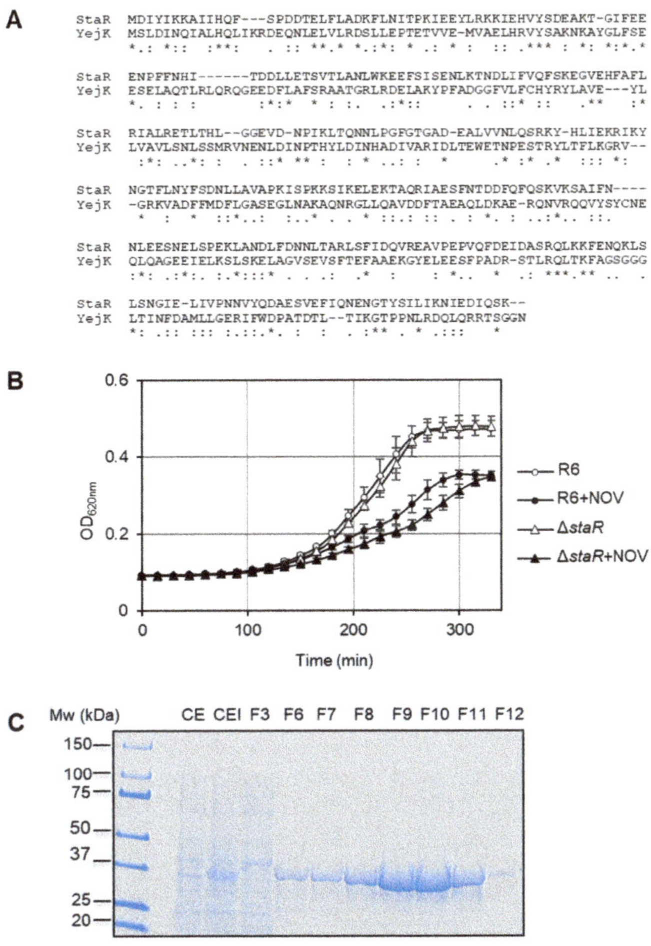

Figure 1. Identification of StaR. (**A**) Alignment between *E.coli* YejK and *S. pneumoniae* StaR using multiple sequence alignment. Asterisks label identical residues between both proteins and colons denote similar residues. (**B**) Absence of *staR* affects NOV susceptibility. Isogenic strains carrying wild-type *staR* (R6) or a deletion of *staR* (Δ*staR*) were grown to $OD_{620\ nm}$ = 0.3, diluted 200-fold in the absence or presence of NOV at 0.5 × MIC and their growth was recorded in a TECAN Infinite 200 PRO reader. Data are the average of three independent replicates ± SEM. (**C**) Purification of H_6-StaR from *E. coli* BL21-CodonPlus (DE3) cells harboring pET28-*staR*. SDS-PAGE of 30 μL of crude extract samples of no-induced (CE) or IPTG-induced (CEI) cultures. Fraction numbers eluted from the AKTA prime system column are indicated; 30 μL of fractions were run in the gel that was stained with Coomassie-blue. Molecular weights are indicated on the left.

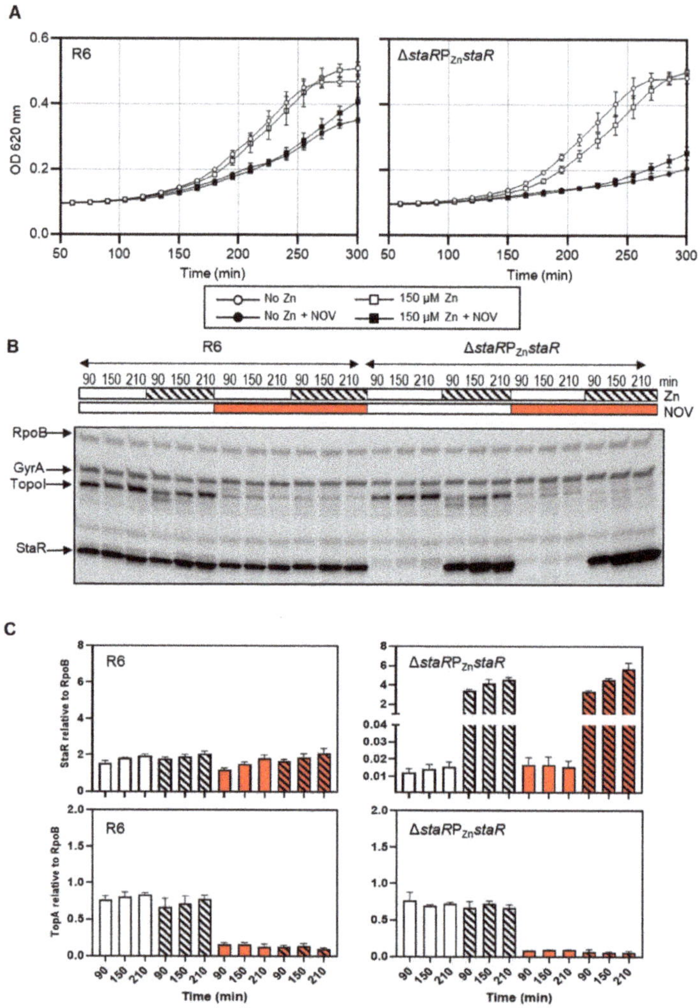

Figure 2. Effect of StaR expression under the control of P_{Zn} promoter. Wild-type R6 and Δ*staR*P_{Zn}*staR* strains were used. Strain Δ*staR*P_{Zn}*staR* carries a deletion of *staR* and a single ectopic copy of *staR* under the control of P_{Zn} promoter. (**A**) Expression of StaR under the control of P_{Zn} impairs growth under NOV treatment. Strains were grown to $OD_{620\,nm}$ = 0.3 in medium without $ZnSO_4$, diluted 200-fold in medium without (empty symbols) or with (black symbols) NOV (0.5 × MIC) together with 150 μM $ZnSO_4$ (squares) or without $ZnSO_4$ (circles) and cultures were grown in a TECAN Infinite 200 PRO reader. (**B**) Western blot assays of one of the three replicates using antibodies against GyrA (1:2000), TopoI (1:500), StaR (1:1000) and RpoB (1:2000) as internal control. Samples from the cultures described in (**A**) were taken at the indicated time points and processed as described in the Materials and Methods section. (**C**) Levels of StaR and TopoI proteins in the absence (empty bars) or in the presence (striped bars) of $ZnSO_4$, and in the absence (white bars) or in the presence (red bars) of NOV, as quantified from Western blots. Protein levels relative to RpoB are shown. Data presented are the average of three independent replicates ± SEM.

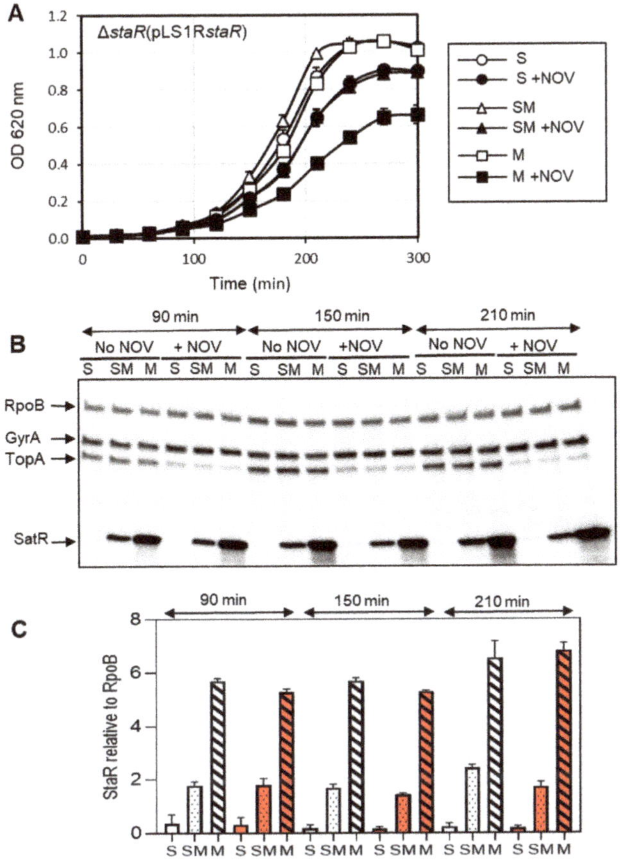

Figure 3. Effect on NOV susceptibility of increasing amounts of StaR. The strain used was Δ*staR* transformed with plasmid pLS1R- *staR*, which carries *staR* under the control of the maltose-regulated promoter P_M. (**A**) Growth curves in the presence of 0.8% sucrose (S), 0.4% sucrose + 0.4% maltose (SM), or 0.8% maltose (M) and in the absence or presence of NOV at 0.5 × MIC. The bacterial culture was grown to OD_{620nm} = 0.3 in medium with S. Cells were washed and diluted 100-fold in media containing either S, SM or M, in the absence or presence of NOV. (**B**) Western blot assays of GyrA, TopoI and StaR using antibodies against GyrA (1:2000), TopoI (1:500), StaR (1:1000) and RpoB (1:500) as internal control. Samples were collected at the indicated time points and treated as described in the Material and Methods section. A representative gel of three replicates is shown. (**C**) StaR levels in cultures grown with S (empty bars), SM (pointed bars), or M (striped bars), and in the absence (white bars) or in the presence (red bars) of NOV, quantified from Western blots. Protein levels relative to RpoB are shown. Data presented are the average of three independent replicates ± SEM.

These results suggest that StaR has a direct role in resolving DNA relaxation stress and that its levels need to be precisely regulated. In fact, *staR* is located in a DOWN domain in the R6 genome, and its transcription decreases more than 2-fold under NOV treatment [7]. This explains the increased NOV susceptibility observed when *staR* is ectopically located in Δ*staR*P_{Zn}*staR* into a locus (*bgaA*) not located in a supercoiling domain, as well as when it is overproduced in the presence of $ZnSO_4$ or maltose.

2.3. The Supercoiling Level Is Affected by StaR under Novobiocin Treatment

As part of the Sc homeostasis response, a decrease in TopoI is produced after NOV treatment (Figure 3), coincidentally with a decrease in *topA* transcription [7]. The effect of StaR levels in DNA Sc was analyzed by measuring Sc density of pLS1 in Δ*staR*P_{Zn}*staR* with or without NOV. We analyzed the topoisomer distribution of the internal plasmid pLS1 by two-dimensional agarose electrophoresis, a suitable approach for studying Sc levels, which is a reflection of nucleoid compaction [7]. This technique allows separation of DNA molecules by mass and shape as an estimation of the chromosomal Sc. As expected, treatment of Δ*staR*P_{Zn}*staR* with NOV resulted in DNA relaxation. In the absence of StaR expression (i.e., no $ZnSO_4$ in the medium), the value of negative DNA Sc density (σ) was -0.056, and it decreased to -0.051 and -0.049 at 0.5 × MIC and 1 × MIC NOV, respectively (Figure 4A). However, under StaR overexpression (150 µM ZnSO4) this variation in σ was higher: from -0.057 to -0.048 and -0.045 at 0.5 × MIC and 1 × MIC NOV, respectively (Figure 4B). Therefore, under inhibitory concentrations of NOV, negative Sc density was higher without StaR ($\sigma = -0.049$) than with a 2-fold overproduction of StaR ($\sigma = -0.045$), $p = 0.01$. These results suggest that StaR increases the level of NOV-induced DNA relaxation.

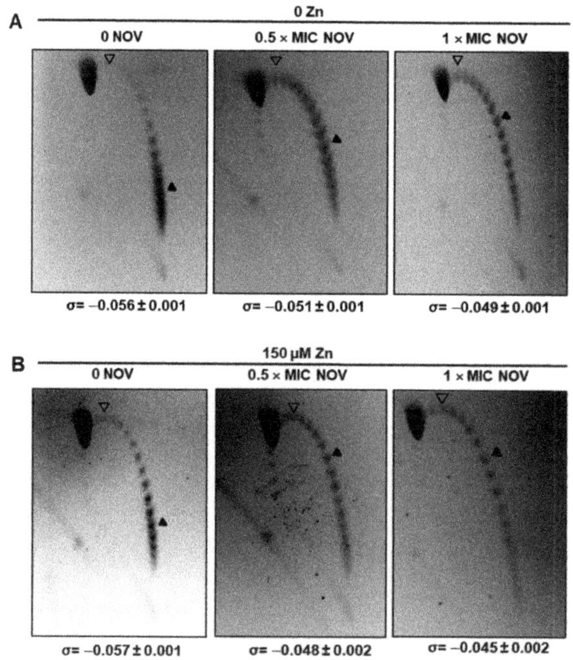

Figure 4. pLS1 topoisomers distribution in strain Δ*staR*P_{Zn}*staR* at the indicated $ZnSO_4$ and NOV concentrations. Cultures were grown to $OD_{620\ nm} = 0.2$, diluted 50-fold in medium without $ZnSO_4$ (**A**) or with $ZnSO_4$ (**B**) and grown to $OD_{620\ nm} = 0.4$. The samples taken at this point corresponded to those not treated with NOV. For NOV-treated samples, NOV was added to the cultures for 30 min so that the final concentration of NOV was 0.5 × MIC or 1 × MIC NOV and the $ZnSO_4$ concentration remained constant. Plasmids were extracted and run in 2D-agarose gels in the presence of 1 and 2 µg/mL chloroquine in the first and second dimensions, respectively. Values are the mean of four independent replicates ± SD. Gels of a typical experiment are shown. Empty arrowheads indicate the topoisomer that migrated with $\Delta Lk = 0$ in the second dimension and has a $\Delta Wr = -14$ during the first dimension (the number of positive supercoils introduced by 2 µg/mL chloroquine). Black arrowheads indicate the most abundant topoisomer.

2.4. *StaR Is a Nucleoid-Associated Protein*

To investigate the possible association of StaR with the nucleoid, we performed Stimulated Emission Depletion (STED) super-resolution microscopy of the wild-type strain (R6). After fixation, the nucleoids were stained with Sytox orange and StaR was immunofluorescently labeled. A total of 66.2% of cells (n = 4912) expressed detectable levels of StaR. In these, co-localization was readily observed, indicating that StaR localizes to the nucleoid (Figure 5A). To confirm co-localization, images of these cells were used to calculate the Pearson's correlation index (Figure 5C). Thresholds corresponding to the average intensity of the green signal of the Δ*staR*P$_{Zn}$*staR* strain (0.042) and a Pearson correlation index of 0.5 were applied. Under these criteria, StaR co-localized with the nucleoid in 51.5% (1674) of the R6 cells analyzed (Figure 5C). As a control, Δ*staR*P$_{Zn}$*staR* cells were imaged under the same growth, fixation and staining conditions in the absence of $ZnSO_4$ to image StaR-depleted cells (Figure 5B). Consistently, the distribution of the background green signal detected drastically differed compared to that of R6. The Δ*staR*P$_{Zn}$*staR* strain showed brilliant green areas without any conserved pattern that could be attributed to nonspecific binding of antibodies, and a low-intensity green signal which could correspond to cellular auto-fluorescence (Figure 5B). Moreover, under the same co-localization criteria that was applied to the wild-type (R6) strain, none of the Δ*staR*P$_{Zn}$*staR* cells analyzed showed any co-localization (Figure 5C). This further reinforces the co-localization analysis and shows that, like its homolog YejK, StaR associates with the nucleoid and must be considered a new NAP in *S. pneumoniae*.

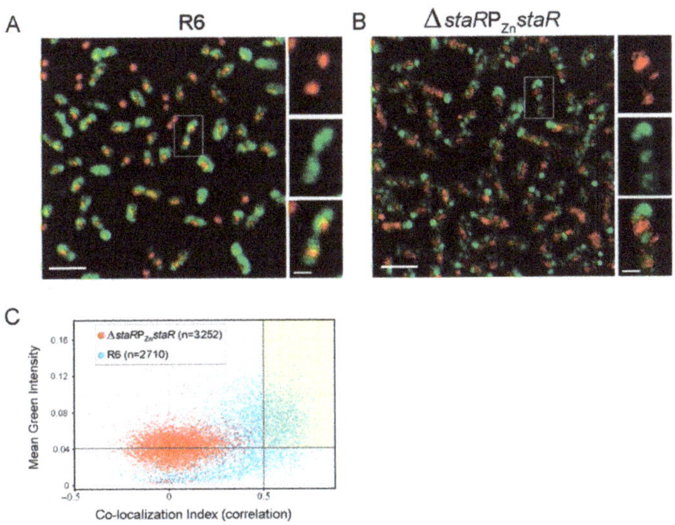

Figure 5. StaR localizes to the nucleoid. Strain R6 and Δ*staR*P$_{Zn}$*staR* were grown in the absence of $ZnSO_4$ to mid-log phase ($OD_{620\ nm}$ = 0.4). Cells were fixed, the nucleoids were stained with Sytox orange and StaR was immunostained using the Abberior Star 488 anti-rabbit secondary antibody as described in the Materials and Methods section. STED microscopy was used to determine the intracellular localization of nucleoids and StaR. (**A**) General view (left panel) and inset of a representative co-localizing cell of R6, showing nucleoids (red) and StaR labeling (green). Scale bar indicates 2 μm in the general views or 0.5 μM in the insets. (**B**) General view (left panel) and inset of a representative co-localizing cell of Δ*staR*P$_{Zn}$*staR*, as a control of StaR depletion. (**C**) Pearson correlation analysis of the strains indicated in (**A**,**B**). y-axis represents the mean intensity of green label (StaR) for Δ*staRPZnstaR* (red points) and R6 (blue points) and the x-axis indicated the Pearson correlation between green and red (nucleoids) labels. Gray lines in the x- and y-axis correspond to the thresholds applied to calculate the number of cells where StaR co-localizes with the nucleoid. A yellow square highlights the R6 population showing StaR–nucleoid co-localization.

2.5. StaR Specifically Activates S. pneumoniae TopoI

In vivo results of Sc variation in the presence or absence of StaR suggest a role of this protein in TopoI activation, since DNA relaxation in the presence of NOV is higher when StaR is overproduced. Alternatively, this protein could further inhibit DNA gyrase leading to the same observations.

The effect of StaR on the enzymatic relaxing activity of TopoI was tested. Results showed that StaR activates the relaxation of pBR322 by TopoI in vitro, depending on the amount of StaR used (Figure 6A). We have quantified the activation of TopoI by measuring both decrease of CCC plasmid form and increase of topoisomers (Figure 6B). Since the gel running conditions do not allow for differentiation between OC and RC, both topoisomers with intermediate linking numbers as well as OC/RC forms were considered for calculating TopoI activity. The results showed the simultaneous decrease of CCC and increase of topoisomers as the StaR concentration increases. The CCC and topoisomers reached 2.5% and 97.5%, respectively, at the highest StaR concentration. Moreover, activation of TopoI by StaR depended on the incubation time (Figure 6C). In contrast, StaR did not show topoisomerase activity by itself at concentrations equal to or higher than those used in TopoI activity assays (Figure 6D). In contrast, no activation of *S. pneumoniae* DNA gyrase activity was observed, since 73% of CCC plus topoisomers was observed at all StaR concentrations tested (from 0.25 to 4 μM). Although an excess of StaR over TopoI is required for activation, these results suggest that StaR is a specific activator of TopoI.

2.6. Physical Interaction between StaR and TopoI

The specific activation of TopoI by StaR suggested a protein–protein interaction between them. To verify this, in vitro co-immunoprecipitation assays (Co-IP) were carried out with the purified proteins. The two proteins were incubated together and immune-precipitated using polyclonal antibodies against either TopoI or StaR. The presence of TopoI in the anti-StaR pull-down fraction and the presence of StaR in the anti-TopoI pull-down fraction (Figure 7A) indicated a direct interaction between these proteins in vitro.

We also performed Co-IP in vivo after formaldehyde crosslinking (Figure 7B). We used extracts from cultures of Δ*staR*P_{Zn}*staR* grown in either the absence or presence of $ZnSO_4$, i.e., in the absence or presence of StaR. Western blots showed that StaR-Ab was able to pull down TopoI. Three bands were observed in the Western blot when the StaR-antibody was used in the Co-IP: a band presumably corresponding to TopoI (79 kDa), the band corresponding to StaR (37 kDa) and another band with an apparent molecular weight of 116 kDa, which may correspond to the TopoI-StaR complex. The last band corresponded to 26% of the total amount of the immunoprecipitated StaR. However, no pulldown of StaR by TopoI-Ab was observed, suggesting that the polyclonal antibody against TopoI dissociates the TopoI-StaR complex.

2.7. No Global Changes in Transcriptome Are Induced by the Absence or Presence of StaR

The transcriptome of the strain Δ*staR*P_{Zn}*staR* was analyzed either in the absence of StaR (no $ZnSO_4$ added) or in its presence (150 μM $ZnSO_4$). In the presence of the P_{Zn} inductor, 34 differentially expressed genes (DEGs) were detected (Table 1), with the same number of genes (17) up- or down-regulated. As expected, genes known to be regulated by high concentration of $ZnSO_4$ [7] were found among these DEGs, with seven up-regulated and six down-regulated. Up-regulated genes included the *psaBCA* operon coding a Mn/Zn transporter, *czcD* coding a cation efflux protein, *prtA* coding a virulence factor, and *adhB* involved in carbon metabolism, while *celA* was down-regulated. Consistent with the increased SatR levels detected in Western blots (280-fold increase, Figure 2) upon $ZnSO_4$ induction, mRNA levels of *staR* also increased about 400-fold.

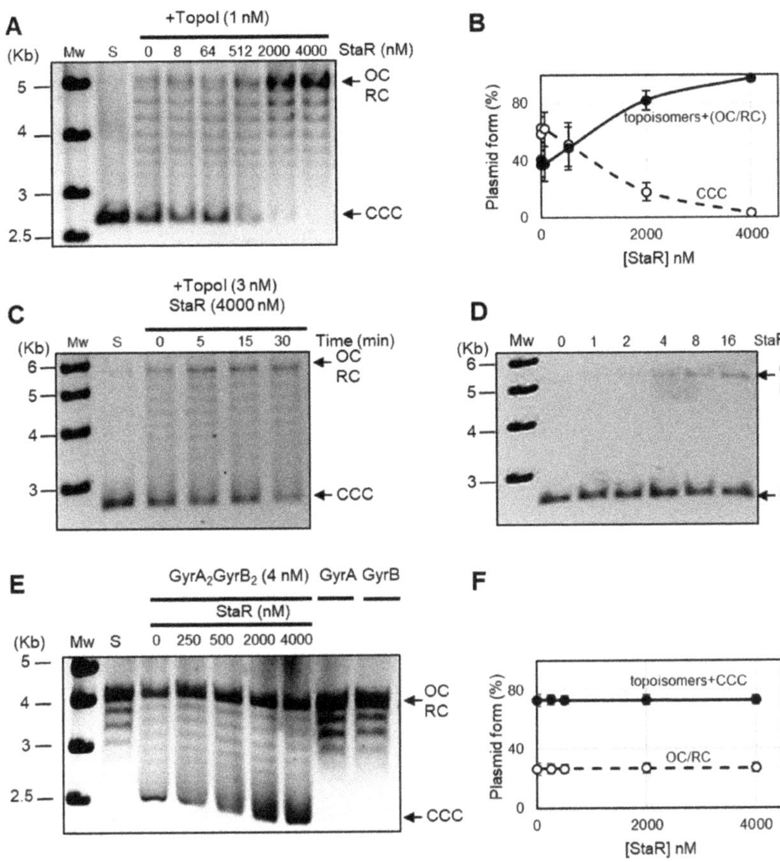

Figure 6. StaR activates the relaxation activity of *S. pneumoniae* TopoI but does not have any effect on gyrase. (**A**) Activation of TopoI as a function of StaR concentration. Plasmid pBR322 was incubated at 37 °C with purified TopoI for 1 h. The indicated StaR concentrations were added simultaneously to the reaction mix. Samples were processed and analyzed as described in the Materials and Methods section. Mw, molecular weight standard (Kb); S, substrate; CCC, covalently closed circles; OC, relaxed open circles; RC, relaxed circular plasmid forms. (**B**) Activity of TopoI determined as increase of the topoisomers (including OC/RC forms) and decrease of CCC form in the presence of StaR. Results are the average ± SEM of three independent replicates. (**C**) Activation as a function of reaction time. Reactions were carried out as in (**A**). (**D**) StaR does not have topoisomerase activity by itself. Reactions were carried out as in A, except that TopoI was not added. Concentrations of StaR are indicated. (**E**) DNA gyrase activity is not affected by StaR. Supercoiling activity over relaxed pBR322 using 4 nM of reconstituted gyrase was assayed in the presence of different StaR concentrations at 37 °C for 1 h. Incubation with GyrA or GyrB alone is included as a control. (**F**) Activity of gyrase determined as percentage of topoisomers +CCC and percentage of OC/RC forms in the presence of StaR. Results are the average ± SEM of three independent replicates.

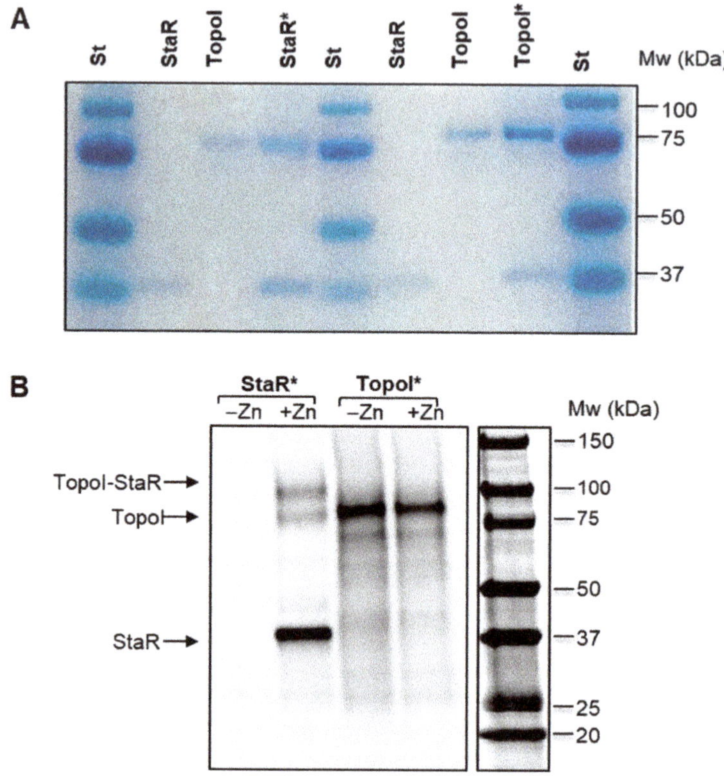

Figure 7. Co-immunoprecipitation of StaR and TopoI. (**A**) In vitro assays. Purified proteins were incubated together in the presence of the antibody against either StaR (StaR*) or TopoI (TopoI*). Samples were treated as described in the Materials and Methods section and run in a 4–20% polyacrylamide gel, which was stained with Coomassie blue. Purified proteins were loaded as controls. (**B**) In vivo assays. Cultures of Δ*staR*P_{Zn}*staR* were grown either in the absence (−Zn) or presence (+Zn) of 150 μM $ZnSO_4$ to $OD_{620\ nm}$ = 0.4, fixed with formaldehyde and treated as described in the Material and Methods section. The samples were run in a 4–20% polyacrylamide gel and Western blots were performed using antibodies against TopoI (1:500 dilution) and StaR (1:1000 dilution). The expected positions of TopoI and StaR proteins and of the TopoI-StaR complex are indicated.

When the effect of NOV on the transcriptome in the absence of StaR (−Zn) was compared with its effect when StaR was overexpressed (+Zn), no important differences were observed (Figure 8A). A total of 324 (110 up-regulated) and 313 (108 up-regulated) DEGs were detected in the absence or presence of StaR, respectively, with 214 DEGs common to both conditions (Figure 8A). However, a slightly differential response was observed, since about one-third of the DEGs under each condition was specific. In the absence of StaR, NOV triggered the expression of 110 genes (67 down-regulated and 43 up-regulated) that did not change in the presence of StaR. Similarly, when StaR was over-expressed, NOV changed the expression of 99 genes (58 down-regulated and 41 up-regulated) that did not change when *staR* was repressed. A comparison of the functional classes of the DEGs did not present significant differences between both conditions. In addition, no difference in the location of DEGs along the genome was detected in the absence or presence of $ZnSO_4$ (Figure 8B).

Table 1. DEGs detected in Δ*staR*P_{Zn}*staR* in the presence of $ZnSO_4$ versus in its absence. Cultures were grown to $OD_{620\ nm}$ = 0.2, diluted 50-fold in medium either without $ZnSO_4$ or with 150 μM $ZnSO_4$ and grown to $OD_{620\ nm}$ = 0.4.

Role o Subrole	Common Name	R6 Locus (Gene) [a]	Relative Fold Variation [b]
Carbohydrate metabolism	Beta-galactosidase	spr0059 (*bgaC*)	−2.2
Hypothetical proteins		spr0066	**3.9**
Transport: Carbohydrates	ABCT membrane permease	spr0082	−2.2
Hypothetical proteins		spr0085	−2.0
Carbohydrate metabolism	6-P-beta-glucosidase	spr0276 (*celA*)	−3.0
Transcription factors	Antiterminator	spr0279 (*bglG*)	−3.2
Transport: Carbohydrates	PTS sugar-specific EII component	spr0280 (*celC*)	−2.3
Hypothetical proteins		spr0281	−2.3
Transcription factors	Antiterminator	spr0504	−2.2
Pathogenesis	LPxTG protein	spr0561 (*prtA*)	**2.3**
Hypothetical proteins		spr0600	−2.0
Hypothetical proteins		spr0601	−2.1
Hypothetical proteins		**spr0929 (*staR*)**	**438.1**
Transport: antibiotics	ABCT ATP-binding/membrane	spr1289	−2.2
Hypothetical proteins		spr1402	**2.3**
Transport: cations	ABCT ATP-binding-Mn	spr1492 (*psaB*)	**4.1**
Transport: cations	ABCT membrane permease -Mn	spr1493 (*psaC*)	**4.7**
Transport: cations	ABCT substrate-binding-Mn	spr1494 (*psaA*)	**4.8**
Carbohydrate metabolism	Galactose-1-P uridylyltransferase	spr1667	**2.1**
Carbohydrate metabolism	Galactokinase	spr1668 (*galK*)	**2.3**
Carbohydrate metabolism	Alcohol dehydrogenase	spr1670 (*adhB*)	**73.7**
Hypothetical proteins		spr1671	**105.3**
Transport: cations	Cation diffusion facilitator	spr1672 (*czcD*)	**507.5**
Carbohydrate metabolism	Dextran glucosidase	spr1698	−3.2
Hypothetical proteins		spr1940	−2.2
Hypothetical proteins		spr1966	**2.6**
Transport: other	Glycerol uptake facilitator protein	spr1988 (*glpF*)	−2.2
Carbohydrate metabolism	Glycerol-3-P dehydrogenase	spr1989	−2.6
Carbohydrate metabolism	Glycerol-3-P dehydrogenase	spr1990	−3.1
Carbohydrate metabolism	Glycerol kinase	spr1991 (*glpK*)	−2.0
Hypothetical proteins		spr2037	**3.1**
Hypothetical proteins		spr2038	**3.4**
Hypothetical proteins		spr2039	**3.4**
Hypothetical proteins		spr2040	**2.0**

[a] Genes shadowed in gray are those known to be regulated by high $ZnSO_4$. [b] The responsive genes from two replicates included those showing a significant fold change (absolute value ≥ 2) and a *p*-value-adjusted ≤ 0.01.

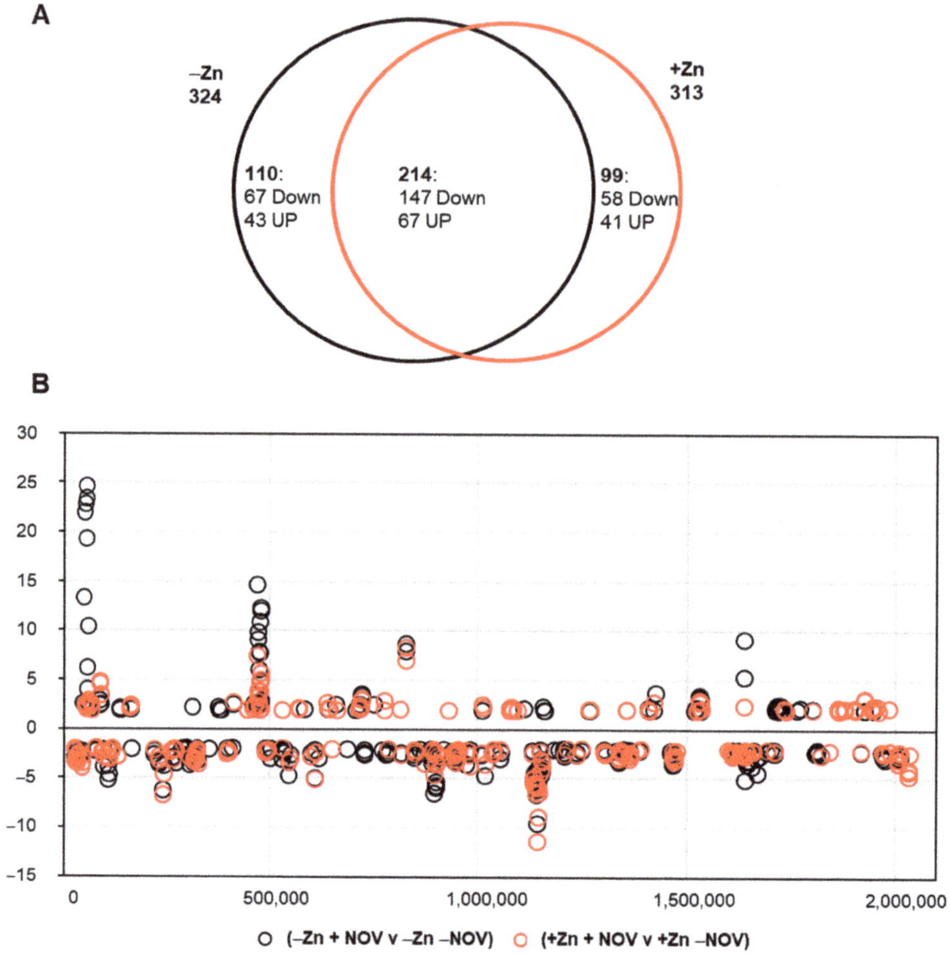

Figure 8. Number and localization of DEGs in the Δ*staR*P_{Zn}*staR* chromosome upon treatment with NOV in the presence or absence of StaR. Cultures were grown to $OD_{620\ nm}$ = 0.2, diluted 50-fold in medium either without or with 150 μM $ZnSO_4$ and grown to $OD_{620\ nm}$ = 0.4. Samples taken at this point were considered as non-treated with NOV. NOV-treated samples were grown for another 30 min with 0.5 × MIC of NOV. Total RNA was isolated and sequenced as described in the Materials and Methods section. (**A**) Venn diagram showing DEGs upon treatment with NOV in the absence or presence of $ZnSO_4$. (**B**) The relative fold variation of each gene represented against the 5′ location of each open reading frame in the *S. pneumoniae* R6 chromosome (bases 1 to 2,038,615). A fold change ≥ 2 (absolute value) and a *p*-value-adjusted ≤0.01 were considered.

These results suggest that the overexpression of StaR does not affect global transcription induced by NOV in a chromosomal-domain manner, and thus discard StaR as a transcriptional regulator.

3. Discussion

The existence of protein networks governing DNA chromosomal topology of pathogens remains an unknown field. Sc levels, nucleoid compaction and viability in *S. pneumoniae* are determined by the balance between gyrase and TopoI activities [7]. This bacterium triggers homeostatic transcriptional responses when Sc density decreases by 25% [8] or when it increases by 40% [16]. These global responses affect transcription of topoisomerases. Relaxation triggers up-regulation (about 2-fold) of gyrase genes (*gyrA* and *gyrB*) and down-regulation of both TopoI (*topA*, around 10-fold) and TopoIV (*parE* and *parC*, around 3-fold) [7]. When Sc increases, *topA* is again down-regulated (about 2-fold), while *gyrA*, *gyrB*, *parE* and *parC* remain unchanged [16]. In this way, the down-regulation of *topA* transcription, although to different levels, allowed cell growth and the recovery of Sc.

In addition, the transcriptomic response of wild-type strain R6 to DNA relaxation, which is triggered by NOV, also involved down-regulation (about 2-fold) of the gene coding StaR (*spr0929*), which is indeed included in a down-regulated domain [8]. A similar decrease in transcription (around 3-fold) was observed for the gene coding HU, the only NAP described so far in *S. pneumoniae*, which is essential for Sc maintenance [32]. However, the down-regulation of HU and StaR only occurred at inhibitory NOV concentrations. No change in the transcription or amount of StaR (Figure 2) was observed at sub-inhibitory NOV concentrations. Nevertheless, the location of StaR in a down-regulated domain is important for cell survival in the presence of NOV (Figure 1). This location is consistent with the homeostatic response to relaxation, which involves the down-regulation of *topA* [7], and to the decrease of TopoI levels detected in Western blots (Figure 2). Therefore, relaxation triggers down-regulation of both *topA* and *staR*. This would contribute to the recovery of supercoiling levels after relaxation, since the levels of StaR, which would normally activate TopoI, would decrease. Accordingly, the overproduction of StaR observed in Δ*staR*P_{Zn}*staR* upon addition of $ZnSO_4$ reduces recovery of Sc at inhibitory (1 × MIC) concentrations of NOV: 21% of Sc variation after 30 min of NOV treatment versus 12.5% without induction (Figure 4). This is presumably due to TopoI activation by StaR. The role of StaR would be the activation of TopoI by protein–protein interaction. Although this activation would hypothetically affect transcription, a global effect was not detected in the in the RNA-Seq experiments carried out in the present paper at the concentration of NOV used.

S. pneumoniae lacks most of the NAPs characterized in *E. coli*, with the exception of HU [32] and SMC [33]. Using DAPI to stain DNA and immunostaining for NAPs, it has been shown in *E. coli* that HU and other NAPs localize to the nucleoid [38]. We have recently used super-resolution confocal microscopy and staining of the bacterial nucleoid with the DNA intercalant Sytox [9]. This technique allowed us to accurately identify and locate nucleoids and determine their level of compaction [9]. In this study, we took advantage of this technique to demonstrate that StaR indeed localizes in the nucleoid (Figure 5). A statistical analysis performed with more than 2500 cells showed co-localization of StaR and the nucleoid in more than 50% of analyzed bacteria. Although we cannot compare this number with previous reports, given that, to our knowledge, no statistical quantification has been performed previously for any NAP, we consider that a co-localization higher than 50% indeed reflects a true co-localization. It is important to notice that cultures were not synchronous, and the localization of StaR in the nucleoid could depend on the cell cycle phase. This is the first demonstration of a NAP localizing to the nucleoid in *S. pneumoniae*.

Therefore, StaR should be considered a new NAP, which regulates the activity of DNA topoisomerases. Other NAPs regulating topoisomerases have been identified in other bacteria. This is the case of GyrI, which inhibits *E. coli* gyrase [39], GapR from *Caulobacter crescentus* that stimulates gyrase and Topo IV to remove (+) Sc during DNA replication [40], and HU that activates Topo I in *M. tuberculosis* [41]. *E. coli* YejK, homologous to StaR, has been shown to inhibit both gyrase and TopoIV activities [42]. We have not found gyrase activity alteration by StaR, which probably reflects the low sequence identity between YejK and StaR (Figure 1). Activation of Topo IV has not been tested, given that *S. pneumoniae* Sc is mostly controlled by the opposing activities of Topo I and gyrase [7,16]. Regulation of the topoisomerase type I activity has also been observed in eukaryotes. It has been described that TOP1 activity is controlled by the phosphorylated RNAPII to modulate elongation during transcription [43]. In *S. pneumoniae*, TopoI is the main topoisomerase involved in the regulation of supercoiling [7,16] and in transcription [14]; therefore, its activity must be finely regulated.

While co-membership to the same Pfam family does not imply sharing exactly the same properties, we have confirmed that StaR is a NAP and has a similar size to *E. coli* YejK, indicating that NA37 is a proper global name for this protein family. In future studies, the structure–function relationship may add support at residue-level to the observations described here. Unfortunately, no homologous structural templates for modeling are currently available. Numerous attempts to resolve the StaR structure from *S. pneumoniae* and *Enterococcus faecalis* were unsuccessful since crystals showed low-resolution diffraction. In all, evidence indicate NA37 proteins are not prone to experimental structural analyses. In contrast, novel ab initio modeling tools based on artificial intelligence such as AlphaFold achieve unprecedented accuracy [44]. Such homology-free models may help to evaluate critical traits of StaR. These include the co-location of basic residues able to interact with DNA, the domain delineation, and the identification of interacting sides for docking studies.

To summarize, StaR is a new NAP of *S. pneumoniae* and is the first characterized TopoI modulator in this bacterium, playing an important role in the genome-relaxation-triggered homeostasis. Therefore, we propose that StaR is involved in the fine-tuning control of the activity of the essential TopoI, contributing to the homeostatic response to Sc changes.

4. Materials and Methods

4.1. Bacterial Strains, Growth Conditions and Transformation

S. pneumoniae strains were grown as static cultures at 37 °C in a casein hydrolysate-based liquid medium (AGCH) containing 0.2% yeast extract and 0.3% sucrose [45] (C + Y). *E. coli* was grown in LB medium at 37 °C with agitation. A strain with a deletion in *staR* was constructed as described [34]. Transformation of pneumococcal strains was performed as previously reported [45] and selection of transformants was in 1 μg/mL tetracycline, 250 μg/mL kanamycin, 2.5 μg/mL chloramphenicol, or 1 μg/mL erythromycin. Growth was followed by measuring the OD at 620 $_{nm}$ either with a UV-visible spectrophotometer (Evolution 201, Thermo Scientific, Waltham, MA, USA) or in a microplate reader (Infinite F200, Tecan, Männedorf, Switzerland). Both measures correlate linearly by means of the equation y = 0.2163x + 0.1151 (y = microplate reader measure, x = spectrophotometer), R^2 = 0.98 [32].

4.2. Construction of Strains and Plasmids

To overproduce and purify StaR, a PCR fragment containing its coding gene was amplified from R6 genomic DNA with primers *spr0929_F* (5′-GCGCGCTAGCATGGATATTTATA TTAAGAAAGCC) and *spr0929_R* (5′-GCGCGGATCCTTATTTACTTTGGATATCCTCG) [34]. Restriction enzyme sites are showed underlined in all oligonucleotides. The amplicon was digested with NheI and BamHI and cloned into pET28a cut with the same enzymes. The plasmid was transferred to BL21-CodonPlus (DE3) (Agilent Technologies, Santa Clara, CA, USA) *E. coli* cells. To test the physiological effects of the overexpression of *staR*, a genetic fusion between the P_{Zn}–ribosome binding site and the *staR* coding region was made to render the Δ*staRP*$_{Zn}$*staR* strain. The *staR* gene was amplified with spr0929_ScIF (5′-: GCGCGAGCTCATGGATATTTATATTAAGAAAGCC) and spr0929_SlIR (5′-GCGCGTCGACTTATTTACTTTGGATATCCTCG) oligonucleotides using R6 chromosomal DNA as template. The amplicon was digested with SacI and SalI enzymes, targets included in the oligonucleotides, and introduced into pZK plasmid cleaved with the same enzymes to render plasmid pZK_0929. The pZK plasmid contains the upstream region of the *czcD* gene that shows up-regulation in the presence of zinc [46]. A PCR fragment containing the Zn-inducible promoter upstream the spr0929 ORF was amplified from pZK_0929 with KmR_B1_Rv (5′-CGCGGGATCCAGGATCCATCGATACAAATTCC) and pZK_Xb1_Rv (5′-GCGCTCTAGACACCATAAAAAATGAACTTGG) oligonucleotides. This construction was introduced into the *spr1806* locus of the Δ*staR* strain. The *spr1806* gene codes for a surface protein, which in laboratory conditions, is not expressed; subsequently, the deletion mutant shows no relevant differences with respect to the wild-type strain [47]. For that, DNA fragments of ~1 Kb corresponding to the exact flanking regions of *spr1806* were amplified with oligonucleotides spr1806_ExF (5′-CATAGGTCTCCACATCGAAA) and spr1806_LiURB1 (5′-GCGCGGATCCAACAGCTAGAAAATTCTATTCT), for the upstream region, and oligonucleotides spr1806_LiDFXb (5′-GCGCTCTAGATACTGAACCATCTGGGAATG) and spr1806_ExR (5′-TTGCCCGTCAAAATGATTTG) for the downstream region. Amplicons were cut with BamHI and XbaI, respectively, and simultaneously ligated to the fragment containing the Zn-inducible promoter and spr0929 ORF. The ligation fragment was introduced into pre-competent Δ*staR* cells by natural transformation and transformants were selected with 50 μg/mL kanamycin. Recombination was checked by PCR.

To test the role of *staR* overproduction, the gene was cloned into pLS1ROM vector under the control of P_{Mal} promoter, which is controlled by MalR transcriptional repressor that is cloned in pLS1ROM under the control of the constitutive P_{tet} promoter [48]. For this purpose, a PCR fragment containing *staR* was obtained from R6 chromosome using oligonucleotides 0929HindF (5′-CGCAAGCTTGGAGGACTTTTATGGATATTTATATTAAGAAAGC) and 0929BamR (5′-GCGGATCCGCAAAAAGGAAGACTGCTAGTACAAG), cut with HindIII and BamHI, and ligated to pLS1ROM cut with the same enzymes. The constructed plasmid pLS1ROMstaR was transformed into Δ*staR* strain and transformants were selected with 1 μg/mL erythromycin.

4.3. Cloning, Expression and Purification of Proteins

H_6-StaR was purified by His-affinity chromatography using HiTrap Chelating HP columns (Amersham-Merck, Madrid, Spain) and an AKTA prime system (Amersham Bioscience, Amersham, UK) from *E. coli* BL21-CodonPlus (DE3) cells harboring pET28-StaR. The strain was grown in LB medium containing 50 μg/mL kanamycin at 37 °C to $OD_{600\ nm}$ = 0.6 and *staR* transcription was induced with 1 mM isopropyl thio-β-D-galactoside (IPTG) for 1 h. The cells were harvested by centrifugation, washed with phosphate-buffered saline (PBS) and suspended in 72 mL of buffer A (50 mM Tris-HCl pH 8.0, 0.5 M NaCl). The cells were sonicated (8 cycles of 20 sec ON/60 sec OFF) and cell extracts were clarified by centrifugation at 100,000× *g* for 1 h at 4 °C. A two-step protein fractionation with ammonium sulfate at 40% pellet and 60% saturation was performed. The pellet from 60% saturation was suspended in 51 mL of buffer A containing 40 mM imidazole and applied to a 5 mL HiTrap column equilibrated buffer A with 40 mM imidazole. Protein elution was achieved with 100 mL of a 100-to-500-mM imidazol gradient in buffer A, and fractions were analyzed by SDS-PAGE. Imidazol was removed from samples by dialysis against 20 mM Tris-HCl pH 7.5, 1 mM EDTA, 0.1 M NaCl, 50% glycerol. Protein concentration was determined with a Qubit 4 fluorimeter using the Qubit® Protein Assay Kit (Invitrogen-ThermoFisher Scientific, Waltham, MA, USA). TopoI, GyrA and GyrB were purified by affinity chromatography in a Ni-NTA (Quiagen, Hilden, Germany) column following manufacturer's instructions as described previously [26].

4.4. Confocal Microscopy

The cells were grown until $OD_{620\ nm}$ = 0.4, centrifuged and suspended in PBS to a number of 8×10^9 cells/mL. The cells were fixed with 2% formaldehyde for at least 16 h. For immunofluorescence microscopy, 8×10^7 cells were incubated with 5 μM SytoxTM Orange Nucleic Acid Stain (Invitrogen) at room temperature for 5 min and extended into a poly-L-lysine coated glass slide. The cells were permeabilized by immersing the slide in methanol at −20 °C for 10 min. The slide was first incubated with 2% BSA, 0.2% Triton X100 in PBS (BSA-PBST), incubated with a rabbit StaR polyclonal antibody at 1/100 dilution in BSA-PBST buffer, washed with PBS and incubated with the anti-rabbit Abberior® STAR 488 antibody (Abberior, Göttingen, Germany) at 1/200 dilution. After a wash with PBS, the slides were mounted with ProLongTM Gold Antifade Mountant (Invitrogen) and sealed. Nucleoids and StaR were imaged as previously described [9] in a confocal microscope STELLARIS 8–FALCON/STED (Leica Microsystems, Wetzlar, Germany) with a HC PL APO 100×/1.40 NA × OIL immersion objective. Super resolution images were acquired by Stimulated Emission Depletion (STED) microscopy using 660 nm depletion laser. Analysis was performed with Cell profiler software v4.2.5, and co-localization-correlation index was calculated for each bacteria with a custom-made routine.

4.5. Western Blot Assays and Antibody Purification

Whole cell lysates (~5×10^5 cells) were obtained by centrifugation of 10 mL cultures ($OD_{620\ nm}$ = 0.4). They were suspended in 400 μL of sample loading buffer (0.3 M Tris-HCl pH 6.8, 10% SDS, 50% *v/v* glycerol and 0.05% bromophenol) supplemented with 0.5 M β-mercaptoethanol and incubated for 5 min at 100 °C. Lysates were separated on Any kD™ Criterion™ TGX Stain-Free™ Protein Gels (Bio-Rad, Hercules, CA, USA). They were transferred to 0.2 μm PVDF membranes with a Trans-Blot Turbo Transfer System (Bio-Rad) at 25 V, 1 A for 30 min. Membranes were blocked with 5% skim milk in Tris-buffered saline for 2 h and incubated with anti-TopoI and anti-GyrA [16], anti-RpoB [14] and anti-StaR (diluted 1:1000). Rabbit polyclonal antibody against StaR was obtained from Davids Biotechnologie from 0.5 mg of protein extracted from SDS-gel following a 28-day SuperFast immunization protocol. Anti-rabbit IgG-peroxidase Ab (Sigma-Aldrich-Merck, Madrid, Spain) was used as the secondary Ab. SuperSignal West Pico chemiluminescent substrate (Thermo-Fisher, Waltham, MA, USA) was used to develop the membranes. Signal was detected with a ChemiDocTM MP system (Bio-Rad). Images were analyzed using Image LabTM software (Bio-Rad). Molecular masses of GyrA and TopoI are 92 kDa and 79 kDa, respectively.

4.6. DNA Topoisomerase Assays

Relaxation reactions of pBR322 by TopoI were carried out exactly as described previously [49]. Reactions of 200 μL contained 0.5 μg of CCC pBR322 in 20 mM Tris-HCl pH 8, 100 mM KCl, 10 mM $MgCl_2$, 1 mM DTT, 50 μg BSA/mL and TopoI at the indicated concentrations. Incubation with TopoI was at 37 °C for 1 h and the reaction was terminated by 2 min incubation at 37 °C with 50 mM EDTA. Then an additional incubation of 1 h at 37 °C with 1% SDS, 100 μg/mL proteinase K was performed. Reaction products were ethanol precipitated, suspended in electrophoresis loading buffer and analyzed by electrophoresis in 1% agarose gels run at 18 V for 18 h. DNA quantification of agarose gels was done by scanning densitometry after electrophoresis and ethidium bromide staining. Quantification of TopoI activity was calculated by gel densitometry using the Image Lab program (Bio-Rad Laboratories, Hercules, CA, USA). To calculate activity, the OC form amount was determined and divided by the total amount of DNA in each well. Enzymatic reactions with gyrase were conducted as described previously [50].

4.7. Co-Immunoprecipitation (Co-IP) Assays

For in vitro Co-IP, the buffer used along the experiment was Co-IP buffer (10 mM sodium phosphate pH 7.4, 50 mM NaCl, 0.5% Nonidet P40). Purified TopoI (60 ng/μL) was dialyzed against the buffer, and StaR (9.2 μg/mL) was diluted 1000-fold in the same buffer. A total of 2 μg of TopoI and StaR, respectively, was mixed into 200 μL of buffer for 2 h at 4 °C with rotation at room temperature. Purified Abs (4 μg) were bound to Dynabeads® Protein G for 2 h in the buffer at room temperature with rotation as described by the manufacturer and were added to the protein mix. After incubation, the mix was washed 3 times and it was collected from the magnetic grid with SDS-loading buffer without β-mercaptoethanol. The samples were boiled for 5 min, loaded in a 4–20% SDS-polyacrylamide gel and run for 1 h at 170 v. Gels were stained with Coomassie blue for about 1 h.

For in vivo Co-IP, 10 mL of fixing buffer (50 mM Tris-HCl pH 8.0, 100 mM NaCl, 0.5 mM EGTA, 1 mM EDTA, 11% (*v*/*v*) formaldehyde) was added to 100 mL of culture grown to $OD_{620\ nm}$ = 0.4 (7×10^9 cells) with or without 150 μM $ZnSO_4$. This mix was incubated for 30 min at room temperature. Crosslinking was stopped by adding 10 mL of cold quenching solution (1.25 M glycine in 50 mM Tris-HCl pH 8.0, 100 mM NaCl, 0.5 mM EGTA, 1 mM EDTA). The mix was transferred to an ice/water bath and rotated for 30 min at 4 °C. Cells were collected by centrifugation at 4 °C and suspended in 50 mL ice-cold PBS.

Washing was repeated twice. Finally, cells were collected by centrifugation and suspended in 1 mL of lysis buffer (50 mM Hepes-KOH pH 7.5, 140 mM NaCl, 1 mM EDTA, 1% (*v*/*v*) Triton X-100, 0.1% (*w*/*v*) sodium deoxicholate, 1 mM PMSF), which contained 100 μg/mL of RNase A. Cells were sonicated using 25 cycles (30 s on/30 s off) in a Bioruptor® Pico sonicator (Diagenode, Ougrée, Belgium). The sonicated suspension was centrifuged at 21,000× *g* for 10 min at 4 °C. 100 μL of the supernatant was kept at −20 °C as whole cell extract control.

TopoI and StaR antibodies were purified as described previously [14]. The antibodies (10 μg) were bound to 50 μL Dynabeads® Protein G. For immunoprecipitation, 500 μL of sonicated suspension (antigens) was added to 200 μL of the magnetic-bead–Ab complex and incubated on a rotating mixer for 4 h at 4 °C. The tube was placed on a magnet and the supernatant was removed. It was washed once with the lysis buffer, once with the lysis buffer containing 500 mM NaCl and once with the lysis buffer containing 250 mM LiCl. The supernatant was removed and proteins bound to Dynabeads were resupended in 30 μL of the sample loading buffer containing 50 mM glycine. Dynabeads® Protein G—antigen was incubated at 65 °C overnight with shaking to elute DNA from the beads and reverse cross-links. Samples were fractionated in SDS/PAGE and proteins were detected by Western blot using TopoI-Ab (1:500 dilution) and StaR-Ab (1:1000 dilution) antibodies.

4.8. Analysis of the Topology of Plasmids

Plasmid DNA topoisomers were analyzed in neutral/neutral two-dimensional agarose gels. The first dimension was run at 1.5 V/cm in a 0.4% agarose (Seakem-Lonza, Basel, Switzerland) gel in 1 × Tris-borate-EDTA (TBE) buffer for 20 h at room temperature. The second dimension was run at 7.5 V/cm in 1% agarose gel in 1 × TBE buffer for 7–9 h at 4 °C. Chloroquine (Sigma-Aldrich-Merck, Madrid, Spain) was added to the TBE buffer in both the agarose and the running buffer. Chloroquine is a DNA intercalating agent that removes negative Sc in bacterial plasmids. Increasing the concentration of chloroquine progressively eliminates negative Sc until the plasmid is relaxed and can then introduce net positive Sc. In this way, the use of adequate concentrations of chloroquine during each dimension in the 2D analysis allows the efficient resolution of the different topoisomers [51]. Chloroquine concentrations used were 1 μg/mL and 2 μg/mL in the first and second dimension, respectively. Gels were stained with 0.5 μg/mL ethidium bromide for 1 h at room temperature. Images were captured in a ChemiDoc Imaging System (Bio-Rad) and analyzed with the Image Lab software (BioRad). DNA supercoiling density (σ) was calculated from the equation $\sigma = \Delta Lk/Lk_0$. Linking number differences (ΔLk) were determined using the equation $\Delta Lk = Lk - Lk_0$, in which $Lk_0 = N/10.5$, being N the size of the molecule in bp (4408) and 10.5 the number of bp per one complete turn in B-DNA. To simplify it, σ = Lk of the most abundant topoisomer/(N/10.5). The Lk of the most abundant topoisomer was calculated taking into account that the topoisomer that migrated with $\Delta Lk = 0$ in the second dimension has a $\Delta Wr = -14$ during the first dimension (the number of positive supercoils introduced by 2 μg/mL chloroquine).

4.9. RNA Library Preparation for RNA-Seq

Total RNA (5 μg) obtained from samples containing 2 to 4 × 10^8 cells with the RNeasy mini kit (Quiagen) was depleted of ribosomal RNA using the Ribo-Zero Magnetic Kit (Bacteria) (Illumina, San Diego, CA, USA). Libraries for RNA sequencing (RNA-Seq) were prepared using the ScriptSeq v2 RNA-Seq system (Illumina). Briefly, 1 μg of RNA samples were chemically fragmented. Using randomly primed cDNA synthesis, cDNAs that were tagged at the 5′ end (equivalent to the 3′ end of the original RNA) were synthesized. The cDNAs were then tagged at the 3′ end using Terminal-Tagging Oligos (TTO). These oligonucleotides randomly annealed to the cDNA and were extended by DNA polymerase. The resulting di-tagged cDNAs were purified with the Ampure bead XP system. The enrichment and barcoding of the purified di-tagged cDNAs were done with 15 cycles of PCR. The library size was determined with a 2100 Bioanalyzer Instrument (Agilent

Technologies). qPCR quantifications were done with a Kapa Library quantification Kit (Kapa Biosystems, Sigma-Aldrich-Merck, Madrid, Spain).

4.10. RNA Seq Data Analysis

Analysis of RNA-Seq data was carried out using the web-based platform Galaxy [14]. The quality of raw sequence data was analyzed with the FASQC tool. Sequencing reads were mapped against the *S. pneumoniae* R6 genome (ASM704v1) using the BWA software package (Galaxy version 0.7.17.4) in simple Illumina mode. The number of reads overlapping each coding gene was obtained using program feature Count. Count tables were used as input in DESeq2 for the analysis of differential expression. A threshold *p*-value-adjusted of 0.01 was considered.

Author Contributions: Conceptualization, A.G.d.l.C., M.A., M.-J.F. and P.H.; methodology, A.A.d.V.J., A.J.M.-G., J.M.T.-V., D.M. and M.-J.F.; software, P.H.; formal analysis, A.G.d.l.C., M.A., M.-J.F. and P.H.; investigation, A.A.d.V.J., J.M.T.-V., M.-J.F., A.J.M.-G., D.M. and M.A.; data curation, P.H.; writing—original draft preparation, A.G.d.l.C.; writing—review and editing, M.A., M.-J.F. and P.H.; supervision, A.G.d.l.C., M.A., M.-J.F. and P.H.; project administration, A.G.d.l.C.; funding acquisition, A.G.d.l.C. All authors have read and agreed to the published version of the manuscript.

Funding: This research and the APC were funded by project PID2021-124738OB-100 to A.G.d.l.C., financed by MCIN/AEI/10.13039/501100011033/FEDER, UE.

Institutional Review Board Statement: Not applicable.

Informed Consent Statement: Not applicable.

Data Availability Statement: The data discussed in this publication have been deposited in NCBI's Gene Expression Omnibus and are accessible through GEO Series accession number GSE225261 (https://www.ncbi.nlm.nih.gov/geo/query/acc.cgi?acc=GSE225261).

Acknowledgments: We thank Miguel Hernández-González (The Francis Crick Institute, London, UK) and Michael McConell (Centro Nacional de Microbiología, ISCIII) for their critical reading and editing.

Conflicts of Interest: The authors declare no conflict of interest. The funders had no role in the design of the study; in the collection, analyses, or interpretation of data; in the writing of the manuscript; or in the decision to publish the results.

References

1. Holmes, V.F.; Cozzarelli, N.R. Closing the ring: Links between SMC proteins and chromosome partitioning, condensation, and supercoiling. *Proc. Natl. Acad. Sci. USA* **2000**, *97*, 1322–1324. [CrossRef] [PubMed]
2. Champoux, J.J. DNA Topoisomerases: Structure, Function, and Mechanism. *Annu. Rev. Biochem.* **2001**, *70*, 369–413. [CrossRef] [PubMed]
3. Gellert, M.; Mizuuchi, K.; O'Dea, M.H.; Nash, H.A. DNA gyrase: An enzyme that introduces superhelical turns into DNA. *Proc. Natl. Acad. Sci. USA* **1976**, *73*, 3872–3876. [CrossRef] [PubMed]
4. Mizuuchi, K.; O'Dea, M.H.; Gellert, M. DNA gyrase: Subunit structure and ATPase activity of the purified enzyme. *Proc. Natl. Acad. Sci. USA* **1978**, *75*, 5960–5963. [CrossRef]
5. DiNardo, S.; Voelkel, K.A.; Sternglanz, R.; Reynolds, A.; Wright, A. Escherichia coli DNA topoisomerase I mutants have compensatory mutations in DNA gyrase genes. *Cell* **1982**, *31*, 43–51. [CrossRef]
6. Pruss, G.J.; Manes, S.H.; Drlica, K. Escherichia coli DNA topoisomerase I mutants: Increased supercoiling is corrected by mutations near gyrase genes. *Cell* **1982**, *31*, 35–42. [CrossRef]
7. Ferrándiz, M.-J.; Martín-Galiano, A.J.; Schvartzman, J.B.; de la Campa, A.G. The genome of Streptococcus pneumoniae is organized in topology-reacting gene clusters. *Nucleic Acids Res.* **2010**, *38*, 3570–3581. [CrossRef]
8. de la Campa, A.G.; Ferrándiz, M.J.; Martín-Galiano, A.J.; García, M.T.; Tirado-Vélez, J.M. The Transcriptome of Streptococcus pneumoniae Induced by Local and Global Changes in Supercoiling. *Front. Microbiol.* **2017**, *8*, 1447. [CrossRef]
9. García-López, M.; Megias, D.; Ferrándiz, M.-J.; de la Campa, A.G. The balance between gyrase and topoisomerase I activities determines levels of supercoiling, nucleoid compaction, and viability in bacteria. *Front. Microbiol.* **2022**, *13*, 1094692. [CrossRef]
10. Wang, X.; Llopis, P.M.; Rudner, D.Z. Organization and segregation of bacterial chromosomes. *Nat. Rev. Genet.* **2013**, *14*, 191–203. [CrossRef]
11. Dorman, C.J. Genome architecture and global gene regulation in bacteria: Making progress towards a unified model? *Nat. Rev. Microbiol.* **2013**, *11*, 349–355. [CrossRef] [PubMed]

12. Liu, L.F.; Wang, J.C. Supercoiling of the DNA template during transcription. *Proc. Natl. Acad. Sci. USA* **1987**, *84*, 7024–7027. [CrossRef] [PubMed]
13. Cheng, B.; Zhu, C.-X.; Ji, C.; Ahumada, A.; Tse-Dinh, Y.-C. Direct Interaction between Escherichia coli RNA Polymerase and the Zinc Ribbon Domains of DNA Topoisomerase I. *J. Biol. Chem.* **2003**, *278*, 30705–30710. [CrossRef]
14. Ferrándiz, M.-J.; Hernández, P.; de la Campa, A.G. Genome-wide proximity between RNA polymerase and DNA topoisomerase I supports transcription in Streptococcus pneumoniae. *PLoS Genet.* **2021**, *17*, e1009542. [CrossRef] [PubMed]
15. Ahmed, W.; Sala, C.; Hegde, S.R.; Jha, R.K.; Cole, S.T.; Nagaraja, V. Transcription facilitated genome-wide recruitment of topoisomerase I and DNA gyrase. *PLoS Genet.* **2017**, *13*, e1006754. [CrossRef]
16. Ferrándiz, M.-J.; Martín-Galiano, A.J.; Arnanz, C.; Camacho-Soguero, I.; Tirado-Vélez, J.-M.; de la Campa, A.G. An increase in negative supercoiling in bacteria reveals topology-reacting gene clusters and a homeostatic response mediated by the DNA topoisomerase I gene. *Nucleic Acids Res.* **2016**, *44*, 7292–7303. [CrossRef]
17. Higgins, N.P.; Yang, X.; Fu, Q.; Roth, J.R. Surveying a supercoil domain by using the gamma delta resolution system in Salmonella typhimurium. *J. Bacteriol.* **1996**, *178*, 2825–2835. [CrossRef]
18. Sinden, R.R.; Pettijohn, D.E. Chromosomes in living Escherichia coli cells are segregated into domains of supercoiling. *Proc. Natl. Acad. Sci. USA* **1981**, *78*, 224–228. [CrossRef]
19. Worcel, A.; Burgi, E. On the structure of the folded chromosome of Escherichia coli. *J. Mol. Biol.* **1972**, *71*, 127–147. [CrossRef]
20. Martín-Galiano, A.J.; Ferrándiz, M.J.; de la Campa, A.G. Bridging Chromosomal Architecture and Pathophysiology of Streptococcus pneumoniae. *Genome Biol. Evol.* **2017**, *9*, 350–361. [CrossRef]
21. Ferrandiz, M.-J.; Arnanz, C.; Martin-Galiano, A.J.; Rodríguez-Martín, C.; De La Campa, A.G. Role of Global and Local Topology in the Regulation of Gene Expression in Streptococcus pneumoniae. *PLoS ONE* **2014**, *9*, e101574. [CrossRef] [PubMed]
22. World Health Organization. Pneumococcal conjugate vaccine for childhood immunization—WHO position paper. *Wkly. Epidemiol. Rec.* **2007**, *82*, 93–104.
23. Jacobs, M.R.; Felmingham, D.; Appelbaum, P.C.; Grüneberg, R.N.; The Alexander Project Group. The Alexander Project 1998-2000: Susceptibility of pathogens isolated from community-acquired respiratory tract infection to commonly used antimicrobial agents. *J. Antimicrob. Chemother.* **2003**, *52*, 229–246. [CrossRef] [PubMed]
24. Domenech, A.; Tirado-Vélez, J.M.; Fenoll, A.; Ardanuy, C.; Yuste, J.; Liñares, J.; de la Campa, A.G. Fluoroquinolone-Resistant Pneumococci: Dynamics of Serotypes and Clones in Spain in 2012 Compared with Those from 2002 and 2006. *Antimicrob. Agents Chemother.* **2014**, *58*, 2393–2399. [CrossRef] [PubMed]
25. Ardanuy, C.; de la Campa, A.G.; García, E.; Fenoll, A.; Calatayud, L.; Cercenado, E.; Pérez-Trallero, E.; Bouza, E.; Liñares, J. Spread of *Streptococcus pneumoniae* Serotype 8-ST63 Multidrug-Resistant Recombinant Clone, Spain. *Emerg. Infect. Dis.* **2014**, *20*, 1848–1856. [CrossRef] [PubMed]
26. Balsalobre, L.; Ferrandiz, M.-J.; de Alba, G.; de la Campa, A.G. Nonoptimal DNA Topoisomerases Allow Maintenance of Supercoiling Levels and Improve Fitness of Streptococcus pneumoniae. *Antimicrob. Agents Chemother.* **2011**, *55*, 1097–1105. [CrossRef] [PubMed]
27. Balsalobre, L.; Ferrándiz, M.J.; Liñares, J.; Tubau, F.; de la Campa, A.G. Viridans Group Streptococci Are Donors in Horizontal Transfer of Topoisomerase IV Genes to Streptococcus pneumoniae. *Antimicrob. Agents Chemother.* **2003**, *47*, 2072–2081. [CrossRef]
28. de la Campa, A.G.; Ardanuy, C.; Balsalobre, L.; Pérez-Trallero, E.; Marimón, J.M.; Fenoll, A.; Liñares, J. Changes in Fluoroquinolone-ResistantStreptococcus pneumoniaafter 7-Valent Conjugate Vaccination, Spain. *Emerg. Infect. Dis.* **2009**, *15*, 905–911. [CrossRef]
29. Alvarado, M.; Martin-Galiano, A.J.; Ferrandiz, M.-J.; Zaballos, Á.; De La Campa, A.G. Upregulation of the PatAB Transporter Confers Fluoroquinolone Resistance to *Streptococcus pseudopneumoniae. Front. Microbiol.* **2017**, *8*, 2074. [CrossRef]
30. Amblar, M.; Zaballos, Á.; de la Campa, A.G. Role of PatAB Transporter in Efflux of Levofloxacin in Streptococcus pneumoniae. *Antibiotics* **2022**, *11*, 1837. [CrossRef]
31. Dillon, S.C.; Dorman, C.J. Bacterial nucleoid-associated proteins, nucleoid structure and gene expression. *Nat. Rev. Microbiol.* **2010**, *8*, 185–195. [CrossRef] [PubMed]
32. Ferrándiz, M.-J.; Carreño, D.; Ayora, S.; de la Campa, A. HU of Streptococcus pneumoniae Is Essential for the Preservation of DNA Supercoiling. *Front. Microbiol.* **2018**, *9*, 493. [CrossRef] [PubMed]
33. Minnen, A.; Attaiech, L.; Thon, M.; Gruber, S.; Jan-Willem Veening, J.-W. SMC is recruited to oriC by ParB and promotes chromosome segregation in Streptococcus pneumoniae. *Mol. Microbiol.* **2011**, *81*, 676–688. [CrossRef] [PubMed]
34. Martín-Galiano, A.J.; Yuste, J.; Cercenado, M.I.; De La Campa, A.G. Inspecting the potential physiological and biomedical value of 44 conserved uncharacterised proteins of Streptococcus pneumoniae. *BMC Genom.* **2014**, *15*, 1–12. [CrossRef] [PubMed]
35. Murphy, L.D.; Rosner, J.L.; Zimmerman, S.B.; Esposito, D. Identification of Two New Proteins in Spermidine Nucleoids Isolated from Escherichia coli. *J. Bacteriol.* **1999**, *181*, 3842–3844. [CrossRef]
36. Gellert, M.; Mizuuchi, K.; O'Dea, M.H.; Itoh, T.; Tomizawa, J.-I. Nalidixic acid resistance: A second genetic character involved in DNA gyrase activity. *Proc. Natl. Acad. Sci. USA* **1977**, *74*, 4772–4776. [CrossRef]
37. Pruss, G.J.; Franco, R.J.; Chevalier, S.G.; Manes, S.H.; Drlica, K. Effects of DNA gyrase inhibitors in Escherichia coli topoisomerase I mutants. *J. Bacteriol.* **1986**, *168*, 276–282. [CrossRef]
38. Azam, T.A.; Hiraga, S.; Ishihama, A. Two types of localization of the DNA-binding proteins within the Escherichia coli nucleoid. *Genes Cells* **2000**, *5*, 613–626. [CrossRef]

39. Nakanishi, A.; Oshida, T.; Matsushita, T.; Imajoh-Ohmi, S.; Ohnuki, T. Identification of DNA Gyrase Inhibitor (GyrI) in *Escherichia coli*. *J. Biol. Chem.* **1998**, *273*, 1933–1938. [CrossRef]
40. Guo, M.S.; Haakonsen, D.L.; Zeng, W.; Schumacher, M.A.; Laub, M.T. A Bacterial Chromosome Structuring Protein Binds Overtwisted DNA to Stimulate Type II Topoisomerases and Enable DNA Replication. *Cell* **2018**, *175*, 583–597.e23. [CrossRef]
41. Ghosh, S.; Mallick, B.; Nagaraja, V. Direct regulation of topoisomerase activity by a nucleoid-associated protein. *Nucleic Acids Res.* **2014**, *42*, 11156–11165. [CrossRef] [PubMed]
42. Lee, C.; Marians, K.J. Characterization of the Nucleoid-associated Protein YejK. *J. Biol. Chem.* **2013**, *288*, 31503–31516. [CrossRef] [PubMed]
43. Baranello, L.; Wojtowicz, D.; Cui, K.; Devaiah, B.N.; Chung, H.-J.; Chan-Salis, K.Y.; Guha, R.; Wilson, K.; Zhang, X.; Zhang, H.; et al. RNA Polymerase II Regulates Topoisomerase 1 Activity to Favor Efficient Transcription. *Cell* **2016**, *165*, 357–371. [CrossRef] [PubMed]
44. Jumper, J.; Evans, R.; Pritzel, A.; Green, T.; Figurnov, M.; Ronneberger, O.; Tunyasuvunakool, K.; Bates, R.; Žídek, A.; Potapenko, A.; et al. Highly accurate protein structure prediction with AlphaFold. *Nature* **2021**, *596*, 583–589. [CrossRef] [PubMed]
45. Lacks, S.A.; López, P.; Greenberg, B.; Espinosa, M. Identification and analysis of genes for tetracycline resistance and replication functions in the broad-host-range plasmid pLS1. *J. Mol. Biol.* **1986**, *192*, 753–765. [CrossRef]
46. Kloosterman, T.G.; Witwicki, R.M.; van der Kooi-Pol, M.M.; Bijlsma, J.J.E.; Kuipers, O.P. Opposite Effects of Mn^{2+} and Zn^{2+} on PsaR-Mediated Expression of the Virulence Genes pcpA, prtA, and psaBCA of Streptococcus pneumoniae. *J. Bacteriol.* **2008**, *190*, 5382–5393. [CrossRef]
47. Escolano-Martínez, M.S.; Domenech, A.; Yuste, J.; Cercenado, M.I.; Ardanuy, C.; Liñares, J.; de la Campa, A.G.; Martin-Galiano, A.J. DiiA is a novel dimorphic cell wall protein of Streptococcus pneumoniae involved in invasive disease. *J. Infect.* **2016**, *73*, 71–81. [CrossRef]
48. Ruiz-Masó, J.A.; López-Aguilar, C.; Nieto, C.; Sanz, M.; Burón, P.; Espinosa, M.; del Solar, G. Construction of a plasmid vector based on the pMV158 replicon for cloning and inducible gene expression in Streptococcus pneumoniae. *Plasmid* **2012**, *67*, 53–59. [CrossRef]
49. García, M.T.; Blázquez, M.A.; Ferrandiz, M.-J.; Sanz, M.J.; Silva-Martín, N.; Hermoso, J.; de la Campa, A.G. New Alkaloid Antibiotics That Target the DNA Topoisomerase I of Streptococcus pneumoniae. *J. Biol. Chem.* **2011**, *286*, 6402–6413. [CrossRef]
50. Fernández-Moreira, E.; Balas, D.; González, I.; DE LA Campa, A.G. Fluoroquinolones Inhibit PreferentiallyStreptococcus pneumoniaeDNA Topoisomerase IV Than DNA Gyrase Native Proteins. *Microb. Drug Resist.* **2000**, *6*, 259–267. [CrossRef]
51. Schvartzman, J.B.; Martínez-Robles, M.-L.; Hernández, P.; Krimer, D.B. Plasmid DNA Topology Assayed by Two-Dimensional Agarose Gel Electrophoresis. In *DNA Electrophoresis*; Methods in Molecular Biology; Humana: Totowa, NJ, USA, 2013; Volume 1054, pp. 121–132. [CrossRef]

International Journal of
Molecular Sciences

MDPI

Review

What's on the Other Side of the Gate: A Structural Perspective on DNA Gate Opening of Type IA and IIA DNA Topoisomerases

Vita Vidmar [1,†], Marlène Vayssières [1,†] and Valérie Lamour [1,2,*]

1 Institut de Génétique et de Biologie Moléculaire et Cellulaire (IGBMC), Université de Strasbourg, CNRS UMR 7104, Inserm U 1258, 67400 Illkirch, France
2 Hôpitaux Universitaires de Strasbourg, 67098 Strasbourg, France
* Correspondence: lamourv@igbmc.fr
† These authors contributed equally to this work.

Abstract: DNA topoisomerases have an essential role in resolving topological problems that arise due to the double-helical structure of DNA. They can recognise DNA topology and catalyse diverse topological reactions by cutting and re-joining DNA ends. Type IA and IIA topoisomerases, which work by strand passage mechanisms, share catalytic domains for DNA binding and cleavage. Structural information has accumulated over the past decades, shedding light on the mechanisms of DNA cleavage and re-ligation. However, the structural rearrangements required for DNA-gate opening and strand transfer remain elusive, in particular for the type IA topoisomerases. In this review, we compare the structural similarities between the type IIA and type IA topoisomerases. The conformational changes that lead to the opening of the DNA-gate and strand passage, as well as allosteric regulation, are discussed, with a focus on the remaining questions about the mechanism of type IA topoisomerases.

Keywords: DNA topoisomerase; DNA supercoiling; DNA relaxation; DNA decatenation; type IA topoisomerase; type IIA topoisomerase; structural biology

Citation: Vidmar, V.; Vayssières, M.; Lamour, V. What's on the Other Side of the Gate: A Structural Perspective on DNA Gate Opening of Type IA and IIA DNA Topoisomerases. *Int. J. Mol. Sci.* **2023**, *24*, 3986. https://doi.org/10.3390/ijms24043986

Academic Editors: Joseph E. Deweese and Neil Osheroff

Received: 11 January 2023
Revised: 9 February 2023
Accepted: 12 February 2023
Published: 16 February 2023

1. Introduction

In all organisms, the intertwined structure of the DNA double helix requires the intervention of DNA topoisomerases in order to maintain the homeostasis of DNA topology during the cell cycle. Topoisomerases are involved in all aspects of DNA metabolism, from replication, DNA repair, and recombination, to gene expression due to their essential role in resolving topological problems [1]. This large family of enzymes can recognise different DNA topologies and catalyse a wide variety of reactions such as DNA relaxation/supercoiling, catenation/decatenation, and knotting/unknotting [2]. All topoisomerases perform the topological transactions by cleaving and resealing DNA, and this reaction involves an intermediate where the enzyme forms a covalent bond with the DNA backbone, forming a transient cleavage complex (Top-cc) [3]. Topoisomerases are divided into two types, depending on whether they introduce single-strand (type I) or double-strand breaks (type II). They are further classified into five subtypes (IA, IB, IC, IIA, or IIB) based on structural homology and reaction mechanisms [4]. Most topoisomerases (types IA, IIA, and IIB) employ a strand passage mechanism in which they transport another segment through the transient break in the DNA. Others (IB and IC) work by a mechanism of controlled rotation, where they let the free end of the cleaved DNA rotate around the intact strand.

The first topoisomerases were identified and characterised in the 1970s as the bacterial ω protein [5] and DNA gyrase [6], which were later classified as type IA and type IIA topoisomerases (Figure 1a). DNA relaxation is one of the major activities of type IA and IIA topoisomerases. However, the bacterial type IIA DNA gyrase and type IA reverse gyrase have the unique ability to introduce negative or positive supercoils, respectively [2]. The type IA topoisomerases relax negatively supercoiled DNA in an ATP-independent manner, whereas the type IIA topoisomerases relax negatively and positively supercoiled DNA in the presence of ATP [2].

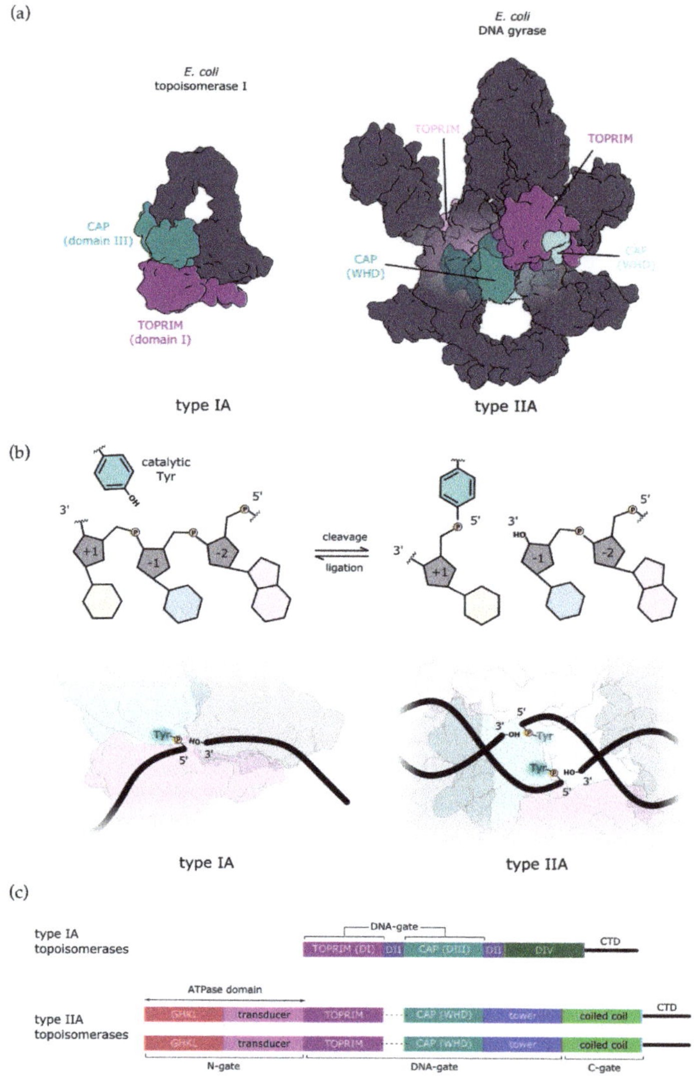

Figure 1. Comparison of type IA and type IIA topoisomerases. (**a**) All type A topoisomerases share catalytic domains for DNA binding and cleavage: CAP domain (dark cyan) bearing the catalytic tyrosine and TOPRIM domain (dark magenta) with a cluster of conserved acidic residues that bind Mg^{2+} which is also required for catalysis. Type IA topoisomerases are monomers with one CAP and one TOPRIM domain, whereas type IIA topoisomerases are dimeric and thus have two pairs of catalytic domains in trans (the second pair is shown in a lighter colour). *E. coli* topoisomerase I is shown without CTD for clarity. (**b**) During DNA cleavage, catalytic tyrosine forms a covalent bond with the 5′-phosphate of the DNA backbone in both types of topoisomerases (top). The reaction is

reversible, with the equilibrium shifting toward ligation. Type IA topoisomerases cleave single-stranded DNA (bottom left), whereas type IIA topoisomerases introduce a transient double-strand break with a stagger of 4 bp (bottom right). (**c**) Schematic representation of domain organisation. Type IA normally consists of a single polypeptide chain encoding four topoisomerase domains, here labelled as DI to DIV. Additionally, they may have a variable CTD. Type IIA topoisomerases are dimers with two copies of each domain. In bacteria, they are heterotetramers (A_2B_2), whereas in eukaryotes, they are homodimers (A_2), where each half of the enzyme is fused into a single subunit (denoted with a dashed line).

The type IA enzymes act on a larger variety of substrates, such as ssDNA, nicked DNA, dsDNA, and RNA, when compared to the type IIA enzymes, which carry out topological transactions solely on dsDNA. In addition to the relaxation activity of supercoils, some topoisomerases preferentially catenate/decatenate, or knot/unknot DNA molecules [2]. For these particular activities, the substrates of the bacterial Topo I enzymes are restricted to ssDNA or nicked DNA. TopIII, another type of IA enzyme, can relax and decatenate DNA and RNA [7,8]. In bacteria and human cells, the types IIA, TopoIV, and TopoIIα, respectively, also possess a robust decatenation activity [2,9,10].

The mechanism of DNA cleavage involves a conserved catalytic Tyr in the active site that participates in a reversible transesterification reaction, forming a transient phosphotyrosine bond with the 5′ ends of the DNA backbone (Figure 1b) [3]. The monomeric structure of the type IA topoisomerase and the dimeric structure of the type IIA or IIB topoisomerases condition the introduction of single- and double-strand breaks respectively (Figure 1b). The mechanism of single- or double-stranded passage that follows the cleavage of a first DNA molecule is tightly regulated to prevent the accumulation of DNA breaks that can lead to cell death.

Structural investigations of the catalytic cycle of the type IA and type IIA enzymes have provided most of the molecular details of the DNA cleavage/re-ligation mechanism and of the conformations that lead to DNA strand passage and DNA relaxation. In particular, the common structural organisation of the cleavage site of the IA and IIA enzymes has been revealed by x-ray crystallography [4]. They contain a common set of CAP domains adopting a Winged-Helix Domain fold (WHD) comprising the conserved tyrosine and a TOPRIM domain with acidic residues chelating two magnesium ions (Figure 1a) [11]. Although the IIB enzymes act with a strand passage mechanism, their primary sequence and global structural organization, down to the CAP and TOPRIM catalytic domains, differ from the IIA enzymes [12–16]. Recent studies have shown that decatenation or unknotting are the preferred activities of the bacterial type IIB topoisomerase VI, which may be related to the fact that they work specifically on certain topologies [17].

In comparison to type IA enzymes, structural studies of type IIA topoisomerases have revealed molecular details for many conformations of the DNA binding/cleavage domain, specifically the closed, partially opened, or fully opened DNA-gate. While structural information is now available to understand how DNA cleavage/re-ligation is catalysed by type IA enzymes, the mechanism of DNA transport through the break and how the structural domains contribute to the opening of the DNA gate at the molecular level remains quite elusive for the type IA family. In this review, we present an overview of the conformational states that condition DNA relaxation and strand passage, focusing on the DNA-gate of the type IA and IIA enzymes, with an emphasis on the remaining questions pertaining to the mechanism and regulation of gate opening in the type IA family.

2. Mechanism of DNA-Gate Opening by Type IIA DNA Topoisomerases

Type IIA topoisomerases are present as dimeric (eukaryotes) or heterotetrameric (prokaryotes and archaea) assemblies (Figure 1c). They contain two copies of evolutionarily conserved domains forming three protein interfaces, also called "gates". The ATPase (GHKL) domains form the N-gate and are responsible for ATP hydrolysis [18]. The DNA gate is composed of the conserved CAP (WHD), TOPRIM, and Tower domains (Figure 1c). They contain the conserved catalytic tyrosine and acidic residues chelating two magnesium ions and are responsible for DNA cleavage [19–21]. The C-gate is composed of two long coil-coiled helices at the C-terminal end of the protein, where a DNA duplex can exit the enzyme interface after strand passage [22–25]. Together, biochemical and structural data have led to the proposal of a "three-gates" mechanism where the protein interfaces successively interact with DNA duplexes at the different steps of the intermolecular reaction of DNA relaxation, catenation, or unknotting [18,19,21–29]. Interestingly, the type IIB family, for which structures of the archaeal TopoVI and eukaryotic Spo11 have been solved, lacks the C-gate, and only structures of the closed conformation of the DNA gate have been determined so far [14,16].

2.1. Structural Rearrangements Leading to DNA-Gate Opening in Type IIA Topoisomerases

The first step of the catalytic cycle is the binding of DNA, the G-segment (for "Gated" DNA), at the dimeric interface of the DNA-gate. Crystal structures of the DNA-binding/cleavage domain revealed conformational changes of the DNA-gate upon DNA binding [19,24,26,30] (Figure 2, step 1). The G-segment binding triggers a ~80° rotation of the TOPRIM domains relative to their initial position, leading to new protein-protein interactions with the CAP and Tower domains from the opposite monomer. This movement creates a positively charged groove in which the double helix of DNA can perfectly fit [24]. Cleavage of DNA is mediated by two metal ions for type IA and type IIA topoisomerases [11] (Figure 2, step 2). After DNA binding, the metal ion binding motif (E-DxD) in the TOPRIM domain is positioned close to the conserved catalytic tyrosine in the CAP domain. The comparison of the active site structures with cleaved or uncleaved DNA has revealed finely regulated conformational changes before, during, and after DNA cleavage [11,25,31–35].

The overall comparison of DNA-gate structures in presence of cleaved or uncleaved DNA suggests that protein interfaces move when they open or close. These movements are carefully regulated to prevent DNA breaks in the genome. ATP hydrolysis at the N-gate triggers allosteric signals controlling the opening or closing of the gates, notably through the positioning of the catalytic tyrosine residues [11,24]. Therefore, once the G-segment is cleaved, the C-gate is kept closed so that the DNA gate can open for the transport of another DNA duplex, called the T-segment (Figure 2, steps 3 and 4). When the G-segment is religated, the DNA-gate closes and the C-gate opens, allowing the T-segment to exit (Figure 2, step 5). The fine regulation of the protein interfaces opening during DNA cleavage is of paramount importance for the unidirectionality of DNA transport (Figure 2, step 4).

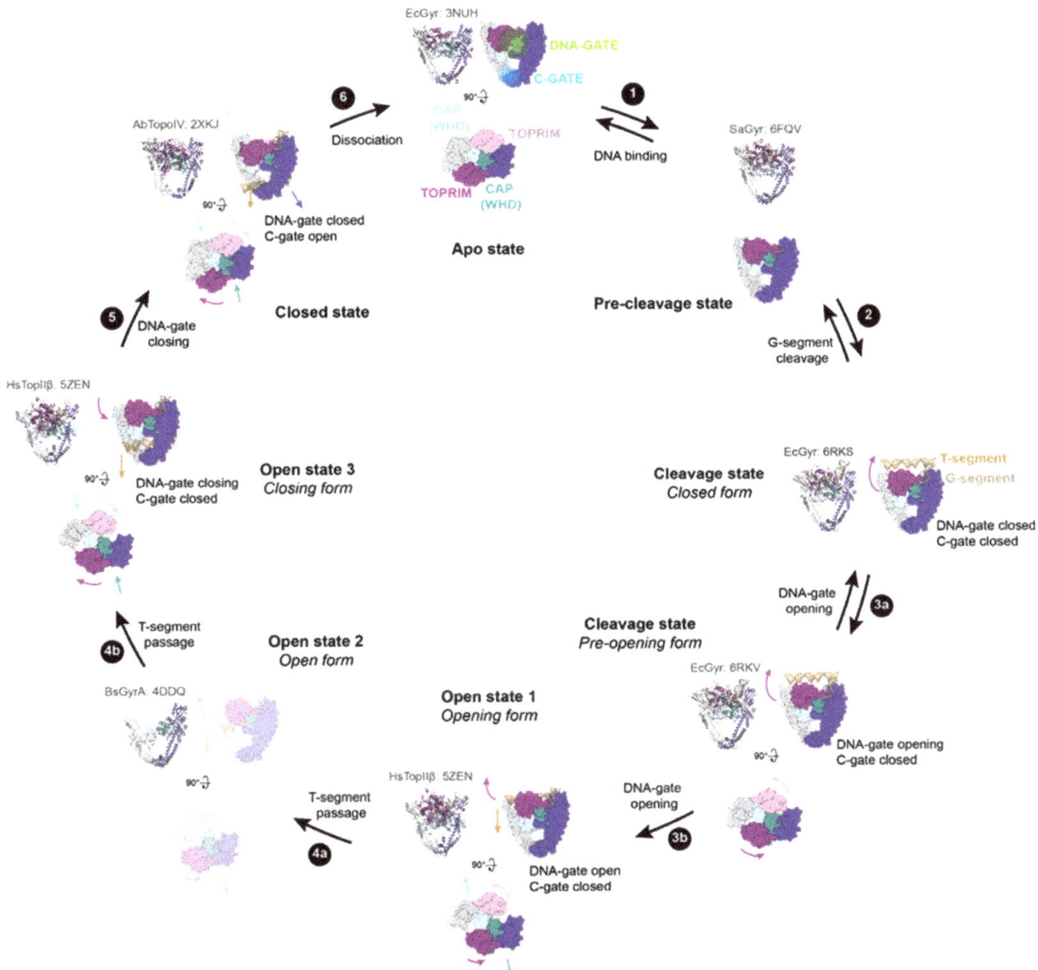

Figure 2. DNA-gate opening mechanism of type IIA enzymes. All the shown intermediates were structurally determined by X-ray crystallography or cryo-electron microscopy (cryo-EM). PDB accession codes are indicated in light grey. Structure representations restricted to the DNA-binding/cleavage domain were generated from the *E. coli* DNA gyrase structure without DNA and the insertion domain for clarity. TOPRIM domains are coloured dark magenta and plum. CAP (WHD) domains are coloured light blue and dark cyan. The other domains are coloured in dark slate blue and grey. The G-segment is shown in tan, and the T-segment (modelled) is in sandy brown. (1) G-segment binds to the DNA binding-groove formed by the TOPRIM-CAP(WHD)-Tower domains. (2) DNA binding triggers a rotation of the TOPRIM domains, resulting in a repositioning of the acidic motif with metal ions close to the catalytic tyrosine residues (CAP/WHD) and cleavage of the G-segment. (3) DNA-gate opening involving sliding and swivelling motions (TOPRIM/WHD (CAP)) of the two halves of the protein against each other. 3a and 3b show the transition to the closed form to the pre-opening form of the DNA-gate. The dimer remains together thanks to interactions in the C-gate. (4) From these movements, a funnel-shaped channel is revealed between the two halves of the DNA-gate allowing T-segment passage to change DNA topology. 4a shows the T-segment passage through the upper part of the DNA-gate. The structure of GyrA in an opened conformation suggests that a wider opening of the DNA gate happens to complete strand passage as illustrated in

step 4b. (5) After the T-segment passage through cleaved G-segment, the DNA-gate is closed and the G-segment is re-ligated. Once the DNA-gate is closed, the C-gate opens to release the T-segment. (6) After a complete reaction cycle, the enzyme dissociates from the DNA and is reset for another cycle.

Type IIA topoisomerases also modify the DNA geometry in order to simplify DNA topology. DNA bending is induced by a conserved isoleucine (Ile 833 for yeast TopII) that intercalates between two base pairs [24]. The high degree of DNA bending generated by Topo IIA may explain its preferential binding to highly bent and flexible DNA regions [36,37].

The first observation of type IIA DNA-gate structures in closed conformation reported twenty years ago led to the proposal that DNA-gate opening is similar to a book-opening movement [19,26]. However, a crystal structure of the DNA-gate in an open conformation has been published recently, and a new opening mechanism has been proposed [34] involving a path for the T-segment through the gate as well as conformational changes that could account for the transition from a closed to an opened state. In this conformation, the dimer is maintained only by protein interactions at the C-gate. The CAP domains move away to create a funnel-shaped channel between the two parts of DNA-gate. This cavity is able to accommodate the T-segment in the DNA-gate. The helical axes of the two parts of the G-segment are shifted ~20° and move in the opposite direction when compared to the initial axis of the G-segment. Therefore, a rocker-switch-type movement of the CAP domains is responsible for DNA-gate opening (Figure 2, step 3b).

The cavity created between the two subunits is increased by local conformational changes that involve an outward movement of the TOPRIM and Tower domains surrounding the G-segment. Separation of the two parts of the G-segment creates a channel in the DNA-gate that can accommodate a B-form DNA duplex (Figure 2, step 4a). Consequently, DNA-gate opening relies on an outward tilting of the two halves of DNA-gate and a sliding/swivelling of the two subunits against each other, breaking the G-segment axis. However, the cavity at the bottom of the funnel is narrower than the DNA duplex. It is therefore necessary to further expand this region for strand passage. The open conformation structure observed in structural studies thus far [34] corresponds to an early intermediate of T-segment passage through DNA-gate. Recent evidence from cryo-EM structures of full-length DNA gyrase and a human type IIA topoisomerase in closed and pre-opening states of the DNA-gate supports the "sliding and swivelling" opening mechanism [38,39]. The pre-opening conformation precedes the open conformation, where the G-segment is still tightly bound to the groove at the DNA-gate interface but the CAP, TOPRIM, and Tower domains have started to undergo conformational changes.

2.2. Regulation of DNA Gate Opening of Type IIA Topoisomerases

DNA-gate structures solved over the past decades have helped to unravel the structural mechanism of type IIA DNA-gate opening. However, many questions remain unanswered, particularly about communication between functional domains in the presence of DNA crossovers, as well as the molecular basis for distinguishing specific DNA topology and the role of the variable C-terminal domain (CTD) in catalytic cycle regulation. In particular, the eukaryotic topoisomerase II possesses a non-conserved CTD, which is predicted to be unfolded and connected to the last helix of the coiled-coil domain. Until now, no structural information is available for this region. The CTD contains a nuclear localization sequence and several post-translational modifications (PTM) [40]. It has been proposed that the PTM may modulate the activity of human topoisomerase IIα and the recruitment of associated proteins [40,41]. The recent cryo-EM structure of human topoisomerase IIα has revealed that the linker connecting to the CTD stimulates DNA cleavage and enhances the catalytic activity of human TopIIα [39]. The CTD of DNA gyrase from different bacterial species, on the other hand, is conserved and adopts a β-pinwheel shape [42–46]. DNA wrapping around this domain is critical for the supercoiling activity of DNA gyrase [43,47]. Topoisomerase IV, the other type of IIA bacterial topoisomerase, possesses a degenerate form of the homologous GyrA CTD [45,46]. This structural difference accounts in part for the distinct supercoiling and decatenation activities of DNA gyrase and TopIV. Indeed, the deletion of DNA gyrase CTD converts it into a TopIV enzyme [47]. In addition, the deletion of the TopIV CTD results in a loss of activity on positively supercoiled DNA or catenanes [45]. An acidic tail containing a stretch of non-conserved negatively charged residues is another structural element typical of *E. coli* DNA gyrase CTD. It has been suggested that this extension may tightly regulate the supercoiling activity of the enzyme through interaction with the non-conserved domain insertion found in the TOPRIM domain of *E. coli* DNA gyrase [48–50]. All together, the functional domains of topoisomerase IIA communicate through allosteric signals, which result in the opening and closing of the three gates to change the chirality of DNA duplex supercoiling.

3. Mechanism of DNA Gate Opening by Type IA DNA Topoisomerases

Although there are some mechanistic parallels with type IIA enzymes, the catalytic cycle of type IA topoisomerases is more poorly characterised. While some steps of the catalytic cycle are firmly established with solid structural and biochemical data, the mechanisms of gate-opening and strand passage steps remain elusive, and the reaction intermediates are still largely hypothetical due to their transient nature.

The catalytic core of type IA topoisomerases consists of four structural domains (I–IV) arranged in a padlock shape with a distinct toroidal hole in the centre [51] (Figure 3a). The CAP (DIII) and TOPRIM (DI) domains work together to form a single DNA-gate for strand passage. The CAP domain is flexibly connected to the rest of the enzyme via domain DII, which mediates gate opening. Domain DIV forms a binding site for the G-strand.

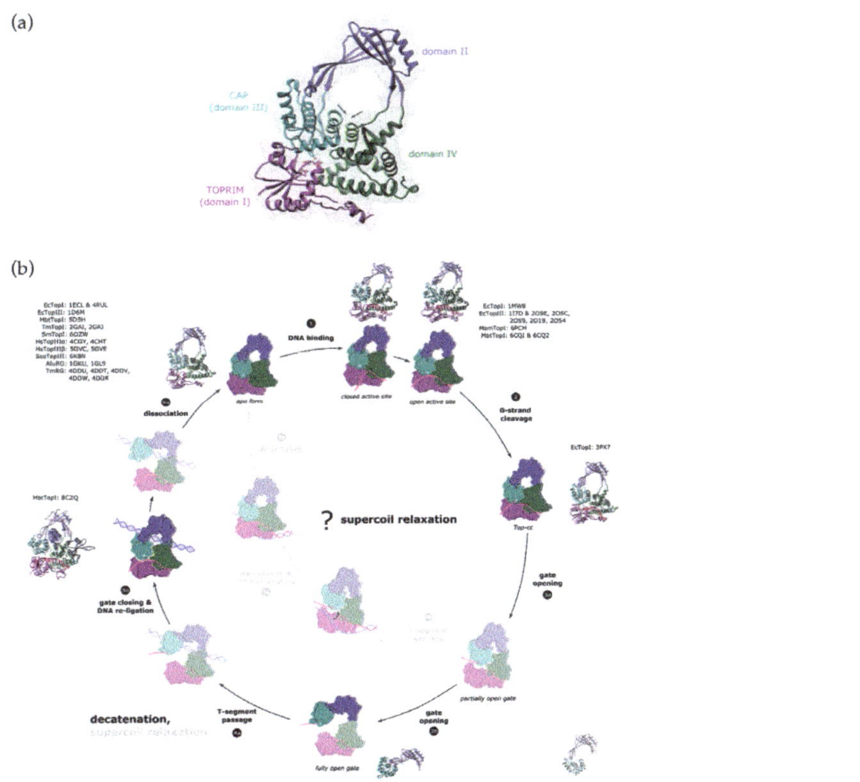

Figure 3. Proposed reaction mechanism of type IA enzymes. (**a**) Domain organisation of type IA topoisomerases. Type IA topoisomerases consist of four domains (labelled I–IV). CAP and TOPRIM

domains contain the active site (the catalytic Tyr is shown in cyan and the residues from the TOPRIM motif are shown in magenta), and they also form the DNA-gate. The interface alpha-helices in domains DIII and DIV that are closing the entrance into the toroid hole are labelled with arrows. PDB ID: 1ECL. (**b**) Topo IA catalytic cycle. Structurally characterised intermediates are shown in full colour with the crystal structures and the PDB accession codes next to them (EcTopI = *Escherichia coli* topoisomerase I, EcTopIII = *Escherichia coli* topoisomerase III, MbtTopI = *Mycobacterium tuberculosis* topoisomerase I, TmTopI = *Thermotoga maritima* topoisomerase I, SmTopI = *Streptococcus mutans* topoisomerase I, HsTopIIIα = human topoisomerase IIIα, HsTopIIIβ= human topoisomerase IIIβ, SsoTopIII = *Sulfolobus solfataricus* topoisomerase III, AfuRG = *Archaeoglobus fulgidus* reverse gyrase, TmRG = *Thermotoga maritima* reverse gyrase, MsmTopI = *Mycobacterium smegmatis* topoisomerase I). Domains I to IV are coloured in dark magenta, dark slate blue, dark cyan, and dark green. The catalytic tyrosine is marked with a cyan star. G-strand is shown in deep pink and the T-segment in blue (either single-stranded or double-stranded). First, the single-stranded G-segment binds to the DNA-binding groove that runs along domain IV. This triggers a conformational change in the enzyme, which results in the active site rearrangement and aligns the scissile bond of the DNA for cleavage (1). Transesterification reaction leads to the formation of the covalent complex (Top-cc), where the 5′-end of the cleaved DNA is covalently bound to the catalytic tyrosine. The 3′-end remains tightly bound within the DNA-binding groove (2). The gate located between the domains I and III opens to form a gap in the G-segment to allow for the strand passage (3). The gate opening might proceed through several mechanistic/kinetic steps [52]. The process could begin with domain sliding that separates the domain III from TOPRIM but retains its interaction with the domain IV (3a, partially open gate) [53]. Breaking the contact with domain IV would result in the opening of the toroidal hole (3b, fully open gate). The T-segment must pass through the nicked G-strand to change the DNA topology. The enzyme is believed to capture the T-segment within the central hole, which is capable of accommodating dsDNA [54,55]. The binding of a double-stranded T-segment is likely accompanied by a conformational change that results in widening of the toroid hole [56] (4a). Alternatively, partial DNA-gate opening might already provide an opening that is big enough for passage of single-stranded DNA, leading to supercoil relaxation [53] (4b). Following the T-segment passage, the gate must close to bring the broken DNA ends back together, and the nick gets religated (5a,b). This process could be potentially facilitated by rewinding the duplex [57]. After the complete reaction cycle, the enzyme dissociates from the DNA (6a,b). The gate would have to re-open to release the T-segment from the toroidal hole. Both supercoil relaxation and decatenation could happen either way, but path 4a→6a with a fully open gate would favour passage of double-stranded segments (decatenation), and the path 4b→6b with a partially open gate the transfer of single-stranded segments (supercoil relaxation).

3.1. Structural Mechanisms Leading to DNA-Gate Opening in Type IA Topoisomerases

Type IA enzymes can act on a wide variety of substrates, including RNA, but have a strict requirement for single-stranded regions. In the apo form, the active site of type IA topoisomerase is buried in the interface between the CAP (DIII) and TOPRIM (DI) domains [51,58,59]. The binding of ssDNA (G-segment) to the DNA-binding groove, which is running along the domain DIV, induces a domain rearrangement, which leads to the formation of the catalytically competent active site and aligns the scissile phosphodiester bond for cleavage [60–62] (Figure 3b, states 1 and 2). Although the G-strand is held in the DNA binding groove mostly via interactions with the DNA backbone [62–64], type IA topoisomerases usually display DNA sequence selectivity for cleavage [65,66]. A cytosine residue at position −four relatives to the cleavage site is required for the bacterial topoisomerase IA enzymes. A recent genome-wide study identified a consensus binding and cleavage motif for *E. coli* topoisomerase I, whose sequence is AT-rich and contains a strictly conserved cytosine residue [67]. A crystal structure of *E. coli* topoisomerase I, a covalent intermediate, revealed that the enzyme has a cavity that specifically accommodates this base [68] and acts as a molecular ruler that helps to precisely position the scissile bond in the active site [65]. Interestingly, it was shown that for human topoisomerase IIIβ, the sequence requirements for RNA cleavage significantly differ from the DNA cleavage sequence [66]. To date, there is no structure of type IA topoisomerase bound to RNA that would explain the molecular basis for the differential selectivity. However, it might be a consequence of different backbone conformations of DNA vs. RNA and thus differences in their interactions with the enzyme.

After the nucleophilic attack by the active site tyrosine residue, the G-strand gets cleaved, leaving its 5′-end covalently bound to the enzyme, whereas the free 3′-end remains non-covalently associated with domain DIV (Figure 1b, state 2). After the strand passage event, the DNA is re-ligated in the reversible reaction, where the free 3′-OH group of the DNA acts as a nucleophile. For this transesterification reaction, divalent cations are required (usually magnesium), which are coordinated by conserved residues of the TOPRIM motif. Based on the evidence for the type IIA topoisomerases, a common two-metal ion mechanism was proposed [11] in which the first metal ion stabilises the transition state, whereas the second is believed to participate in DNA binding. Although type IA has homologous catalytic domains with conserved residues [69], their mechanisms might be slightly different. It is not entirely clear if cleavage needs a metal ion or not, because type IA topoisomerases have an extra lysine residue in the active site that could replace the cation in the cleavage reaction [11]. Re-ligation, on the other hand, strictly depends on the metal ions, which are needed to precisely align the DNA ends for re-joining. In those few crystal structures where the ion is bound to the active site, only one could be identified [61,70,71], making it uncertain whether one ion is sufficient for the reaction or whether there should be two ions, like in type IIA topoisomerases, with the second metal ion bound only transiently.

While the single-strandedness of the G-segment is an absolute requirement for cleavage, there seems to be almost no limitation for the nature of the T-segment. Type IA topoisomerases are capable of performing a wide range of topological reactions: they can relax (negative) supercoils [5], decatenate hemicatenated as well as fully catenated substrates [72], and untangle RNA [73,74]. For supercoil relaxation and manipulation of RNA topology, transfer of a single strand generally needs to occur. However, decatenation of fully catenated substrates requires several catalytic cycles (strand passage steps) and involves hemi-catenated intermediates. A topoisomerase can decatenate two DNA molecules either in two steps, separately transferring a whole duplex through two single-stranded nicks, or in four steps, passing only one strand at a time (two strands through two gaps) [75,76]. Moreover, it was shown that plasmids containing R-loops are excellent substrates for supercoil relaxation by bacterial TopI [77]. Although not formally demonstrated, this implies that the topoisomerase could also use a DNA/RNA hybrid as the T-segment. All this raises the question of how one enzyme can carry out so many topological reactions.

Three decades ago, when the first crystal structure of the type IA topoisomerase (the catalytic core of the *E. coli* TopI) was solved, an elegant mechanism unifying all these 'complicated' strand transfers was proposed: after cleavage, the toroid structure of the enzyme opens up and transiently captures the T-segment inside its central cavity [51]. The enzyme's clamp-like shape would allow it to be opened, and the size of the toroid hole in the closed state is approximately 25 Å, which is large enough to accommodate either single-stranded or double-stranded DNA (a diameter ~ 20 Å). Additionally, the inner side of the hole is lined with positively charged residues, which would enable non-specific DNA binding via electrostatic interactions with its backbone. It was later shown that both single-stranded and double-stranded DNA can be captured inside the central hole by two DNA decatenating enzymes, *E. coli* TopIII [54] and *M. smegmatis* TopI [55]. Each group used a different approach to stabilise the topoisomerase in the closed state after the DNA capture. Li et al. created a disulfide bond between domains DIII (CAP) and DIV to keep the central cavity closed, whereas Leelaram et al. used an inhibitory antibody to keep the topoisomerase in the closed state. Both have also shown that preventing the topoisomerase from opening inhibits its catalytic activity (DNA relaxation). However, there are two considerations with these experiments. First, the capture of DNA inside the central hole was not a result of the topoisomerase reaction (e.g., a trapped intermediate of the DNA relaxation), since for their DNA-trapping experiments both groups used a catalytically dead enzyme, where the catalytic tyrosine was mutated into phenylalanine. Although this undisputedly proves that the toroidal hole can open and accept a DNA molecule, it raises the question as to whether this is a part of the catalytic cycle or just a result of stochastic DNA trapping by the opening and closing of the enzyme in solution. In that regard, it was shown that topoisomerases are very flexible enzymes that readily undergo conformational changes even in the absence of DNA [78]. Second, the stabilisation of the enzyme in the closed conformation might not only affect gate opening but also other steps in the catalytic cycle. The inherent flexibility of topoisomerases is crucial for their function, since mutations or factors that impair conformational changes also affect earlier stages of the catalytic cycle, e.g., DNA binding or cleavage [55,79], which makes it very difficult to uncouple their effects from the subsequent steps.

A crystal structure of *M. tuberculosis* TopI bound to the G-segment and the DNA trapped within the toroid hole was recently published and proposed to represent the state immediately following the G-strand re-ligation [56] (Figure 3b, state 5a). However, since only one DNA oligonucleotide was used to form the complex, the trapped T-segment is a non-canonical parallel duplex that formed as a result of crystal packing. In this complex, TopI only interacts with one of the strands, making sequence-independent contacts with the DNA backbone, which would be consistent with the T-segment being either single- or double-stranded.

Another question pertaining to the generally accepted hypothesis on the reaction mechanism is whether all type IA topoisomerases can bind DNA within their central hole. Reverse gyrases appear to have a smaller toroid hole than the other topoisomerases (~16 Å vs. ~25 Å), which can only hold single-stranded DNA but not a double helix [80,81] (Figure 4). Admittedly, this could be in line with their specialisation in positively supercoiled DNA, which would not require strand passage of the duplex. It was also suggested that a smaller hole size could contribute to the thermostability of the enzyme [80]. In type IA enzymes, the amino acid sequence of domain DII, which comprises the largest part of the toroid hole, is less conserved compared to the rest of the catalytic core [82]. Even though they generally do have positively charged residues facing the inner side of the central cavity, their number and distribution vary depending on species and subtypes [56,82]. This points to the absence of specific protein-DNA interactions within the hole. Accordingly, it was proposed that the DNA, which is trapped inside the central hole, could slide [54,83]. The lack of strong interactions would presumably facilitate rapid strand transfer into and/or out of the hole. Something similar has been proposed for the T-segment passage through the DNA-gate of type IIA DNA gyrase, except that the residues interacting with the DNA are well-conserved in this case [84]. In addition, some topoisomerases have loops inserted at different positions in the domain DII that are protruding into the hole [59,85], or present surfaces that can bind to interaction partners [70,71]. Both would result, at least to some degree, in a reduction of the effective size of the central cavity. Currently, it is unknown what the roles of these loops are, but being flexible elements, they might regulate the strand passage and/or gate opening [56,85]. The available structural as well as biophysical data suggested that type IA enzymes have sufficient plasticity to widen the central hole to accept double-stranded DNA [56,83].

Over a dozen full-length or partial crystal structures of IA topoisomerases from various organisms with or without bound DNA have been solved to date, but these structures have not provided information on the conformational changes that open the DNA-gate and promote T-segment passage [51,56,58–64,68,70,71,80,81,85,86]. One exception might be the structure of the isolated fragment of the DII/DIII domains from *E. coli* TopI [87] (Figure 3b, state 3b). This fragment adopts a conformation where a 'break' in the β-sheet of domain DII would move domain DIII away from the domains DI and DIV, thus forming an entrance into the central hole of the enzyme. The size of the opening would be sufficient to allow the passage of the DNA duplex. The structure revealed the conformational flexibility of domain DII, which is believed to mediate gate opening. However, considerations of the conformation of this fragment alone do not account for the interactions with the rest of the protein. It is worth pointing out that in the same asymmetric unit of the crystal structure, the DII/DIII fragment can be found in another conformation, which would be incompatible with the context of the full-length TopI due to a major clash with domain DIV (Figure 3b, bottom right).

Some very useful insights on DNA-gate opening come from single-molecule experiments that compared two of the most studied IA topoisomerases from *E. coli*: TopI, as an example of an enzyme that excels in supercoil relaxation but is less efficient at decatenation, and TopIII, which has the opposite properties. As it turns out, the catalytic activity of TopIII outperforms that of TopI in both DNA relaxation and decatenation. The only thing that makes TopI a better DNA-relaxing topoisomerase is a shorter waiting time between the relaxation cycles [76,88].

Mills et al. directly measured the kinetic parameters of gate opening for *E. coli* TopI and TopIII using magnetic tweezers [52]. The authors measured the distance between the DIII and TOPRIM domains as the DNA-gate opened by monitoring the DNA extension caused by DNA cleavage, which directly corresponds to the distance between the cleaved DNA ends. The measured distance was the same for both enzymes, but, interestingly, it was shown that the gate remained open much longer for TopIII than for TopI. Apart from the differences in their C-terminal domains, TopIII possesses a small insertion in domain IV, called the decatenation loop. The second difference lies in the sequence of the hinges that connect domain DII with domains DIII and DIV [89]. The molecular dynamics simulation that Mills et al. also performed suggested that, in the open conformation, the decatenation loop would make stabilising contacts with another loop in domain DII, both loops being missing in TopI [52]. This observation and the impact of the hinge sequences on the mechanism remain to be experimentally confirmed. This study also proposed that gate opening might proceed through several mechanistic/kinetic steps.

Bakx et al. performed a similar study on the human TopIIIα-RMI1-RMI2 complex, which has a primary decatenating activity, and on the *E. coli* TopI [83]. Although not at the single-molecule level, they simultaneously monitored DNA-gate opening by measuring the extension of the DNA caused by cleavage and the binding of topoisomerase to the G- and T-segments during DNA catenation, using a combination of dual-trap optical tweezers and fluorescence microscopy. Consistently, they detected a two-step change associated with DNA-gate opening, where they interpreted the second step as a widening of the gate. Furthermore, they have shown that transfer of double-stranded T-segment results in the expansion of the DNA-gate, whereas a single-stranded T-strand did not affect the gate size.

A different single-molecule study by Gunn et al. [53] monitored the relaxation of supercoils by *E. coli* TopI using magnetic tweezers, while simultaneously observing the concomitant conformational change of the protein by Total Internal Reflection Fluorescence (TIRF) Microscopy. They obtained seemingly contradictory conclusions about the topoisomerase mechanism. First of all, they proposed that domain DIII might not move away from the main body of the enzyme but instead slide closer to create a gate and capture the passing strand. Such movement could potentially be possible, but it would not result in the opening of the central hole because this would require the complete separation of domain DIII. As previously stated, two α-helices, one from domain DIII and one from DIV, form a major interface between these domains, keeping the toroid hole closed (Figure 3a). One possible explanation is that the DNA-gate only partially opens, disrupting contacts between domains DIII (CAP) and DI (TOPRIM) but not with DIV. Furthermore, Gunn et al. [53] suggested that the topoisomerase I catalytic cycle involves a sub-cycle in which the enzyme is trying to capture the T-strand, moving between closed and open conformations, with only successful events resulting in strand passage and DNA relaxation.

What could be the cause of the discrepancy between the studies by Mills et al. [52] and Gunn et al. [53], and where do their conclusions meet? One obvious answer is the experimental setup. By applying high forces on the DNA, Mills et al. likely measured the largest extent of the gate opening, whereas the assay by Gunn et al., which was monitoring supercoil relaxation, was probably closer to the native conditions in terms of the forces that act on the DNA, thus complementing the experiments by the former. Mills et al. observed that it was difficult to capture TopI in the open state due to its fast closing rate, and the interpretation of the results by Gunn et al. points in the same direction: that the gate of TopI does not readily open, but in spite of this 'shortcoming', it still efficiently relaxes supercoils. It was also noted by Bakx et al. that TopI does not efficiently bind and thus transfer double-stranded T-segments, which could be potentially related to the DNA-gate opening [83]. In contrast, Topo III enzymes open the gate much more easily, which would be favourable for the passage of more bulky substrates during decatenation.

Why would such discrimination between the relaxation of supercoils and decatenation be necessary? What role does the T-segment have in the catalytic cycle other than being transported? Unlike their big siblings, type IIA enzymes, the type IA topoisomerases are not powered by ATP hydrolysis, and they depend entirely on the energy stored in the DNA to drive strand passage. While the passage is unidirectional in type IIA enzymes, type IA enzymes can, in principle, transfer the strand in both directions. TopIII from *Sulfolobus solfataricus*, a hyperthermophilic archaeon, can perform two topologically opposite reactions [90]. At low temperatures, it can link single-stranded circles into double-stranded DNA (increasing the linking number of the substrate), whereas at low temperatures, it produces single-stranded circles (decreasing the linking number). Apparently, this enzyme can perform the strand passage reaction in either direction, which is determined by the temperature. Additionally, *E. coli* TopI, which normally relaxes negative supercoils, can also relax positive supercoils provided that the substrate contains a single-stranded region (e.g., a mismatched bubble or a bulge) [91]. Such substrates are often used in single molecule studies to constrain the topoisomerase to a single region on the DNA [53,88,92]. Strand passage in IA topoisomerases seems to work well in either direction, but this raises another question: given the asymmetry of the enzyme, are both directions mechanistically equivalent, or is one direction preferred over the other? Not limited exclusively to supercoiling, it appears that the geometry of the DNA matters for different IA topoisomerases. Single-molecule experiments on *E. coli* TopI and TopIII comparing supercoil relaxation and decatenation indicated that TopIII might be particularly sensitive to the type of substrate, making it inefficient for relaxing supercoils but good at decatenation, particularly with large DNA crossover angles [76,88].

The transferred segment might have more than just a passive role in the catalytic cycle—a possibility that has not been explored yet. A study on *Sulfolobus solfataricus* TopIII showed that annealing of the complementary strand facilitated re-ligation of the cleaved strand [57]. If this finding could be generally applied to other type IA topoisomerases, this would have important consequences for negative supercoil relaxation. Apart from shifting the reaction equilibrium towards re-ligation, it might also contribute to DNA rewinding after strand passage and a faster turnover of the enzyme through the different affinities for double-stranded versus single-stranded DNA. Indeed, in the crystal structures of the topoisomerases bound to the G-strand, the DNA is in a B-like conformation, which could presumably prepare it for duplex formation [62–64].

Taken together, it has become clear that different types of IA enzymes work in different ways. Relatively subtle structural differences result in strong preferences for different reactions (e.g., supercoil relaxation versus decatenation), although it is difficult to precisely pinpoint the molecular basis for this specificity. One possibility is that decatenation and relaxation of supercoils by the type IA enzymes might use two slightly different pathways, which are not mutually exclusive (Figure 3b). The DNA-gate must be opened for T-strand passage after the initial steps of G-segment binding and cleavage. Even though the conformational changes are mostly made up, it seems likely that there are steps in between. As a result, the DNA-gate may open to a greater or lesser extent, though it is unclear whether this necessarily entails opening the central hole. Regardless, the size of the gap would determine if the T-segment could be transported, and single-stranded segments might be able to slip through the gate before it is fully open. This would explain the observed differences, but we need more experimental evidence at this point.

3.2. Structural Regulation of DNA Gate Opening and Strand Passage in Type IA Topoisomerases

Although in vitro data show that the DNA-gate can open spontaneously in solution regardless of the enzymatic activity [54,55], gate opening during the catalytic cycle must be coordinated with the cleavage and strand passage steps. To improve their catalytic efficiency, type IA enzymes often have 'helpers', either as additional domains in the enzyme itself or through other proteins that associate with them.

The type IA topoisomerases greatly differ in their CTDs, which often contain a variable number of Zn fingers [93] or some other type of DNA-binding domain [59]. The eukaryotic Topo III enzymes additionally have RGG domains for RNA-binding [94]. These domains on the C-terminus bind DNA, usually with high affinity [62,95], forming additional stabilising interactions with the substrate that enhance the activity of the core enzyme [66,96]. Though the exact role of CTD is unknown, it is essential for catalytic activity in some topoisomerases but not in others, and its deletion only impairs enzyme processivity [59,66,86,96–99]. For example, *E. coli* and mycobacterial TopI absolutely require the CTD for their DNA relaxation activity, and it was proposed that the CTD binds the T-strand to promote its passage [62,85]. The CTDs of both enzymes are structurally unrelated, with *E. coli* TopI having Zn finger and Zn finger-like domains, and mycobacterial TopI's C-terminal consisting of structural repeats with a unique fold [59]. *E. coli* TopI has an additional α-helix in its second Zn finger domain that interacts with the hinge region in domain DII and may be involved in DNA-gate opening regulation by pushing or pulling at the hinges [85] (Figure 4). *Sulfolobus solfataricus* TopIII has a single Zn finger of a different type on its CTD, which also contacts domain DII at the hinges, possibly with a similar role [86] (Figure 4).

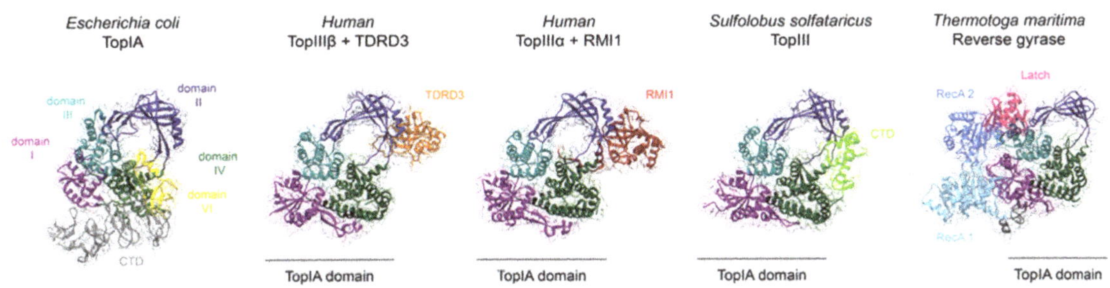

Figure 4. Structural elements, partners, or associated domains regulate topoisomerase IA activity. The domains I to IV of topoisomerase IA are coloured dark magenta, dark slate blue, dark cyan, and dark green. Additional domains of *E. coli* TopI (PDB 4RUL) located at the C-terminal end are coloured grey, and the Zn finger of domain VI interacting with domain DII appears in yellow. The catalytic domain of topoisomerase III and reverse gyrase adopts the typical topoisomerase IA toroidal fold. TDRD3 (orange) or RMI1 (red) are accessory proteins interacting mainly with the toroidal region of human TopIIIβ (PDB 5GVE) and TopIIIα (PDB 4CGY), respectively. Similarly, the C-terminal domain (CTD in light green) of TopIII from *S. solfataricus* (PDB 6KN8) interacts with the toroid domain. Reverse gyrase (PDB 4DDU) is composed of two RecA helicases domains (light and medium blue) separated from the Topo IA domain by the latch domain (deep pink).

The activity of topoisomerases can also be regulated by their interacting partners. Topo III enzymes are often associated with helicases that provide single-stranded DNA regions, but there is also evidence for direct regulation. In yeast, TopIII is part of a larger complex that participates in the resolution of double Holliday junctions and stalled replication forks. It consists of the Sgs1 helicase, Rmi1 protein, RPA (replication protein A), which binds single-stranded DNA, and TopIII [75]. Rmi1 was shown to redirect the preferential activity of TopIII from supercoil relaxation to decatenation, by stabilising the cleavage intermediate. Furthermore, Sgs1 and RPA seemed to additionally stimulate TopIII not just through their primary activities but also through specific interactions. In contrast with the human homologous complex of BLM helicase with TopIIIα-RMI1-RMI2, it was suggested that BLM reduces DNA-gate flexibility [83].

The crystal structure of a human TopIIIα in a complex with the interacting domain from RMI1 (OB fold) revealed that RMI1 inserts a loop into the toroid hole of the TopIIIα [70] (Figure 4). This long loop contacts the hinges in domain DII, likely directly affecting gate opening. Eukaryotic Topo III enzymes do not have a decatenation loop, and RMI1 provides this element in trans to enable efficient decatenation. This hypothesis fits rather well with the observations made on *E. coli* TopI and TopIII discussed earlier, suggesting that stabilisation of the open gate conformation is favourable for decatenation, whereas this is not required for supercoil relaxation.

Human TopIIIβ has a different interaction partner, the TDRD3 protein, which interacts with the topoisomerase in a similar way as RMI1 does with TopIIIα, but with a shorter loop [71]. TDRD3 also binds to domain DII of TopIIIβ at the same position, interacting with the hinges and possibly regulating gate opening (Figure 4). Yang et al. demonstrated that TDRD3 enhances DNA relaxation by TopIIIβ and proposed that their interaction could reduce the flexibility of the DNA-gate to facilitate T-strand passage and/or G-strand re-ligation [66].

Reverse gyrases are equipped with their own RecA helicase domain on the N-terminal end. An element called latch in the helicase domain interacts with domain DIII (CAP) of the topoisomerase to prevent the gate from opening on its own [80,81] (Figure 4). The conformational change of the helicase domain induced by the ATP binding is thought to release the latch [100]. When compared to reverse gyrase, most Topo IA enzymes have their associated domains or protein partners on the opposite side of the toroid hole. This could mean that positive supercoiling uses a different way to control how the DNA gate opens.

4. Conclusions

Over the past twenty years, a lot of information has been learned about how the Topo IA and IIA families simplify their DNA topology. Biochemical, kinetics, and structural biology experiments have provided some clues on how DNA cleavage and re-ligation are coupled to conformational changes of the conserved functional domains forming the DNA gate of the type IA and IIA DNA topoisomerases.

Advances in structural biology have led to the determination of several conformations of the large nucleoprotein complexes of type IIA topoisomerases, revealing some of the key allosteric mechanisms of DNA gate opening. Although the molecular basis for specific DNA topology resolution remains to be elucidated, the fact that type IIA intermolecular reactions involve solely DNA duplexes as primary substrates at the DNA gate somehow facilitates our understanding of these mechanisms.

In contrast, the diversity of substrates and reactions that can be recognised and catalysed by type IA enzymes represents an additional complexity that has to be considered for experimental designs. However, this has been partly compensated by several important studies highlighting the dynamic and kinetic properties of the type IA enzymes.

Together, more information is needed to fully understand the key steps of strand passage and the subsequent conformational changes for both families. Recent studies have highlighted that the combination of different techniques, including single-molecule approaches, is key to getting a comprehensive view of these complex mechanisms. Structural studies of the type IA and type IIA DNA topoisomerases have proven challenging, sometimes due to their relatively large size but mostly owing to their modular and highly flexible architecture. Stabilizing other reaction intermediates or building up larger DNA topology substrates to approach physiological conditions, will now be necessary to get insights into the critical steps that occur beyond the DNA gate.

Author Contributions: Writing—original draft preparation, V.V., M.V. and V.L.; writing—review and editing, V.V., M.V. and V.L.; funding acquisition, V.L. All authors have read and agreed to the published version of the manuscript.

Funding: This work was funded by the Region Grand-Est, the grants from the Agence Nationale de la Recherche ANR-19-CE11-0001-01 and ANR-10-LABX-0030-INRT (managed by the under the frame program Investissements d'Avenir ANR-10-IDEX-0002-02). The authors acknowledge the support and the use of resources of the French Infrastructure for Integrated Structural Biology FRISBI ANR-10-INBS-05 and of Instruct-ERIC. The APC was funded by the grant ANR-19-CE11-0001-01.

Institutional Review Board Statement: Not applicable.

Informed Consent Statement: Not applicable.

Data Availability Statement: Data sharing not applicable.

Acknowledgments: We thank Robert Drilien for proofreading the manuscript.

Conflicts of Interest: The authors declare no conflict of interest.

References

1. McKie, S.J.; Neuman, K.C.; Maxwell, A. DNA Topoisomerases: Advances in Understanding of Cellular Roles and Multi-Protein Complexes via Structure-Function Analysis. *BioEssays* **2021**, *43*, 2000286. [CrossRef]
2. Maxwell, A.; Bush, N.G.; Evans-Roberts, K. DNA Topoisomerases. *EcoSal Plus* **2015**, *6*. [CrossRef]
3. Tse, Y.C.; Kirkegaard, K.; Wang, J.C. Covalent Bonds between Protein and DNA. Formation of Phosphotyrosine Linkage between Certain DNA Topoisomerases and DNA. *J. Biol. Chem.* **1980**, *255*, 5560–5565. [CrossRef]
4. Schoeffler, A.J.; Berger, J.M. DNA Topoisomerases: Harnessing and Constraining Energy to Govern Chromosome Topology. *Q. Rev. Biophys.* **2008**, *41*, 41–101. [CrossRef] [PubMed]
5. Wang, J.C. Interaction between DNA and an Escherichia Coli Protein ω. *J. Mol. Biol.* **1971**, *55*, 523-IN16. [CrossRef] [PubMed]
6. Gellert, M.; Mizuuchi, K.; O'Dea, M.H.; Nash, H.A. DNA Gyrase: An Enzyme That Introduces Superhelical Turns into DNA. *Proc. Natl. Acad. Sci. USA* **1976**, *73*, 3872–3876. [CrossRef]
7. DiGate, R.J.; Marians, K.J. Molecular Cloning and DNA Sequence Analysis of Escherichia Coli TopB, the Gene Encoding Topoisomerase III. *J. Biol. Chem.* **1989**, *264*, 17924–17930.
8. DiGate, R.J.; Marians, K.J. Escherichia Coli Topoisomerase III-Catalyzed Cleavage of RNA. *J. Biol. Chem.* **1992**, *267*, 20532–20535.
9. LIU, L.; LIU, C.; ALBERTS, B. Type II DNA Topoisomerases: Enzymes That Can Unknot a Topologically Knotted DNA Molecule via a Reversible Double-Strand Break. *Cell* **1980**, *19*, 697–707. [CrossRef]
10. Kreuzer, K.N.; Cozzarelli, N.R. Formation and Resolution of DNA Catenanes by DNA Gyrase. *Cell* **1980**, *20*, 245–254. [CrossRef]
11. Schmidt, B.H.; Burgin, A.B.; Deweese, J.E.; Osheroff, N.; Berger, J.M. A Novel and Unified Two-Metal Mechanism for DNA Cleavage by Type II and IA Topoisomerases. *Nature* **2010**, *465*, 641–644. [CrossRef] [PubMed]
12. Nichols, M.D. Structure and Function of an Archaeal Topoisomerase VI Subunit with Homology to the Meiotic Recombination Factor Spo11. *EMBO J.* **1999**, *18*, 6177–6188. [CrossRef] [PubMed]
13. Corbett, K.D. Structure of the Topoisomerase VI-B Subunit: Implications for Type II Topoisomerase Mechanism and Evolution. *EMBO J.* **2003**, *22*, 151–163. [CrossRef] [PubMed]
14. Corbett, K.D.; Benedetti, P.; Berger, J.M. Holoenzyme Assembly and ATP-Mediated Conformational Dynamics of Topoisomerase VI. *Nat. Struct. Mol. Biol.* **2007**, *14*, 611–619. [CrossRef]
15. Vrielynck, N.; Chambon, A.; Vezon, D.; Pereira, L.; Chelysheva, L.; De Muyt, A.; Mézard, C.; Mayer, C.; Grelon, M. A DNA Topoisomerase VI–like Complex Initiates Meiotic Recombination. *Science* **2016**, *351*, 939–943. [CrossRef]
16. Claeys Bouuaert, C.; Tischfield, S.E.; Pu, S.; Mimitou, E.P.; Arias-Palomo, E.; Berger, J.M.; Keeney, S. Structural and Functional Characterization of the Spo11 Core Complex. *Nat. Struct. Mol. Biol.* **2021**, *28*, 92–102. [CrossRef]
17. McKie, S.J.; Desai, P.R.; Seol, Y.; Allen, A.M.; Maxwell, A.; Neuman, K.C. Topoisomerase VI Is a Chirally-Selective, Preferential DNA Decatenase. *Elife* **2022**, *11*, e67021. [CrossRef]
18. Wigley, D.B.; Davies, G.J.; Dodson, E.J.; Maxwell, A.; Dodson, G. Crystal Structure of an N-Terminal Fragment of the DNA Gyrase B Protein. *Nature* **1991**, *351*, 624–629. [CrossRef]
19. Berger, J.M.; Wang, J.C. Recent Developments in DNA Topoisomerase II Structure and Mechanism. *Curr. Opin. Struct. Biol.* **1996**, *6*, 84–90. [CrossRef]
20. Aravind, L.; Leipe, D.D.; Koonin, E. V Toprim–a Conserved Catalytic Domain in Type IA and II Topoisomerases, DnaG-Type Primases, OLD Family Nucleases and RecR Proteins. *Nucleic Acids Res.* **1998**, *26*, 4205–4213. [CrossRef]
21. Cabral, J.H.M.; Jackson, A.P.; Smith, C.V.; Shikotra, N.; Maxwell, A.; Liddington, R.C. Crystal Structure of the Breakage–Reunion Domain of DNA Gyrase. *Nature* **1997**, *388*, 903–906. [CrossRef]
22. Roca, J.; Wang, J.C. DNA Transport by a Type II DNA Topoisomerase: Evidence in Favor of a Two-Gate Mechanism. *Cell* **1994**, *77*, 609–616. [CrossRef] [PubMed]

23. Roca, J.; Wang, J.C. The Probabilities of Supercoil Removal and Decatenation by Yeast DNA Topoisomerase II. *Genes Cells* **1996**, *1*, 17–27. [CrossRef] [PubMed]
24. Dong, K.C.; Berger, J.M. Structural Basis for Gate-DNA Recognition and Bending by Type IIA Topoisomerases. *Nature* **2007**, *450*, 1201–1205. [CrossRef] [PubMed]
25. Wendorff, T.J.; Schmidt, B.H.; Heslop, P.; Austin, C.A.; Berger, J.M. The Structure of DNA-Bound Human Topoisomerase II Alpha: Conformational Mechanisms for Coordinating Inter-Subunit Interactions with DNA Cleavage. *J. Mol. Biol.* **2012**, *424*, 109–124. [CrossRef]
26. Fass, D.; Bogden, C.E.; Berger, J.M. Quaternary Changes in Topoisomerase II May Direct Orthogonal Movement of Two DNA Strands. *Nat. Struct. Biol.* **1999**, *6*, 322–326. [CrossRef]
27. Roca, J.; Wang, J.C. The Capture of a DNA Double Helix by an ATP-Dependent Protein Clamp: A Key Step in DNA Transport by Type II DNA Topoisomerases. *Cell* **1992**, *71*, 833–840. [CrossRef]
28. Bjergbæk, L.; Jensen, S.; Westergaard, O.; Andersen, A.H. Using a Biochemical Approach to Identify the Primary Dimerization Regions in Human DNA Topoisomerase IIα. *J. Biol. Chem.* **1999**, *274*, 26529–26536. [CrossRef]
29. Harkins, T.T.; Lewis, T.J.; Lindsley, J.E. Pre-Steady-State Analysis of ATP Hydrolysis by Saccharomyces Cerevisiae DNA Topoisomerase II. 2. Kinetic Mechanism for the Sequential Hydrolysis of Two ATP. *Biochemistry* **1998**, *37*, 7299–7312. [CrossRef]
30. Schmidt, B.H.; Osheroff, N.; Berger, J.M. Structure of a Topoisomerase II-DNA-Nucleotide Complex Reveals a New Control Mechanism for ATPase Activity. *Nat. Struct. Mol. Biol.* **2012**, *19*, 1147–1154. [CrossRef]
31. Germe, T.; Vörös, J.; Jeannot, F.; Taillier, T.; Stavenger, R.A.; Bacqué, E.; Maxwell, A.; Bax, B.D. A New Class of Antibacterials, the Imidazopyrazinones, Reveal Structural Transitions Involved in DNA Gyrase Poisoning and Mechanisms of Resistance. *Nucleic Acids Res.* **2018**, *46*, 4114–4128. [CrossRef] [PubMed]
32. Bax, B.D.; Chan, P.F.; Eggleston, D.S.; Fosberry, A.; Gentry, D.R.; Gorrec, F.; Giordano, I.; Hann, M.M.; Hennessy, A.; Hibbs, M.; et al. Type IIA Topoisomerase Inhibition by a New Class of Antibacterial Agents. *Nature* **2010**, *466*, 935–940. [CrossRef] [PubMed]
33. Wohlkonig, A.; Chan, P.F.; Fosberry, A.P.; Homes, P.; Huang, J.; Kranz, M.; Leydon, V.R.; Miles, T.J.; Pearson, N.D.; Perera, R.L.; et al. Structural Basis of Quinolone Inhibition of Type IIA Topoisomerases and Target-Mediated Resistance. *Nat. Struct. Mol. Biol.* **2010**, *17*, 1152–1153. [CrossRef] [PubMed]
34. Chen, S.-F.; Huang, N.-L.; Lin, J.-H.; Wu, C.-C.; Wang, Y.-R.; Yu, Y.-J.; Gilson, M.K.; Chan, N.-L. Structural Insights into the Gating of DNA Passage by the Topoisomerase II DNA-Gate. *Nat. Commun.* **2018**, *9*, 3085. [CrossRef] [PubMed]
35. Wang, Y.-R.; Chen, S.-F.; Wu, C.-C.; Liao, Y.-W.; Lin, T.-S.; Liu, K.-T.; Chen, Y.-S.; Li, T.-K.; Chien, T.-C.; Chan, N.-L. Producing Irreversible Topoisomerase II-Mediated DNA Breaks by Site-Specific Pt(II)-Methionine Coordination Chemistry. *Nucleic Acids Res.* **2017**, *45*, 10920. [CrossRef] [PubMed]
36. Zechiedrich, E.L.; Osheroff, N. Eukaryotic Topoisomerases Recognize Nucleic Acid Topology by Preferentially Interacting with DNA Crossovers. *EMBO J.* **1990**, *9*, 4555–4562. [CrossRef] [PubMed]
37. Vologodskii, A.V.; Zhang, W.; Rybenkov, V.V.; Podtelezhnikov, A.A.; Subramanian, D.; Griffith, J.D.; Cozzarelli, N.R. Mechanism of Topology Simplification by Type II DNA Topoisomerases. *Proc. Natl. Acad. Sci. USA* **2001**, *98*, 3045–3049. [CrossRef]
38. Vanden Broeck, A.; Lotz, C.; Ortiz, J.; Lamour, V. Cryo-EM Structure of the Complete E. Coli DNA Gyrase Nucleoprotein Complex. *Nat. Commun.* **2019**, *10*, 4935. [CrossRef]
39. Vanden Broeck, A.; Lotz, C.; Drillien, R.; Haas, L.; Bedez, C.; Lamour, V. Structural Basis for Allosteric Regulation of Human Topoisomerase IIα. *Nat. Commun.* **2021**, *12*, 2962. [CrossRef]
40. Bedez, C.; Lotz, C.; Batisse, C.; Broeck, A.V.; Stote, R.H.; Howard, E.; Pradeau-Aubreton, K.; Ruff, M.; Lamour, V. Post-Translational Modifications in DNA Topoisomerase 2α Highlight the Role of a Eukaryote-Specific Residue in the ATPase Domain. *Sci. Rep.* **2018**, *8*, 9272. [CrossRef]
41. Dougherty, A.C.; Hawaz, M.G.; Hoang, K.G.; Trac, J.; Keck, J.M.; Ayes, C.; Deweese, J.E. Exploration of the Role of the C-Terminal Domain of Human DNA Topoisomerase IIα in Catalytic Activity. *ACS Omega* **2021**, *6*, 25892–25903. [CrossRef] [PubMed]
42. Hsieh, T.-J.; Yen, T.-J.; Lin, T.-S.; Chang, H.-T.; Huang, S.-Y.; Hsu, C.-H.; Farh, L.; Chan, N.-L. Twisting of the DNA-Binding Surface by a β-Strand-Bearing Proline Modulates DNA Gyrase Activity. *Nucleic Acids Res.* **2010**, *38*, 4173–4181. [CrossRef]
43. Ruthenburg, A.J.; Graybosch, D.M.; Huetsch, J.C.; Verdine, G.L. A Superhelical Spiral in the Escherichia Coli DNA Gyrase A C-Terminal Domain Imparts Unidirectional Supercoiling Bias. *J. Biol. Chem.* **2005**, *280*, 26177–26184. [CrossRef] [PubMed]
44. Corbett, K.D.; Shultzaberger, R.K.; Berger, J.M. The C-Terminal Domain of DNA Gyrase A Adopts a DNA-Bending β-Pinwheel Fold. *Proc. Natl. Acad. Sci. USA* **2004**, *101*, 7293–7298. [CrossRef]
45. Corbett, K.D.; Schoeffler, A.J.; Thomsen, N.D.; Berger, J.M. The Structural Basis for Substrate Specificity in DNA Topoisomerase IV. *J. Mol. Biol.* **2005**, *351*, 545–561. [CrossRef] [PubMed]
46. Bouige, A.; Darmon, A.; Piton, J.; Roue, M.; Petrella, S.; Capton, E.; Forterre, P.; Aubry, A.; Mayer, C. Mycobacterium Tuberculosis DNA Gyrase Possesses Two Functional GyrA-Boxes. *Biochem. J.* **2013**, *455*, 285–294. [CrossRef] [PubMed]
47. Kampranis, S.C.; Maxwell, A. Conversion of DNA Gyrase into a Conventional Type II Topoisomerase. *Proc. Natl. Acad. Sci. USA* **1996**, *93*, 14416–14421. [CrossRef]
48. Tretter, E.M.; Berger, J.M. Mechanisms for Defining Supercoiling Set Point of DNA Gyrase Orthologs: I. A Nonconserved Acidic C-Terminal Tail Modulates Escherichia Coli Gyrase Activity. *J. Biol. Chem.* **2012**, *287*, 18636–18644. [CrossRef]

49. Tretter, E.M.; Berger, J.M. Mechanisms for Defining Supercoiling Set Point of DNA Gyrase Orthologs: II. The Shape of the GyrA Subunit C-Terminal Domain (CTD) Is Not a Sole Determinant for Controlling Supercoiling Efficiency. *J. Biol. Chem.* **2012**, *287*, 18645–18654. [CrossRef]
50. Lanz, M.A.; Klostermeier, D. The GyrA-Box Determines the Geometry of DNA Bound to Gyrase and Couples DNA Binding to the Nucleotide Cycle. *Nucleic Acids Res.* **2012**, *40*, 10893–10903. [CrossRef]
51. Lima, C.D.; Wang, J.C.; Mondragón, A. Three-Dimensional Structure of the 67K N-Terminal Fragment of E. Coli DNA Topoisomerase I. *Nat. 1994 3676459* **1994**, *367*, 138–146. [CrossRef] [PubMed]
52. Mills, M.; Tse-Dinh, Y.-C.; Neuman, K.C. Direct Observation of Topoisomerase IA Gate Dynamics. *Nat. Struct. Mol. Biol.* **2018**, *25*, 1111–1118. [CrossRef] [PubMed]
53. Gunn, K.H.; Marko, J.F.; Mondragón, A. An Orthogonal Single-Molecule Experiment Reveals Multiple-Attempt Dynamics of Type IA Topoisomerases. *Nat. Struct. Mol. Biol.* **2017**, *24*, 484–490. [CrossRef] [PubMed]
54. Li, Z.; Mondragón, A.; DiGate, R.J. The Mechanism of Type IA Topoisomerase-Mediated DNA Topological Transformations. *Mol. Cell* **2001**, *7*, 301–307. [CrossRef]
55. Leelaram, M.N.; Bhat, A.G.; Godbole, A.A.; Bhat, R.S.; Manjunath, R.; Nagaraja, V. Type IA Topoisomerase Inhibition by Clamp Closure. *FASEB J.* **2013**, *27*, 3030–3038. [CrossRef]
56. Ferdous, S.; Dasgupta, T.; Annamalai, T.; Tan, K.; Tse-Dinh, Y.-C. The Interaction between Transport-Segment DNA and Topoisomerase IA—Crystal Structure of MtbTOP1 in Complex with Both G- and T-Segments. *Nucleic Acids Res.* **2022**, *51*, 349–364. [CrossRef]
57. Chen, L.; Huang, L. Oligonucleotide Cleavage and Rejoining by Topoisomerase III from the Hyperthermophilic Archaeon Sulfolobus Solfataricus: Temperature Dependence and Strand Annealing-Promoted DNA Religation. *Mol. Microbiol.* **2006**, *60*, 783–794. [CrossRef] [PubMed]
58. Mondragón, A.; DiGate, R. The Structure of Escherichia Coli DNA Topoisomerase III. *Structure* **1999**, *7*, 1373–1383. [CrossRef]
59. Tan, K.; Cao, N.; Cheng, B.; Joachimiak, A.; Tse-Dinh, Y.-C. Insights from the Structure of Mycobacterium Tuberculosis Topoisomerase I with a Novel Protein Fold. *J. Mol. Biol.* **2016**, *428*, 182–193. [CrossRef] [PubMed]
60. Changela, A.; DiGate, R.J.; Mondragón, A. Structural Studies of E. Coli Topoisomerase III-DNA Complexes Reveal A Novel Type IA Topoisomerase-DNA Conformational Intermediate. *J. Mol. Biol.* **2007**, *368*, 105. [CrossRef]
61. Cao, N.; Tan, K.; Annamalai, T.; Joachimiak, A.; Tse-Dinh, Y.C. Investigating Mycobacterial Topoisomerase I Mechanism from the Analysis of Metal and DNA Substrate Interactions at the Active Site. *Nucleic Acids Res.* **2018**, *46*, 7296–7308. [CrossRef] [PubMed]
62. Cao, N.; Tan, K.; Zuo, X.; Annamalai, T.; Tse-Dinh, Y.C. Mechanistic Insights from Structure of Mycobacterium Smegmatis Topoisomerase I with SsDNA Bound to Both N- and C-Terminal Domains. *Nucleic Acids Res.* **2020**, *48*, 4448. [CrossRef] [PubMed]
63. Perry, K.; Mondragón, A. Structure of a Complex between E. Coli DNA Topoisomerase I and Single-Stranded DNA. *Structure* **2003**, *11*, 1349–1358. [CrossRef] [PubMed]
64. Changela, A.; DiGate, R.J.; Mondragón, A. Crystal Structure of a Complex of a Type IA DNA Topoisomerase with a Single-Stranded DNA Molecule. *Nature* **2001**, *411*, 1077–1081. [CrossRef]
65. Narula, G.; Tse-Dinh, Y.-C. Residues of E. Coli Topoisomerase I Conserved for Interaction with a Specific Cytosine Base to Facilitate DNA Cleavage. *Nucleic Acids Res.* **2012**, *40*, 9233–9243. [CrossRef]
66. Yang, X.; Saha, S.; Yang, W.; Neuman, K.C.; Pommier, Y. Structural and Biochemical Basis for DNA and RNA Catalysis by Human Topoisomerase 3β. *Nat. Commun.* **2022**, *13*, 4656. [CrossRef]
67. Sutormin, D.; Galivondzhyan, A.; Musharova, O.; Travin, D.; Rusanova, A.; Obraztsova, K.; Borukhov, S.; Severinov, K. Interaction between Transcribing RNA Polymerase and Topoisomerase I Prevents R-Loop Formation in E. Coli. *Nat. Commun.* **2022**, *13*, 4524. [CrossRef]
68. Zhang, Z.; Cheng, B.; Tse-Dinh, Y.-C. Crystal Structure of a Covalent Intermediate in DNA Cleavage and Rejoining by Escherichia Coli DNA Topoisomerase I. *Proc. Natl. Acad. Sci. USA* **2011**, *108*, 6939–6944. [CrossRef]
69. Corbett, K.D.; Berger, J.M. Structure, Molecular Mechanisms, and Evolutionary Relationships in DNA Topoisomerases. *Annu. Rev. Biophys. Biomol. Struct.* **2004**, *33*, 95–118. [CrossRef]
70. Bocquet, N.; Bizard, A.H.; Abdulrahman, W.; Larsen, N.B.; Faty, M.; Cavadini, S.; Bunker, R.D.; Kowalczykowski, S.C.; Cejka, P.; Hickson, I.D.; et al. Structural and Mechanistic Insight into Holliday Junction Dissolution by Topoisomerase IIIα and RMI1. *Nat. Struct. Mol. Biol.* **2014**, *21*, 261. [CrossRef]
71. Goto-Ito, S.; Yamagata, A.; Takahashi, T.S.; Sato, Y.; Fukai, S. Structural Basis of the Interaction between Topoisomerase IIIβ and the TDRD3 Auxiliary Factor. *Sci. Rep.* **2017**, *7*, 42123. [CrossRef] [PubMed]
72. Brown, P.O.; Cozzarelli, N.R. Catenation and Knotting of Duplex DNA by Type 1 Topoisomerases: A Mechanistic Parallel with Type 2 Topoisomerases. *Proc. Natl. Acad. Sci. USA* **1981**, *78*, 843–847. [CrossRef] [PubMed]
73. Ahmad, M.; Shen, W.; Li, W.; Xue, Y.; Zou, S.; Xu, D.; Wang, W. Topoisomerase 3β Is the Major Topoisomerase for MRNAs and Linked to Neurodevelopment and Mental Dysfunction. *Nucleic Acids Res.* **2016**, *45*, gkw1293. [CrossRef] [PubMed]
74. Ahmad, M.; Xue, Y.; Lee, S.K.; Martindale, J.L.; Shen, W.; Li, W.; Zou, S.; Ciaramella, M.; Debat, H.; Nadal, M.; et al. RNA Topoisomerase Is Prevalent in All Domains of Life and Associates with Polyribosomes in Animals. *Nucleic Acids Res.* **2016**, *44*, 6335–6349. [CrossRef]
75. Cejka, P.; Plank, J.L.; Dombrowski, C.C.; Kowalczykowski, S.C. Decatenation of DNA by the S. Cerevisiae Sgs1-Top3-Rmi1 and RPA Complex: A Mechanism for Disentangling Chromosomes. *Mol. Cell* **2012**, *47*, 886–896. [CrossRef]

76. Terekhova, K.; Marko, J.F.; Mondragón, A.; Mondragón, M. Single-Molecule Analysis Uncovers the Difference between the Kinetics of DNA Decatenation by Bacterial Topoisomerases I and III. *Nucleic Acids Res.* **2014**, *42*, 11657–11667. [CrossRef]
77. Phoenix, P.; Raymond, M.A.; Massé, É.; Drolet, M. Roles of DNA Topoisomerases in the Regulation of R-Loop Formation in Vitro. *J. Biol. Chem.* **1997**, *272*, 1473–1479. [CrossRef]
78. Jeanne Dit Fouque, K.; Garabedian, A.; Leng, F.; Tse-Dinh, Y.-C.; Fernandez-Lima, F. Microheterogeneity of Topoisomerase IA/IB and Their DNA-Bound States. *ACS Omega* **2019**, *4*, 3619–3626. [CrossRef]
79. Cheng, B.; Feng, J.; Gadgil, S.; Tse-Dinh, Y.-C.C. Flexibility at Gly-194 Is Required for DNA Cleavage and Relaxation Activity of Escherichia Coli DNA Topoisomerase I. *J. Biol. Chem.* **2004**, *279*, 8648–8654. [CrossRef]
80. Rodríguez, A.C.; Stock, D. Crystal Structure of Reverse Gyrase: Insights into the Positive Supercoiling of DNA. *EMBO J.* **2002**, *21*, 418–426. [CrossRef]
81. Rudolph, M.G.; del Toro Duany, Y.; Jungblut, S.P.; Ganguly, A.; Klostermeier, D. Crystal Structures of Thermotoga Maritima Reverse Gyrase: Inferences for the Mechanism of Positive DNA Supercoiling. *Nucleic Acids Res.* **2013**, *41*, 1058–1070. [CrossRef] [PubMed]
82. Duguet, M.; Serre, M.C.; Bouthier de La Tour, C. A Universal Type IA Topoisomerase Fold. *J. Mol. Biol.* **2006**, *359*, 805–812. [CrossRef] [PubMed]
83. Bakx, J.A.M.; Biebricher, A.S.; King, G.A.; Christodoulis, P.; Sarlós, K.; Bizard, A.H.; Hickson, I.D.; Wuite, G.J.L.; Peterman, E.J.G. Duplex DNA and BLM Regulate Gate Opening by the Human TopoIIIα-RMI1-RMI2 Complex. *Nat. Commun.* **2022**, *13*, 584. [CrossRef] [PubMed]
84. Soczek, K.M.; Grant, T.; Rosenthal, P.B.; Mondragón, A. CryoEM Structures of Open Dimers of Gyrase A in Complex with DNA Illuminate Mechanism of Strand Passage. *Elife* **2018**, *7*, e41215. [CrossRef]
85. Tan, K.; Zhou, Q.; Cheng, B.; Zhang, Z.; Joachimiak, A.; Tse-Dinh, Y.-C.C. Structural Basis for Suppression of Hypernegative DNA Supercoiling by E. Coli Topoisomerase I. *Nucleic Acids Res.* **2015**, *43*, 11031–11046. [CrossRef]
86. Wang, H.; Zhang, J.; Gao, Z.; Zheng, X.; Zhu, K.; Zhang, Z.; Zhang, Z.; Dong, Y.; Huang, L.; Gong, Y. Crystal Structure of Sulfolobus Solfataricus Topoisomerase III Reveals a Novel Carboxyl-Terminal Zinc Finger Domain Essential for Decatenation Activity. *bioRxiv* **2021**. [CrossRef]
87. Feinberg, H.; Lima, C.D.; Mondragon, A. Conformational Changes in E. Coli DNA Topoisomerase I. *Nat. Struct. Biol.* **1999**, *6*, 918–922. [CrossRef]
88. Terekhova, K.; Gunn, K.H.; Marko, J.F.; Mondragón, A. Bacterial Topoisomerase I and Topoisomerase III Relax Supercoiled DNA via Distinct Pathways. *Nucleic Acids Res.* **2012**, *40*, 10432–10440. [CrossRef]
89. Viard, T.; de la Tour, C.B. Type IA Topoisomerases: A Simple Puzzle? *Biochimie* **2007**, *89*, 456–467. [CrossRef]
90. Bizard, A.H.; Yang, X.; Débat, H.; Fogg, J.M.; Zechiedrich, L.; Strick, T.R.; Garnier, F.; Nadal, M. TopA, the Sulfolobus Solfataricus Topoisomerase III, Is a Decatenase. *Nucleic Acids Res.* **2018**, *46*, 861–872. [CrossRef]
91. Kirkegaard, K.; Wang, J.C. Bacterial DNA Topoisomerase I Can Relax Positively Supercoiled DNA Containing a Single-Stranded Loop. *J. Mol. Biol.* **1985**, *185*, 625–637. [CrossRef] [PubMed]
92. Dekker, N.H.; Rybenkov, V.V.; Duguet, M.; Crisona, N.J.; Cozzarelli, N.R.; Bensimon, D.; Croquette, V. The Mechanism of Type IA Topoisomerases. *Proc. Natl. Acad. Sci. USA* **2002**, *99*, 12126–12131. [CrossRef] [PubMed]
93. Viard, T.; Lamour, V.; Duguet, M.; Bouthier de la Tour, C. Hyperthermophilic Topoisomerase I from Thermotoga Maritima. *J. Biol. Chem.* **2001**, *276*, 46495–46503. [CrossRef] [PubMed]
94. Xu, D.; Shen, W.; Guo, R.; Xue, Y.; Peng, W.; Sima, J.; Yang, J.; Sharov, A.; Srikantan, S.; Yang, J.; et al. Top3β Is an RNA Topoisomerase That Works with Fragile X Syndrome Protein to Promote Synapse Formation HHS Public Access Author Manuscript. *Nat. Neurosci.* **2013**, *16*, 1238–1247. [CrossRef] [PubMed]
95. Ahumada, A.; Tse-Dinh, Y.C. The Zn(II) Binding Motifs of E. Coli DNA Topoisomerase I Is Part of a High-Affinity DNA Binding Domain. *Biochem. Biophys. Res. Commun.* **1998**, *251*, 509–514. [CrossRef] [PubMed]
96. Viard, T.; Cossard, R.; Duguet, M.; de La Tour, C.B. Thermotoga Maritima-Escherichia Coli Chimeric Topoisomerases. *J. Biol. Chem.* **2004**, *279*, 30073–30080. [CrossRef]
97. Ahumada, A.; Tse-Dinh, Y.-C. The Role of the Zn(II) Binding Domain in the Mechanism of E. Coli DNA Topoisomerase I. *BMC Biochem.* **2002**, *3*, 13. [CrossRef]
98. Strzałka, A.; Szafran, M.J.; Strick, T.; Jakimowicz, D. C-Terminal Lysine Repeats in Streptomyces Topoisomerase I Stabilize the Enzyme–DNA Complex and Confer High Enzyme Processivity. *Nucleic Acids Res.* **2017**, *45*, 11908–11924. [CrossRef]
99. Ahmed, W.; Bhat, A.G.; Leelaram, M.N.; Menon, S.; Nagaraja, V. Carboxyl Terminal Domain Basic Amino Acids of Mycobacterial Topoisomerase I Bind DNA to Promote Strand Passage. *Nucleic Acids Res.* **2013**, *41*, 7462–7471. [CrossRef]
100. Yang, X.; Garnier, F.; Débat, H.; Strick, T.R.; Nadal, M. Direct Observation of Helicase-Topoisomerase Coupling within Reverse Gyrase. *Proc. Natl. Acad. Sci. USA* **2020**, *117*, 10856–10864. [CrossRef]

International Journal of
Molecular Sciences

Article

Role of the Water–Metal Ion Bridge in Quinolone Interactions with *Escherichia coli* Gyrase

Hannah E. Carter [1], Baylee Wildman [1], Heidi A. Schwanz [2], Robert J. Kerns [2] and Katie J. Aldred [1,*]

1 Biology Department, University of Evansville, Evansville, IN 47722, USA
2 Department of Pharmaceutical Sciences and Experimental Therapeutics, University of Iowa, Iowa City, IA 42232, USA
* Correspondence: ka59@evansville.edu

Abstract: Fluoroquinolones are an important class of antibacterials, and rising levels of resistance threaten their clinical efficacy. Gaining a more full understanding of their mechanism of action against their target enzymes—the bacterial type II topoisomerases gyrase and topoisomerase IV—may allow us to rationally design quinolone-based drugs that overcome resistance. As a step toward this goal, we investigated whether the water–metal ion bridge that has been found to mediate the major point of interaction between *Escherichia coli* topoisomerase IV and *Bacillus anthracis* topoisomerase IV and gyrase, as well as *Mycobacterium tuberculosis* gyrase, exists in *E. coli* gyrase. This is the first investigation of the water–metal ion bridge and its function in a Gram-negative gyrase. Evidence suggests that the water–metal ion bridge does exist in quinolone interactions with this enzyme and, unlike the Gram-positive *B. anthracis* gyrase, does use both conserved residues (serine and acidic) as bridge anchors. Furthermore, this interaction appears to play a positioning role. These findings raise the possibility that the water–metal ion bridge is a universal point of interaction between quinolones and type II topoisomerases and that it functions primarily as a binding contact in Gram-positive species and primarily as a positioning interaction in Gram-negative species. Future studies will explore this possibility.

Keywords: quinolone; quinolone resistance; water–metal ion bridge; gyrase; topoisomerase

Citation: Carter, H.E.; Wildman, B.; Schwanz, H.A.; Kerns, R.J.; Aldred, K.J. Role of the Water–Metal Ion Bridge in Quinolone Interactions with *Escherichia coli* Gyrase. *Int. J. Mol. Sci.* **2023**, *24*, 2879. https://doi.org/10.3390/ijms24032879

Academic Editors: Joseph E. Deweese and Neil Osheroff

Received: 15 December 2022
Revised: 27 January 2023
Accepted: 31 January 2023
Published: 2 February 2023

1. Introduction

Escherichia coli is a Gram-negative bacillus that is a common pathogen, as well as a model organism. Treatment for ailments caused by *E. coli* often includes a fluoroquinolone antibacterial [1]. The most commonly prescribed fluoroquinolones are ciprofloxacin ("Cipro") and levofloxacin ("Levaquin") [2,3]. Like many antibacterial agents, the clinical efficacy of the fluoroquinolone class is threatened by increasing rates of resistance [4–9].

Fluoroquinolones kill bacteria by poisoning the activity of type II topoisomerases—essential enzymes that regulate DNA topology [5,8–11]. Most bacterial species, including *E. coli*, contain two type II topoisomerases, gyrase and topoisomerase IV. Gyrase primarily functions ahead of the replication fork to relieve supercoiling that arises due to unwinding of the DNA helix. Meanwhile, topoisomerase IV works primarily behind the fork to resolve knots and tangles that arise in the genome. Topoisomerase IV also functions as an efficient decatenase to separate daughter chromosomes following replication [4,12–16].

Both type II topoisomerases must introduce double-strand DNA breaks into the genome in order to carry out their essential functions [12,13,16]. Briefly, the enzymes cut both strands of the "gate-segment" of DNA and covalently attach to the newly generated termini, creating a structure known as a "cleavage complex". Then, they pass a "transfer-segment" of DNA through the break. This is followed by religation of the break and release of both DNA helices. Fluoroquinolones take advantage of this double-strand break mechanism and insert into the cut to prevent the enzymes from repairing the damage they created, thereby increasing the number of cleavage complexes in the cell. As a result of this blockage and build-up of cleavage complexes, double-strand breaks accumulate in the cell and overwhelm repair systems, ultimately leading to cell death [2,8,9,17–20].

Typically, specific mutations must occur in both gyrase and topoisomerase IV in order to generate clinically relevant levels of resistance. The most common mutations in both enzymes across a range of species have been found to be at the positions equivalent to Ser83 and Asp87 in *E. coli* gyrase [1,4,7,10,11,21–27]. In *E. coli* gyrase, the specific mutations most commonly observed are S83L and D87N or D87Y [28]. Previous biochemical work with *E. coli* topoisomerase IV [29], *Bacillus anthracis* topoisomerase IV [28,30], *Mycobacterium tuberculosis* gyrase [31], and *B. anthracis* gyrase [32] have indicated that these residues coordinate a water–metal ion bridge interaction (first suggested by an x-ray crystallography structure of a cleavage complex formed between *Acinetobacter baumannii* topoisomerase IV and moxifloxacin [33]) that serves as the main interaction point between the drug and enzyme. In the water–metal ion bridge (Figure 1), the C3/C4 keto acid of the quinolone coordinates a Mg^{2+} ion. The hydration sphere of this ion is filled out by four water molecules, two of which are coordinated to the enzyme through hydrogen bonds to Ser83 and Asp87 [28,33]. Interestingly, this bridge has been found to provide different functions in different enzymes. In *B. anthracis* topoisomerase IV [28,30] and gyrase [32], it functions primarily as a binding contact, while in *E. coli* topoisomerase IV [29] it functions primarily as a positioning contact. Based on these limited examples, it appears that the water–metal ion bridge coordinated by "Ser83" and "Asp87" could be universal and that it could serve different functions based on the Gram classification of the bacterial species in question.

In the aforementioned biochemical studies that tested for the presence and function of the water–metal ion bridge interaction [28–32], a quinazolinedione (which lacks the C3/C4 keto acid found in quinolones) has been used as the comparison drug due to its apparent metal-ion-independent function and its ability to overcome resistance caused by mutations at the conserved serine and acidic residues [34]. Based on a crystal structure of a cleavage complex formed by *Streptococcus pneumoniae* topoisomerase IV in the presence of the quinazolinedione PD0305970, quinazolinediones do not interact with the enzyme through either conserved residue, nor do they use a metal ion in any manner. Thus, they serve as a valuable comparator when examining the effects of metal ion variation on drug activity against the target enzymes and when examining the effects of enzyme mutations on drug activity. In this study, the quinolone that was used was ciprofloxacin (Figure 2), which differs from moxifloxacin (used in the crystallographic study; see Figure 1A, inset) only at the C-7 position. In addition, the quinazolinedione that was used here (Figure 2) differs from PD0305970 (used in the crystallographic study; see Figure 1B, inset) only in the absence vs. presence of a methyl group on the C-7 substituent. In both cases, these substituents are not predicted to play a role in drug–enzyme interaction based on the crystallographic studies, nor are they implicated to be involved based on the biochemical data from other topoisomerases in which the water–metal ion bridge has been investigated.

Here, the universality of the water–metal ion bridge and its possible split of function as a binding vs. positioning contact in Gram-positive vs. Gram-negative species will be further explored. Specifically, *E. coli* gyrase was examined, as there is yet to be a published example of a Gram-negative gyrase that addresses the existence and function of the water–metal ion bridge. We found that the water–metal ion bridge likely does exist as the major point of interaction between quinolones and *E. coli* gyrase, is likely anchored by the conserved serine and acidic residues, and appears to function as a positioning contact. Future studies aim to explore additional common pathogenic species of both Gram-positive and Gram-negative bacteria to further address the seeming universality of the bridge interaction and its apparent split of function down Gram-positive and Gram-negative lines.

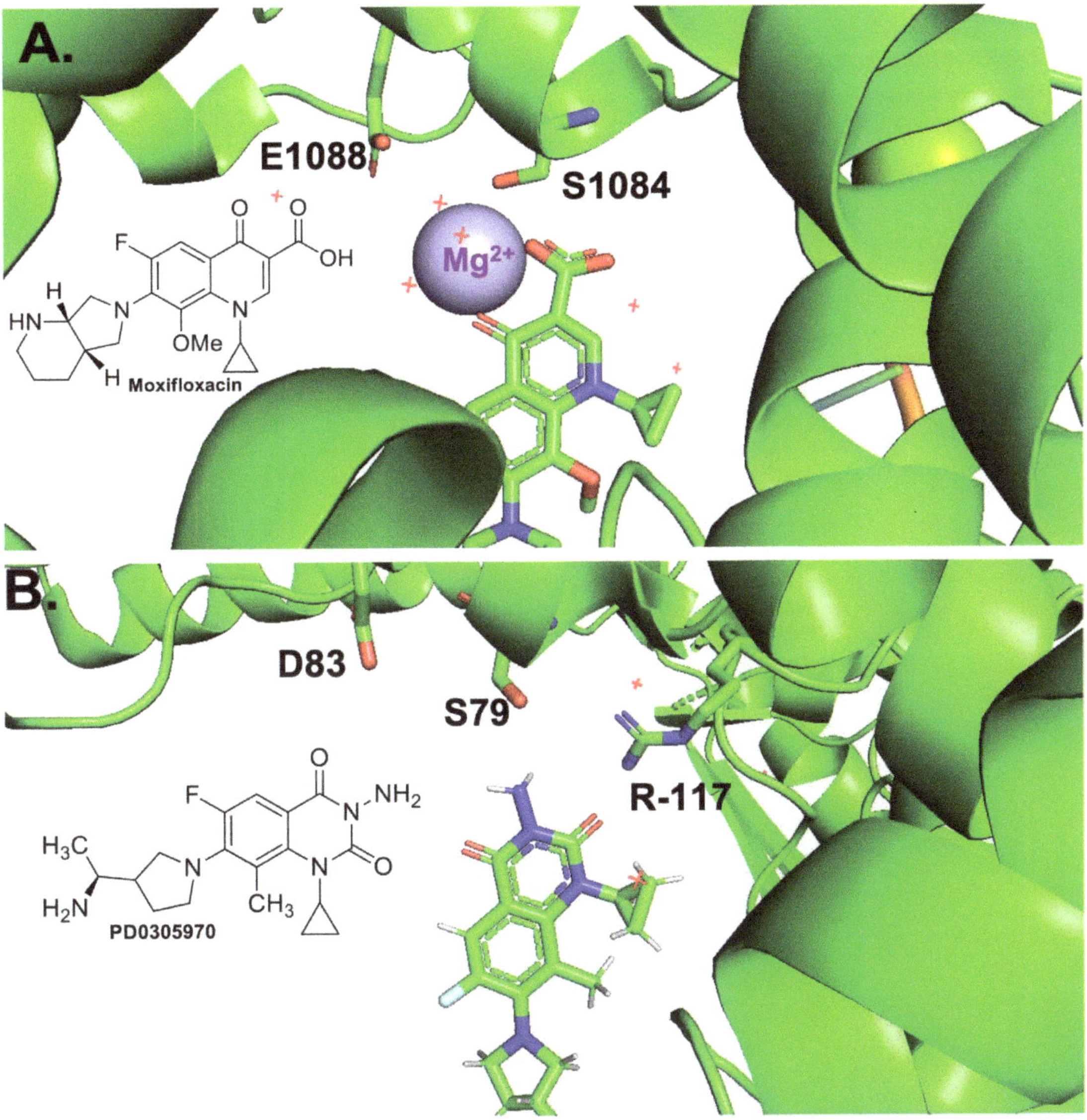

Figure 1. Crystal structures of a quinolone and a quinazolinedione interacting with topoisomerase IV-DNA cleavage complexes. (**A**) Crystal structure of moxifloxacin, DNA, and *A. baumannii* topoisomerase IV. Fluoroquinolone and key residues displayed as sticks, water molecules displayed as red +, divalent magnesium ion displayed in lavender. Adapted from RCSB PDB: 2XKK, visualized with The PyMOL Molecular Graphics System, Version 2.5.2 Schrödinger, LLC. Chemical structure of Moxifloxacin is shown in the inset. (**B**) Crystal structure of PD0305970, DNA, and *S. pneumoniae* topoisomerase IV. Quinazoline-2,4-dione displayed as sticks. Adapted from RCSB PDB: 3RAF, visualized with The PyMOL Molecular Graphics System, Version 2.5.2 Schrödinger, LLC. Chemical structure of PD0305970 is shown in the inset.

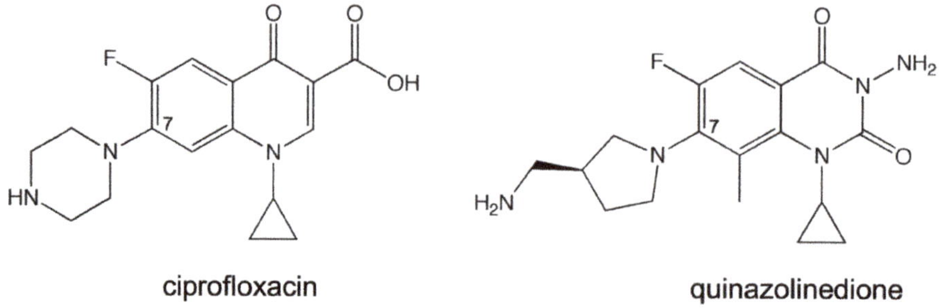

Figure 2. Structures of the quinolone (ciprofloxacin) and quinazolinedione used in this study. The C-7 position is labeled.

2. Results and Discussion

The first step toward determining whether the water–metal ion bridge exists in quinolone interactions with *E. coli* gyrase was to confirm that the purified wild-type enzyme was sensitive to ciprofloxacin and that the mutant enzymes were resistant. As seen in Figure 3 (left panel), wild-type shows increasing levels of DNA cleavage in the presence of increasing concentrations of ciprofloxacin, with maximum cleavage reached at 10 μM. In contrast, the single mutant enzymes $GyrA^{S83L}$ and $GyrA^{D87N}$ show significantly reduced sensitivity to the quinolone, and even at 500 μM ciprofloxacin (left panel, inset) only about 50% of the wild-type level of cleavage is reached. The double mutant gyrase $GyrA^{S83L/D87N}$ shows essentially no sensitivity to ciprofloxacin, even at 500 μM. These findings suggest that these mutations do not cause resistance simply by decreasing binding between the drug and enzyme as high levels of the drug cannot overcome the mutations to reach wild-type levels.

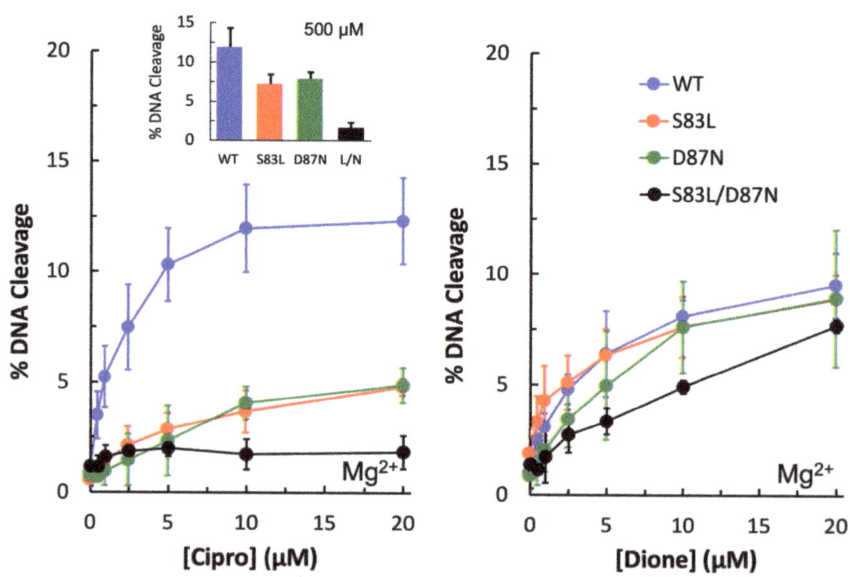

Figure 3. Plasmid DNA cleavage induced by wild-type (blue), $GyrA^{S83L}$ (red), $GyrA^{D87N}$ (green), and $GyrA^{S83L/D87N}$ (black) gyrase enzymes in the presence of 0–20 μM ciprofloxacin (**left**) or quinazolinedione (**right**). Mg^{2+} was the metal ion. The inset in the left panel shows the level of cleavage induced by the enzymes in the presence of 500 μM ciprofloxacin. Error bars represent the standard deviation of three or more independent experiments.

3-amino-7-[(3S)-3-(aminomethyl)-1-pyrrolidin-yl]-1-cyclopropyl-6-fluoro-8-methyl-2,4-(1H,3H)-quinazoline-dione ("quinazolinedione") has previously been shown to be a metal-ion-independent drug and overcome quinolone resistance caused mutations at these conserved residues in a number of type II topoisomerases [28–32,34]. As expected, wild-type as well as all three mutant enzymes were sensitive to the quinazolinedione and showed increasing levels of cleavage in the presence of increasing drug concentrations (Figure 3, right panel).

Next, Mn^{2+} was substituted for Mg^{2+} in cleavage reactions with ciprofloxacin and quinazolinedione with wild-type and single mutant enzymes to determine whether a metal ion plays a role in the function of the quinolone against this Gram-negative gyrase enzyme. Under these conditions, both WT and mutant enzymes induced higher levels of quinazolinedione-induced DNA cleavage than they did in the presence of Mg^{2+} (Figure 4, right panel). This was also seen with wild-type in the presence of ciprofloxacin (Figure 4, left panel). However, with the GyrA^{S83L} and GyrA^{D87N} mutant enzymes, this was not the case. With GyrA^{D87N} gyrase, ciprofloxacin-induced cleavage was approximately equal regardless of which metal ion was used. With GyrA^{S83L} gyrase, little to no increase in cleavage was observed in the presence of ciprofloxacin when Mn^{2+} was the metal ion. These findings suggest that a metal ion is indeed involved in the interaction between quinolones and the enzyme and that these residues play a role in the interaction.

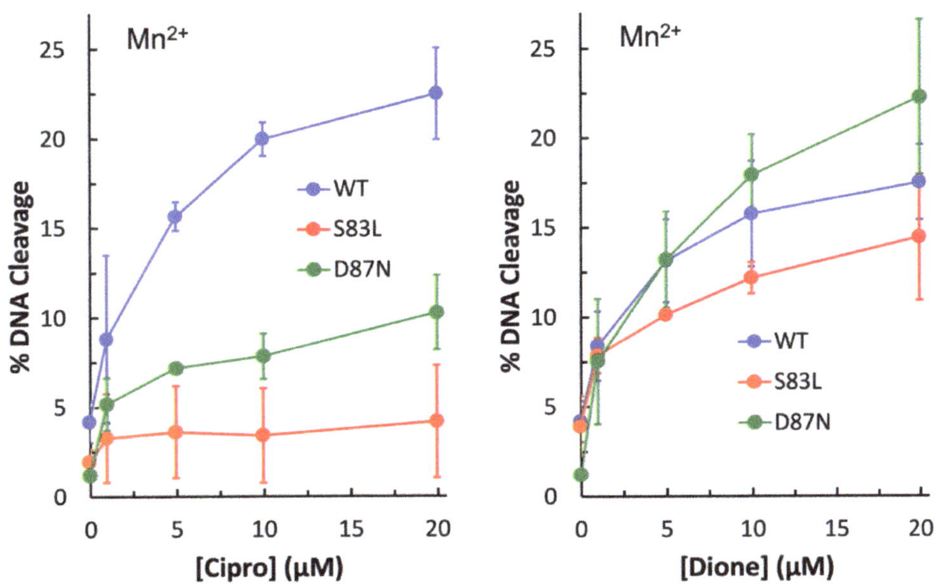

Figure 4. Plasmid DNA cleavage induced by wild-type (blue), GyrA^{S83L} (red), and GyrA^{D87N} (green) gyrase enzymes in the presence of 0–20 μM ciprofloxacin (**left**) or quinazolinedione (**right**). Mn^{2+} was the metal ion. Error bars represent the standard deviation of three or more independent experiments.

To further investigate the metal ion requirement for quinolone vs. quinazolinedione activity against gyrase, Mg^{2+} ion titrations were conducted with wild-type, GyrA^{S83L}, and GyrA^{D87N} gyrase enzymes. As seen in Figure 5 (top left panel), the wild-type enzyme required equivalent amounts of Mg^{2+} regardless of the drug used. However, with GyrA^{D87N} (bottom left panel), there is a right shift of the curve in the presence of the quinolone, indicating that quinolone activity requires higher concentrations of the ion. When coupled with the data shown in Figures 3 and 4, this supports the conclusion that the water–metal ion bridge is the major point of interaction between quinolones and *E. coli* gyrase. Interestingly, there is little difference in Mg^{2+} requirement for ciprofloxacin and quinazolinedione with the GyrA^{S83L} mutant at most concentrations tested (Figure 5, top right panel). However, the quinolone does show a sort of plateau in the induced cleavage level at low levels of Mg^{2+} before rapidly increasing between 1 and 2 mM. This difference in shape between the quinolone and quinazolinedione curves is consistent with the metal ion playing a role in quinolone interactions with the enzyme and the quinazolinedione being a metal-ion-independent drug.

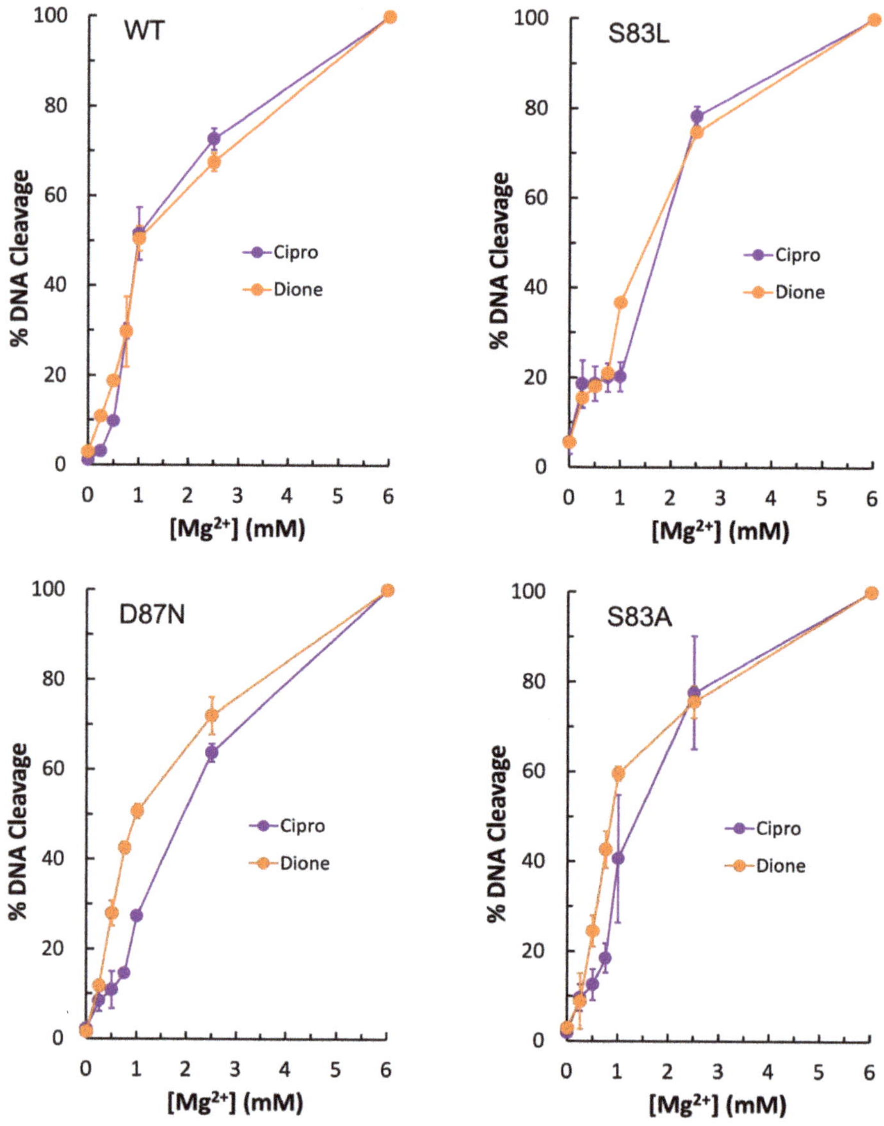

Figure 5. Levels of Mg^{2+} required for wild-type and mutant gyrase enzymes to induce plasmid DNA cleavage in the presence of 50 μM ciprofloxacin (purple) or quinazolinedione (orange). Wild-type is shown in the top left, GyrA^{S83L} in the top right, GyrA^{D87N} in the bottom left, and GyrA^{S83A} in the bottom right. Error bars represent the standard deviation of either two (WT, S83L, and D87N) or three (S83A) independent experiments.

Due to the unexpected shape of the GyrA^{S83L} Mg^{2+} titration curves, it was concluded that if the water–metal ion bridge exists in *E. coli* gyrase–quinolone interactions then the bulky leucine residue in place of the more compact serine could disrupt the quinolone-enzyme interaction beyond simply preventing coordination of the water molecules that are part of the water–metal ion bridge. For this reason, GyrA^{S83A} gyrase was generated. Unexpectedly, this mutant enzyme did not display resistance to ciprofloxacin in the presence of either Mg^{2+} or Mn^{2+} (Figure 6, left panel and right panel, respectively). In fact, much like the wild-type enzyme, GyrA^{S83A} gyrase showed increased levels of ciprofloxacin-induced DNA cleavage with Mn^{2+} as compared to Mg^{2+} (compare Figures 4–6). However, cleavage induced by the quinazolinedione was nearly identical with this mutant regardless of which divalent ion was present. Because changing the metal ion affects quinolone-induced but not quinazolinedione-induced DNA cleavage, this is another piece of evidence supporting the conclusion that a metal ion is important for the interaction between quinolones and *E. coli* gyrase. Moreover, despite not causing quinolone-resistance, the GyrA^{S83A} mutant enzyme required an increased concentration of Mg^{2+} to reach maximal DNA cleavage with the quinolone as compared to the quinazolinedione (Figure 5, bottom right panel). This finding is also consistent with the water–metal ion bridge existing and Ser83 playing a role in coordinating it. Thus, it appears that the water–metal ion bridge exists in *E. coli* gyrase–quinolone interactions and that Ser83 and Asp87 act as the anchors.

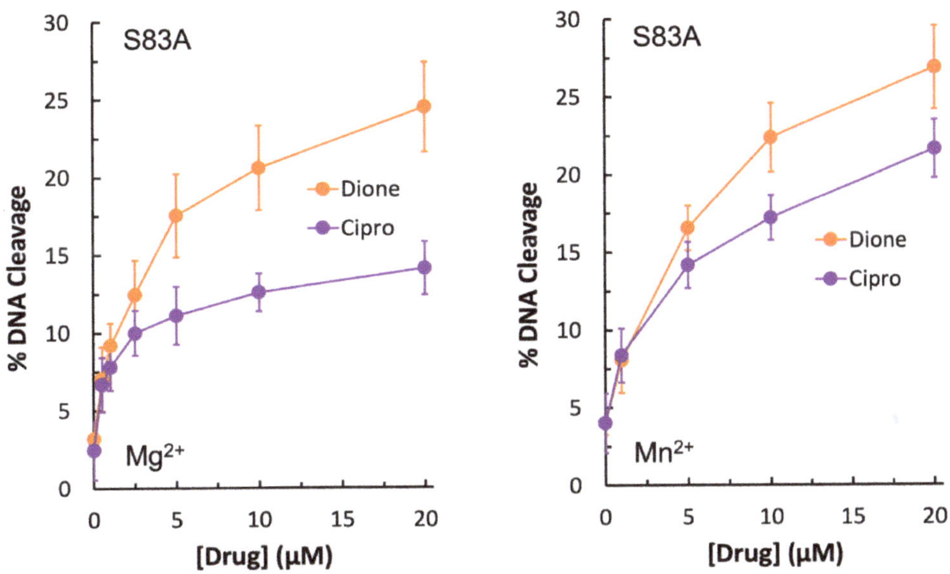

Figure 6. Plasmid DNA cleavage induced by GyrA^{S83A} gyrase in the presence of 0–20 µM ciprofloxacin (purple) or quinazolinedione (orange) with 6 mM Mg^{2+} (**left**) or 5 mM Mn^{2+} (**right**) as the metal ion. Error bars represent the standard deviation of three or more independent experiments.

In order to determine the function of the apparent water–metal ion bridge in the interaction between *E. coli* gyrase and ciprofloxacin, a competition assay was carried out. In this assay, 50 µM quinazolinedione was used to induce cleavage by the double mutant GyrA$^{S83L/D87N}$ enzyme in the presence of increasing concentrations of ciprofloxacin. Because ciprofloxacin did not induce DNA cleavage by this mutant enzyme (see Figure 3, left panel), then any decrease in DNA cleavage would be due to the quinolone competing out the quinazolinedione. As shown in Figure 7, ciprofloxacin was effective at competing out the quinazolinedione and decreasing DNA cleavage. Thus, it appears that the water–metal ion bridge functions primarily as a positioning interaction as the decrease in cleavage with equal concentrations of the two drugs indicates that they do not differ greatly in their ability to bind to the cleavage complex. This conclusion is consistent with the inability of high concentrations of ciprofloxacin to induce high levels of cleavage when the conserved residues are mutated (see Figure 3, left panel). Furthermore, this conclusion is consistent with the data gathered with the GyrA^{S83A} and GyrA^{S83L} mutant enzymes: it appears that the smaller alanine residue does not disrupt the bridge that is coordinated and stabilized by the remaining aspartic acid residue, while the bulkier leucine causes steric hindrance and disrupts this interaction.

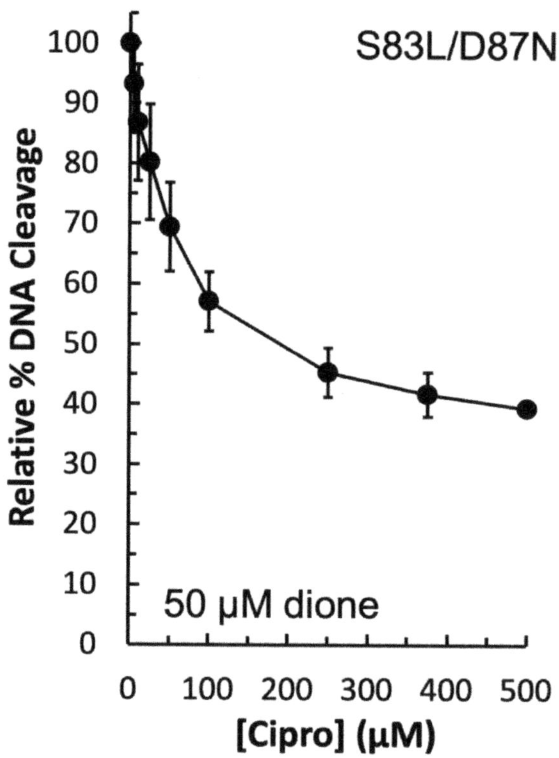

Figure 7. Competition between ciprofloxacin and the quinazolinedione to induce plasmid DNA cleavage by GyrA$^{S83L/D87N}$ gyrase. An amount of 50 μM quinazolinedione (the concentration at which maximal cleavage was achieved with this drug–enzyme combination) was included simultaneously in all reactions with 0–500 μM ciprofloxacin to determine the ability of the quinolone to compete with the quinazolinedione for interaction with the enzyme. The level of cleavage seen in the presence of the quinazolinedione alone was set to 100%. Error bars represent the standard deviation of three independent experiments.

3. Materials and Methods

3.1. Enzymes and Materials

Wild-type GyrA and GyrB genes from *E. coli* were PCR amplified and cloned into pET16b (Novagen, Madison, WI, USA), which added an N-terminal 10x His-tag. GyrA mutants—S83L, D87N, S83F/D87N, and S83A—were generated using a QuikChange Lightning Site-Directed Mutagenesis Kit (Agilent, Santa Clara, CA, USA). All clones, both wild-type and mutant, were sequenced prior to expression to confirm accuracy. Following expression in Rosetta 2 (DE3) pLysS *E. coli* (Novagen), cells were pelleted, resuspended in 20 mM TRIS-HCl (pH 7.9), 10% glycerol, 10 mM imidazole, 500 mM NaCl, and protease inhibitors (cOmplete, EDTA-free protease inhibitor cocktail; Roche), and lysed via sonication on ice. The lysate was cleared by centrifugation before being loaded onto a HisTrap HP column (Ge Healthcare, Chicago, IL, USA). The column was first washed with cold wash buffer (20 mM TRIS-HCl pH 7.9, 30 mM imidazole, 1 M NaCl, and protease inhibitors), then cold wash buffer containing 500 mM NaCl, and finally cold wash buffer containing 500 mM NaCl and 60 mM imidazole. Subunits were then eluted in 20 mM TRIS-HCl pH 7.9, 500 mM imidazole, 500 mM NaCl, and protease inhibitors. Finally, they were buffer exchanged into 50 mM TRIS-HCl pH 7.5, 200 mM NaCl, and 20% glycerol using Amicon Ultra-15 centrifugal filters (30K; Millipore, Burlington, MA, USA). Purity was checked via Coomassie staining, and quantification was via A_{260} reading on a NanoDrop 1000 (Thermo, Waltham, MA, USA). The resulting proteins were stored at −80 °C and used as a 1:2 GyrA:GyrB mixture in all assays. Dilution immediately prior to use was into 50 mM TRIS-HCl pH 7.9, 5 mM DTT, 30% glycerol, 125 mM potassium glutamate, and 200 mM KCl.

The negatively supercoiled pBR322 DNA substrate was purified using a Plasmid Mega Kit (Qiagen, Hilden, Germany) as described by the manufacturer.

Ciprofloxacin (SigmaAldrich, St. Louis, MO, USA) was kept at 4 °C as a 40 mM stock solution dissolved in 0.1 N NaOH. It was diluted 5x into 10 mM TRIS-HCl pH 7.9 immediately prior to use. The quinazolinedione 3-amino-7-[(3S)-3-(aminomethyl)-1-pyrrolidin-yl]-1-cyclopropyl-6-fluoro-8-methyl-2,4-(1H,3H)-quinazoline-dione was synthesized as previously described [35]. For simplicity, it is referred to simply as "quinazolinedione" throughout this paper. It was kept at 4 °C as a 20 mM stock solution dissolved in 100% DMSO.

3.2. Cleavage Assays

Cleavage assays were carried out as previously described [36]. Drug titrations were conducted to test for resistance. In these assays, 100 nM gyrase was combined with 0.6 μg pBR322 and various amounts of drug (as indicated in the figures) in cleavage buffer (10 mM TRIS-HCl pH 7.9, 20 mM NaCl, 5% glycerol, 6 mM $MgCl_2$, 1 mM DTT, and 1 mM ATP). For cleavage assays carried out in the presence of Mn^{2+}, $MnCl_2$ was substituted for $MgCl_2$ in the cleavage buffer at a concentration of 5 mM. For Mg^{2+} titrations, the drug was added to the cleavage buffer while the metal ion was omitted to allow titration of $MgCl_2$ from 0–6 mM as indicated in the figures. For competition assays, reactions contained a constant amount of quinazolinedione and 0–500 μM ciprofloxacin, as indicated in the figures. In all cases, the resulting 20 μL reaction was incubated at 37 °C for 10 min. Reactions were terminated and cleavage complexes were trapped by addition of 2 μL of 5% SDS. After addition of 1 μL of 250 mM Na_2EDTA (pH 8.0) and 2 μL of 0.8 mg/mL proteinase K, samples were incubated at 45 °C for 45 min. Then, 2 μL of agarose gel loading buffer was added to each reaction, which was incubated at 45 °C for 5 min before electrophoresis on a 1% agarose gel made in 40 mM Tris-acetate (pH 8.3) and 2 mM Na_2EDTA containing 0.5 μg/mL ethidium bromide. DNA bands were visualized using the FOTO/Analyst Luminary FX system (Fotodyne) and quantified using the stand-alone AlphaEase program (Alpha Innotech).

4. Conclusions

When the data presented above are viewed holistically, it is likely that the water–metal ion bridge does exist in the quinolone-*E. coli* gyrase interaction and that the conserved Ser and acidic residue 4 positions downstream (on helix 4 of the GyrA subunit) act as the bridge anchors. As seen above, metal ion substitution affected quinolone-induced but not quinazolinedione-induced DNA cleavage mediated by the target enzyme implicating a role for the metal ion in the quinolone interaction. In addition, because the mutant enzymes required higher concentrations of metal ions to reach maximal DNA cleavage in the presence of the quinolone but not the quinazolinedione, it appears that the conserved residues anchor the bridge. All of these data and conclusions are consistent with both crystal structures presented in Figure 1 in which the quinolone interacts with the enzyme via a water–metal ion bridge that is anchored by the conserved Ser and acidic residues and that the quinazolinedione is metal-ion-independent and does not utilize these anchor residues to interact with the enzyme. These findings are also consistent with the biochemical data gathered with *E. coli* topoisomerase IV [29], *B. anthracis* topoisomerase IV [28,30], and *M. tuberculosis* gyrase [31] (which is unusual because it is the only type II topoisomerase in that species and has efficient decatenase activity). However, it is in contrast to *B. anthracis* gyrase, which is a "typical" gyrase that occurs alongside topoisomerase IV and has inefficient decatenase activity, where the water–metal ion bridge appears to be anchored solely by the Ser residue [32]. Studies on the water–metal ion bridge in other "typical" gyrase enzymes, as well as topoisomerase IV enzymes, will reveal whether the bridge anchors in most gyrases are consistent with those in topoisomerase IV or whether "typical" gyrases most often rely on only one residue anchor, which would make this enzyme—*E. coli* gyrase—an outlier.

Moreover, the water–metal ion bridge appears to play a positioning, rather than a binding, role because high concentrations of the quinolone were unable to induce high levels of DNA cleavage in the presence of the mutant enzymes and the quinolone was able to compete out the quinazolinedione and lower cleavage seen when both drugs were present as compared to when only the quinazolinedione was present. A positioning role for the water–metal ion bridge in *E. coli* gyrase is consistent with the function of the bridge in *E. coli* topoisomerase IV [29]. In contrast, previous work with *B. anthracis* topoisomerase IV [28,30] and gyrase [32] found the water–metal ion bridge to function as a binding interaction in that Gram-positive species. In addition, in *M. tuberculosis*, an acid-fast bacterium that contains only gyrase, the water–metal ion bridge was also found to play a binding role [31]. Taken together, these findings raise the possibility that the water–metal ion bridge has a split of function down Gram classification lines, with it functioning primarily as a binding interaction in Gram-positive type II topoisomerases and primarily as a positioning interaction in Gram-negative type II topoisomerases. We are currently working to test for the presence and the function of the water–metal ion bridge in other common pathogenic Gram-positive and Gram-negative bacterial species to determine whether the water–metal ion bridge has the same function in both type II topoisomerases from a given species, and more importantly, whether the function of the bridge in a given species can be predicted based on Gram classification. If there is indeed a split of function along Gram classification lines, then it may be necessary to approach designing a quinolone-based drug that overcomes resistance to the most common mutations (i.e., those that facilitate formation of the bridge), from two different angles in order to solve the two different issues—binding and positioning.

Author Contributions: Conceptualization, K.J.A. and R.J.K.; Formal Analysis, H.E.C., B.W. and H.A.S.; Investigation, H.E.C., B.W. and H.A.S.; Writing—Original Draft Preparation, K.J.A.; Writing—Review and Editing, H.E.C., B.W. and R.J.K.; Supervision, K.J.A. and R.J.K.; Funding Acquisition, H.E.C., K.J.A. and R.J.K. All authors have read and agreed to the published version of the manuscript.

Funding: This research was funded by the UExplore Undergraduate Research Program at the University of Evansville with fall and spring semester grants to H.E.C. and a summer grant to K.J.A., H.A.S. participated in this research as a trainee under grant T32 GM008365 from the National Institutes of Health, and contributions by R.J.K. were supported by National Institutes of Health research grant AI87671.

Institutional Review Board Statement: Not applicable.

Informed Consent Statement: Not applicable.

Data Availability Statement: Data are contained within the article.

Conflicts of Interest: The authors declare no conflict of interest.

References

1. Morgan-Linnell, S.K.; Becnel Boyd, L.; Steffen, D.; Zechiedrich, L. Mechanisms accounting for fluoroquinolone resistance in *Escherichia coli* clinical isolates. *Antimicrob. Agents Chemother.* **2009**, *53*, 235–241. [CrossRef]
2. Emmerson, A.M.; Jones, A.M. The quinolones: Decades of development and use. *J. Antimicrob. Chemother.* **2003**, *51* (Suppl. 1), 13–20. [CrossRef] [PubMed]
3. Linder, J.A.; Huang, E.S.; Steinman, M.A.; Gonzales, R.; Stafford, R.S. Fluoroquinolone prescribing in the United States: 1995 to 2002. *Am. J. Med.* **2005**, *118*, 259–268. [CrossRef] [PubMed]
4. Anderson, V.E.; Osheroff, N. Type II topoisomerases as targets for quinolone antibacterials: Turning Dr. Jekyll into Mr. Hyde. *Curr. Pharm. Des.* **2001**, *7*, 337–353. [CrossRef] [PubMed]
5. Aldred, K.J.; Kerns, R.J.; Osheroff, N. Mechanism of quinolone action and resistance. *Biochemistry* **2014**, *53*, 1565–1574. [CrossRef]
6. Jacoby, G.A. Mechanisms of resistance to quinolones. *Clin. Infect. Dis.* **2005**, *41* (Suppl. 2), S120–S126. [CrossRef]
7. Drlica, K.; Hiasa, H.; Kerns, R.; Malik, M.; Mustaev, A.; Zhao, X. Quinolones: Action and resistance updated. *Curr. Top. Med. Chem.* **2009**, *9*, 981–998. [CrossRef]
8. Bush, N.G.; Diez-Santos, I.; Abbott, L.R.; Maxwell, A. Quinolones: Mechanism, Lethality and Their Contributions to Antibiotic Resistance. *Molecules* **2020**, *25*, 5662. [CrossRef]
9. Hooper, D.C.; Jacoby, G.A. Mechanisms of drug resistance: Quinolone resistance. *Ann. N. Y. Acad. Sci.* **2015**, *1354*, 12–31. [CrossRef]
10. Hooper, D.C. Mode of action of fluoroquinolones. *Drugs* **1999**, *58* (Suppl. 2), 6–10. [CrossRef]
11. Hooper, D.C. Mechanisms of action of antimicrobials: Focus on fluoroquinolones. *Clin. Infect. Dis.* **2001**, *32* (Suppl. 1), S9–S15. [CrossRef] [PubMed]
12. Champoux, J.J. DNA topoisomerases: Structure, function, and mechanism. *Annu. Rev. Biochem.* **2001**, *70*, 369–413. [CrossRef] [PubMed]
13. Levine, C.; Hiasa, H.; Marians, K.J. DNA gyrase and topoisomerase IV: Biochemical activities, physiological roles during chromosome replication, and drug sensitivities. *Biochim. Biophys. Acta* **1998**, *1400*, 29–43. [CrossRef]
14. Forterre, P.; Gribaldo, S.; Gadelle, D.; Serre, M.C. Origin and evolution of DNA topoisomerases. *Biochimie* **2007**, *89*, 427–446. [CrossRef] [PubMed]
15. Forterre, P.; Gadelle, D. Phylogenomics of DNA topoisomerases: Their origin and putative roles in the emergence of modern organisms. *Nucleic Acids Res.* **2009**, *37*, 679–692. [CrossRef] [PubMed]
16. Pommier, Y.; Leo, E.; Zhang, H.; Marchand, C. DNA topoisomerases and their poisoning by anticancer and antibacterial drugs. *Chem. Biol.* **2010**, *17*, 421–433. [CrossRef] [PubMed]
17. Mitscher, L.A. Bacterial topoisomerase inhibitors: Quinolone and pyridone antibacterial agents. *Chem. Rev.* **2005**, *105*, 559–592. [CrossRef]
18. Stein, G.E. The 4-quinolone antibiotics: Past, present, and future. *Pharmacotherapy* **1988**, *8*, 301–314. [CrossRef]
19. Andersson, M.I.; MacGowan, A.P. Development of the quinolones. *J. Antimicrob. Chemother.* **2003**, *51* (Suppl. 1), 1–11. [CrossRef]
20. Andriole, V.T. The quinolones: Past, present, and future. *Clin. Infect. Dis.* **2005**, *41* (Suppl. 2), S113–S119. [CrossRef]
21. Fournier, B.; Zhao, X.; Lu, T.; Drlica, K.; Hooper, D.C. Selective targeting of topoisomerase IV and DNA gyrase in *Staphylococcus aureus*: Different patterns of quinolone-induced inhibition of DNA synthesis. *Antimicrob. Agents Chemother.* **2000**, *44*, 2160–2165. [CrossRef] [PubMed]
22. Price, L.B.; Vogler, A.; Pearson, T.; Busch, J.D.; Schupp, J.M.; Keim, P. In vitro selection and characterization of *Bacillus anthracis* mutants with high-level resistance to ciprofloxacin. *Antimicrob. Agents Chemother.* **2003**, *47*, 2362–2365. [CrossRef] [PubMed]
23. Drlica, K.; Zhao, X. DNA gyrase, topoisomerase IV, and the 4-quinolones. *Microbiol. Mol. Biol. Rev.* **1997**, *61*, 377–392. [PubMed]
24. Li, Z.; Deguchi, T.; Yasuda, M.; Kawamura, T.; Kanematsu, E.; Nishino, Y.; Ishihara, S.; Kawada, Y. Alteration in the GyrA subunit of DNA gyrase and the ParC subunit of DNA topoisomerase IV in quinolone-resistant clinical isolates of Staphylococcus epidermidis. *Antimicrob. Agents Chemother.* **1998**, *42*, 3293–3295. [CrossRef] [PubMed]
25. Yang, J.; Luo, Y.; Li, J.; Ma, Y.; Hu, C.; Jin, S.; Ye, L.; Cui, S. Characterization of clinical *Escherichia coli* isolates from China containing transferable quinolone resistance determinants. *J. Antimicrob. Chemother.* **2010**, *65*, 453–459. [CrossRef] [PubMed]
26. Lautenbach, E.; Metlay, J.P.; Mao, X.; Han, X.; Fishman, N.O.; Bilker, W.B.; Tolomeo, P.; Wheeler, M.; Nachamkin, I. The prevalence of fluoroquinolone resistance mechanisms in colonizing *Escherichia coli* isolates recovered from hospitalized patients. *Clin. Infect. Dis.* **2010**, *51*, 280–285. [CrossRef]
27. Bansal, S.; Tandon, V. Contribution of mutations in DNA gyrase and topoisomerase IV genes to ciprofloxacin resistance in *Escherichia coli* clinical isolates. *Int. J. Antimicrob. Agents* **2011**, *37*, 253–255. [CrossRef]
28. Aldred, K.J.; McPherson, S.A.; Turnbough, C.L., Jr.; Kerns, R.J.; Osheroff, N. Topoisomerase IV-quinolone interactions are mediated through a water-metal ion bridge: Mechanistic basis of quinolone resistance. *Nucleic Acids Res.* **2013**, *41*, 4628–4639. [CrossRef]
29. Aldred, K.J.; Breland, E.J.; Vlckova, V.; Strub, M.P.; Neuman, K.C.; Kerns, R.J.; Osheroff, N. Role of the water-metal ion bridge in mediating interactions between quinolones and *Escherichia coli* topoisomerase IV. *Biochemistry* **2014**, *53*, 5558–5567. [CrossRef]
30. Aldred, K.J.; McPherson, S.A.; Wang, P.; Kerns, R.J.; Graves, D.E.; Turnbough, C.L., Jr.; Osheroff, N. Drug interactions with *Bacillus anthracis* topoisomerase IV: Biochemical basis for quinolone action and resistance. *Biochemistry* **2012**, *51*, 370–381. [CrossRef]

31. Aldred, K.J.; Blower, T.R.; Kerns, R.J.; Berger, J.M.; Osheroff, N. Fluoroquinolone interactions with *Mycobacterium tuberculosis* gyrase: Enhancing drug activity against wild-type and resistant gyrase. *Proc. Natl. Acad. Sci. USA* **2016**, *113*, E839–E846. [CrossRef] [PubMed]
32. Ashley, R.E.; Lindsey, R.H., Jr.; McPherson, S.A.; Turnbough, C.L., Jr.; Kerns, R.J.; Osheroff, N. Interactions between Quinolones and *Bacillus anthracis* Gyrase and the Basis of Drug Resistance. *Biochemistry* **2017**, *56*, 4191–4200. [CrossRef] [PubMed]
33. Wohlkonig, A.; Chan, P.F.; Fosberry, A.P.; Homes, P.; Huang, J.; Kranz, M.; Leydon, V.R.; Miles, T.J.; Pearson, N.D.; Perera, R.L.; et al. Structural basis of quinolone inhibition of type IIA topoisomerases and target-mediated resistance. *Nat. Struct. Mol. Biol.* **2010**, *17*, 1152–1153. [CrossRef]
34. Oppegard, L.M.; Streck, K.R.; Rosen, J.D.; Schwanz, H.A.; Drlica, K.; Kerns, R.J.; Hiasa, H. Comparison of in vitro activities of fluoroquinolone-like 2,4- and 1,3-diones. *Antimicrob. Agents Chemother.* **2010**, *54*, 3011–3014. [CrossRef]
35. Malik, M.; Marks, K.R.; Mustaev, A.; Zhao, X.; Chavda, K.; Kerns, R.J.; Drlica, K. Fluoroquinolone and quinazolinedione activities against wild-type and gyrase mutant strains of Mycobacterium smegmatis. *Antimicrob. Agents Chemother.* **2011**, *55*, 2335–2343. [CrossRef]
36. Fortune, J.M.; Osheroff, N. Merbarone inhibits the catalytic activity of human topoisomerase IIalpha by blocking DNA cleavage. *J. Biol. Chem.* **1998**, *273*, 17643–17650. [CrossRef] [PubMed]

International Journal of *Molecular Sciences*

MDPI

Review

A Mini Review of Novel Topoisomerase II Inhibitors as Future Anticancer Agents

Cosmas O. Okoro * and Toluwase Hezekiah Fatoki

Department of Chemistry, Tennessee State University, Boswell Science Complex, Nashville, TN 37209-1561, USA
* Correspondence: cokoro@tnstate.edu

Abstract: Several reviews of inhibitors of topoisomerase II have been published, covering research before 2018. Therefore, this review is focused primarily on more recent publications with relevant points from the earlier literature. Topoisomerase II is an established target for anticancer drugs, which are further subdivided into poisons and catalytic inhibitors. While most of the topoisomerase II-based drugs in clinical use are mostly topoisomerase II poisons, their mechanism of action has posed severe concern due to DNA damaging potential, including the development of multi-drug resistance. As a result, we are beginning to see a gradual paradigm shift towards non-DNA damaging agents, such as the lesser studied topoisomerase II catalytic inhibitors. In addition, this review describes some novel selective catalytic topoisomerase II inhibitors. The ultimate goal is to bring researchers up to speed by curating and delineating new scaffolds as the leads for the optimization and development of new potent, safe, and selective agents for the treatment of cancer.

Keywords: mini; topoisomerase II; catalytic inhibitors; poisons; DNA; salicylate; cardiotoxicity; MD simulation

Citation: Okoro, C.O.; Fatoki, T.H. A Mini Review of Novel Topoisomerase II Inhibitors as Future Anticancer Agents. *Int. J. Mol. Sci.* **2023**, *24*, 2532. https://doi.org/10.3390/ijms24032532

Academic Editors: Joseph E. Deweese and Neil Osheroff

Received: 2 November 2022
Revised: 15 January 2023
Accepted: 16 January 2023
Published: 28 January 2023

1. Introduction

Cancer is a word that creates deep-seated fear because we immediately associate it with grave illness and high mortality rate. Almost all of us know someone whose life has been blighted by cancer diagnosis and who has suffered the prolonged pain of the illness. Cancer patients are forced to tolerate a tough treatment regime with all the accompanying side effects. Few are fortunate to escape the distress of cancer over their lifetime since the frightening statistics suggest that the vast majority of us will either experience it firsthand or have a loved one afflicted by it. As a result of the above facts, and in order to improve the survival and quality of life of cancer patients, medicinal chemists are actively searching for novel effective, safe, and selective anticancer drugs. This review provides an overview of the chemical structures and bioactivities of recent agents that target the nuclear enzyme, topoisomerase. Essential nucleic acid functions, such as DNA replication and recombination, generate knots and tangles within the double helix. DNA knots are known to impair the ability to separate the two strands. In addition, intermolecular DNA tangles prevent the segregation of chromosomes during mitosis. It is now recognized that topoisomerases regulate the topological structure of DNA. If the above topological obstacles are not removed, they become lethal to cells. Topoisomerases fall into two major classes, topoisomerase I and topoisomerase II. The classification is based upon the linking number changed by the enzyme: type 1 changes the linking number by one while type II changes the linking number by two. Topoisomerase I breaks one strand of the double helix and its functions includes regulation of the levels of DNA supercoiling. On the other hand, topoisomerase II breaks both strands of the double helix and regulates the super-helical density and removal of knots and tangles in the duplex DNA. To maintain genomic integrity during the required DNA cleavage event, all topoisomerases form covalent bond between the active-site tyrosyl residues and the DNA termini generated during the reaction.

This review focuses on topoisomerase II (topo II). There are two isoforms of human topo II, topo IIα and topo IIβ. Both isoforms are encoded by separate genes but appear to share approximately 70% of amino acid sequence identity. However, both isoforms have distinct patterns of expression. Topo IIα has the following characteristics: (1) it is essential for the survival of proliferating cells; (2) its protein levels rise dramatically during periods of cell growth; (3) it is regulated over the cell cycle with the concentration peaking in the G2/M phase; (4) it helps alleviate the torsional stress that accumulates ahead of the replication forks and transcription complexes; and (5) it is required for proper chromosome condensation, cohesion, and segregation. Topo IIβ has the following features: (1) it is dispensable at the cellular level; (2) its concentration is independent of cell cycle; (3) high levels of Topo IIβ are found in most cell types regardless of proliferation status; and (4) it dissociates from chromosome during mitosis. The physiological functions of topo IIβ are yet to be fully understood.

Topoisomerase (Topo) is an established target for anticancer drugs and is known to be responsible for regulating the topological constraints in DNA, such as under-winding, over-winding, knotting, and tangling. In particular, under-winding (negatively supercoiled) is important because the two strands of DNA must be separated in order for replication to start. Thus, one of the physiological roles of topoisomerase is to relax both positive and negative supercoiled DNA [1]. Topoisomerase falls into two major classes in human eukaryotes, namely type I and type II. Type I topoisomerase has the subfamilies type IA (Topo IIIα and Topo IIIβ) and type IB (Topo I and mtTopo 1), while type II topoisomerase has the subfamily type IIA (Topo IIα and Topo IIβ) [2]. Human Topo I is a monomeric protein that does not need cofactors for biological activity. It regulates the topological state of DNA by breaking a single strand and re-ligating it after the strand passage reaction. Human Topo II is a homodimer that requires Mg^{2+} and ATP for its catalytic activity. Topo II breaks double strand and passes an intact DNA through the break.

Topo II inhibitors are classically divided into catalytic inhibitors and topo II poisons, according to their mechanism of action. Topo II catalytic inhibitors destroy cancer cells through the inhibitions of Topo II enzymatic activities, thus preventing the formation of Topo II–DNA complex without increasing DNA cleavage, via the mechanisms of action that include interfering with DNA binding, inhibiting cleavage of the DNA molecule, ATP hydrolysis, and binding to the ATP binding site. Topo II poisons destroy cancer cells by increasing the amount of covalent Topo II–DNA complexes and preventing the religation of the cleaved DNA strands, thus forming unwanted double strand breaks that are toxic to the cells, and, subsequently, leading to apoptosis. Based on the type of DNA binding strategies, Topo II poisons could be DNA-intercalating agents, which have a weak interaction with DNA and function by insnaring Topo II–DNA complexes, or non-DNA-intercalating agents, which can reversibly incorporate themselves into the DNA base pairs and hinder the activity of enzymes responsible for DNA replication and transcription processes. Topo II poisons include drugs such as etoposide **1**, doxorubicin **2**, and m-amsacrine **3** [3–5], as shown in Figure 1. Most of the first-line agents for treating cancer are Topo II poisons, such as etoposide (non-intercalator), doxorubicin, and m-amsacrine (intercalator) [4–8]. However, a study by Ketron et al. [9] indicated that the activity and specificity of m-amsa lies in the head group 4′-amino-methanesulfon-m-anisidide. The DNA intercalation of m-amsa is used primarily to enhance the affinity of the drug for Topo II–DNA cleavage complex. However, due to side effects, such as risk of cardiotoxicity and secondary malignancies, that are often encountered during the use of DNA poisonous drugs, research is now shifting towards the discovery of Topo II catalytic inhibitors, which have good pharmacokinetics profiles.

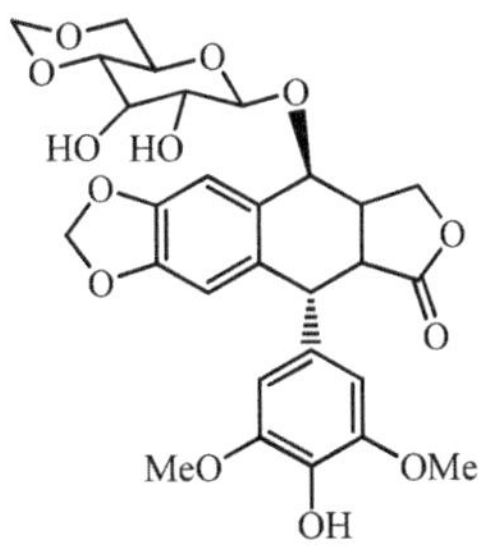

Etoposide **1**

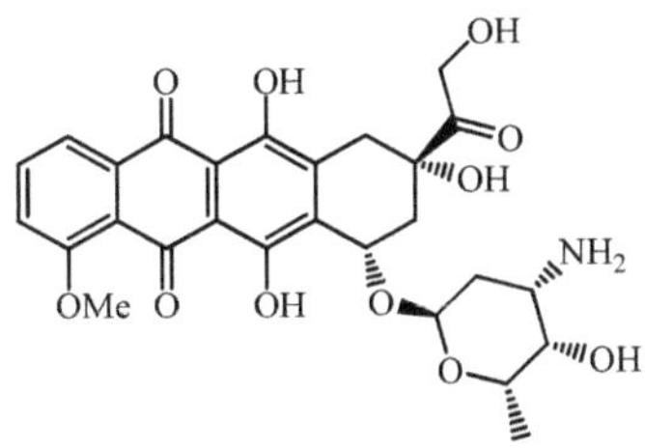

Doxorubicin **2**

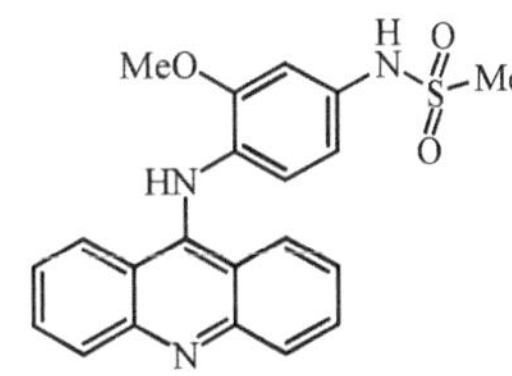

m-Amsacrine **3**

Figure 1. Structure of topoisomerase II poisons.

Topo II is further divided into Topo IIαand Topo IIβ. Topo IIα is highly expressed in proliferating cells, while Topo IIβ is dispensable during proliferation. Although the two isoforms are structurally similar, the precise role of the β-isoform is unclear. There is speculation that the Β-isoform is responsible for the occurrence of acute MLL in patients treated with a topoisomerase inhibitor. Topoisomerase II has four domains: (i) an N gate, (ii) a DNA gate, (iii) a C gate, and (iv) a C-terminal domain that is responsible for DNA recognition [10,11].

Topoisomerase II catalytic cycle consists of six sequential steps as follows: (i) recognition and binding of the enzyme to DNA helix 1; (ii) trapping of DNA helix 2 through the ATPase domain dimerization; (iii) double-strand break that results in covalent bond formation between the enzyme and the 5′ phosphodiester of DNA; (iv) another segment of DNA is passed through the break facilitated by ATP hydrolysis; (v) religation of the DNA strand break mediated by the release of ADP; and (vi) the enzyme and DNA are each restored and ready to start another catalytic cycle. It is worthy of note that DNA cleavage by either Topo I or Topo II is transient and rate determining, while the religation process is fast and well tolerated by cells. The ability of small organic molecules to modulate topoisomerase activity is an effective method for identifying new cancer therapeutics. On the following pages, we delineate recent inhibitors that are still in various stages of development.

A study that used virtual high-throughput screening (VHTS) of the ZINC database showed that four zinc compounds, **4–7**, could be potent inhibitors of TopoIIα based on their better docking score than the standard drug etoposide, as well as their suitable predicted ADME/Tox properties [12]. Similarly, an in silico study conducted in Nigeria by Adeniran et al. [13], which used VHTS, three-dimensional quantitative structure activity and relationship (3D-QSAR), and molecular docking approaches, reported the potential of 20-betaecdysone **8** and andropanoside **9** as better inhibitors of topoisomerase IIα (TopoIIα) than the standard drug, etoposide. This study needs further investigation as these compounds are phytochemicals and could be structurally optimized to deliver efficient anticancer activity. Additionally, in Slovenia, Skok et al. [14] used in silico screening of bacterial topoisomerase inhibitors with in vitro assay to identify ATP-competitive inhibitors of human DNA TopoIIα, and they reported *N-(4-Carbamoyl-2-isopropoxyphenyl)-3,4-dichloro-5-methyl-1H-pyrrole-2-carboxamide* as a potential active inhibitor of TopoIIα. Further investigation of these computationally screened compounds is necessary to validate their biological activity as anticancer agents.

Ellipticine, an alkaloid from *Ochrosia elliptica* labil, has been previously indicated for the treatment of metastatic breast cancer, and several carbazole derivatives based on Ellipticine have shown inhibitory activities against TopoI and TopoII [15–17]. In Italy, Saturnino et al. [18] reported *2-(4-((3-Chloro-9H-carbazol-9-yl)pentyl)piperazin-1-yl)-N,N,N-trimethylethanammonium iodide* **10** as a good inhibitor of TopoII and that it showed antiproliferative activity on breast cancer cells, causing apoptosis by activating the caspase pathway. Additionally, *7-((2-(dimethylamino)ethyl)amino)indolo[2,1-b]quinazoline-6,12-dione*

26, which was derived from trypthantrin, a natural alkaloidal compound containing a basic indoloquinazoline moiety, has been reported for its inhibitory activity against TopoII, and has shown properties such as high water solubility and antiproliferative activity on acute leukemia, colon, and breast cancer cell lines [19]. Garcinol, a polyisoprenylated benzophenone isolated from the *Garcinia* genus, has been reported to inhibit eukaryotic TopoI and TopoII at concentrations comparable to that of the standard drug etoposide [20]. Mai et al. [21] in China investigated *9-bromo-2,3-diethylbenzo[de]chromene-7,8-dione* (MSN54) (**16**), a derivative of Mansonone F (MsF) [22], and they found it to be a non-intercalative catalytic inhibitor of TopoIIα. MsF (**17**) is a phytoalexin obtained from *Helicteres Angustifolia* L. with a rare sesquiterpene o-quinone structure and possesses anti-tumor activity, and its derivatives have shown strong inhibitory activity on TopoII [23].

A study in Italy that made use of biochemical assays along with molecular docking and molecular dynamic methods reported that the derivatives of garcinol acted as catalytic inhibitors of TopoII via a mixed inhibition of ATP hydrolysis, in which guttiferone M showed the most significant effects against TopoIIα and TopoIIβ, whereas oxyguttiferones K and M showed activity against TopoIIβ that was slightly better than that of the parent garcinol compound [24]. A collaborative work by Zidar et al. [25], in Italy and Slovenia, reported that *N-phenoxylypropyl-3-methyl-2-phenyl-1H-indoles* **20** and **21** had potent antiproliferative and anti-topoisomerase II activities against three selected human tumor cell lines, including cervix adenocarcinoma (HeLa), ovarian carcinoma (A2780), and biphasic mesothelioma (MSTO-211H), and they were capable of inducing the apoptosis pathway. These compounds have structural similarities with some naturally occurring flavonoids that inhibit topoisomerase II (TopoII), such as quercetin and luteolin. Thus, they are scientifically acceptable and could have broader application in anticancer drug development.

Zhou et al. [26], in China, studied novel perimidine *o*-quinone derivatives and found that *2-(4-Chlorophenyl)-1-methyl-1H-perimidine-5,6-dione* showed the best antiproliferative activity (IC50 $\leq$ 1 μM) against four cancer cell lines (HL-60, Huh7, Hct116, and Hela) by inducing apoptosis in a dose-dependent manner, while exhibiting potent topoisomerase IIα (TopoIIα) inhibitory activity (IC_{50} = 7.54 μM). They provided evidence that compound 21 did not intercalate into DNA and suggested that it might act as an ATP competitive inhibitor by blocking the ATP-binding site of the enzyme; this was also tested by molecular docking of compound 21 in the ATP-binding domain of human TopoIIα (PDB code: 1ZXM). Overall, this study is robust and may provide advanced opportunities for the design and development of new chemotherapeutic agents.

Sakr et al. [27], in Egypt, reported that *N-Cyclohexyl-2-(3-methyl-[1,2,4]-triazolo[3,4-a]phthalazin-6-yl)-hydrazine-1-carboxamide* **11**, a derivative of triazolophthalazine, showed slightly high cytotoxic activity than doxorubicin when tested against human cancer cell line (HepG2, MCF-7, and HCT-116 cells), induced apoptosis in HepG2 cells and G2/M phase cell cycle arrest, and showed Topo II inhibitory activity. However, they reported that compound 22 showed TopoII poisoning effects at 2.5 μM and Topo II catalytic inhibitory effects at 5 and 10 μM; these results indicate that this compound could serve as a two-edge chemotherapeutic agent, thus requiring further validation. A previous study reported that a compound that contained the 1,4-diaminobenzo[g]phthalazine nucleus had a promising binding affinity against DNA by intercalation [28]. In addition, Arencibia et al. [29] reported *6-Hydroxy-4-oxo-1,3-diphenyl-2-thioxo-N-(3-(trifluoromethoxy)phenyl)-1,2,3,4-tetrahydropyrimidine-5-carboxamide* **24** as the most promising drug-like candidate that acted as a TopoII poison and exhibited good solubility, metabolic (microsomal) stability, and promising cytotoxicity in three cancer cell lines (DU145, HeLa, and A549).

Additionally, four compounds from a new series of [1,2,4]triazolo[4,3-*a*]quinoxaline and bis([1,2,4]triazolo)[4,3-*a*:3′,4′-*c*]quinoxaline derivatives showed cytotoxic activities against three tumor cell lines (HePG-2, Hep-2, and Caco-2), of which *2-(Bis[1,2,4]triazolo[4,3-a:3′,4′-c]quinoxalin-3-ylsulfanyl)-N-(4-fluorophenyl) acetamide* was the most effective inhibitor against TopoII, intercalated DNA, caused cell cycle arrest at the G2/M phase, and induced apoptosis in Caco-2 cells [30]. Moreover, El-Adl et al. [31], in Egypt, worked on twenty four novel [1,2,4]triazolo[4,3-*a*]quinoxaline derivatives, and they reported that *2-{4-([1,2,4]Triazolo[4,3-a]quinoxalin-4-ylamino)benzoyl}-N-cyclohexylhydrazine-1-carboxamide* **12**, *2-{4-([1,2,4]Triazolo[4,3-a]quinoxalin-4-ylamino)benzoyl}-N-cyclohexylhydrazine-1-carbothioamide*, and *4-(Diethylamino)-[1,2,4]triazolo[4,3-a]quinoxaline-1-thiol* **11b** were the most potent derivatives against the tested HepG2, HCT116, and MCF-7 cancer cell lines, and that these three compounds also displayed very-good-to-moderate DNA-binding affinities and exhibited very good inhibitory activities against TopoII enzyme. However, no in vitro differentiation was made between TopoIIα and TopoIIβ, although molecular docking was carried out on DNA-Topo II receptor (PDB code: 4G0U). Previous studies had reported that several quinoxaline compounds, such as *1-(2-Bromoethyl)-1,4-dihydroquinoxaline-2,3-dione, 4-Amino-N′-(3-chloroquinoxalin-2-yl) benzohydrazide, N′-(3-chloroquinoxalin-2-yl)-isonicotinohydrazide*, and *3-mercaptoquinoxalin-2-yl carbamimidothioate*, from the series of quinoxaline derivatives were DNA intercalators, effective inhibitors of TopoII, and showed antiproliferative activities against HePG-2, MCF-7, and HCT-116 cell lines [32–35].

Bruno et al. [36], in USA, explored the CRISPR datasets (19Q3 DepMap Public data) together with biochemical and cell biological assays and showed that CX-5461 **25**, which is structurally similar to ciprofloxacin and voreloxin, exerts its primary cytotoxic activity through topoisomerase II poisoning. This study was holistic in its approach, and it could be applied to other investigational small molecules. The authors attempted to rethink the verdict by Lin et al. [37], which stated that off-target toxicity is a common mechanism of action of cancer drugs undergoing clinical trials. They further suggested that the mechanism of cell death induced by CX-5461 is critical for rational clinical development in patients with relapse/refractory hematopoietic tumors based on its previous clinical indication as an inhibitor of RNA polymerase I [38,39]. Recent study has suggested that CX-5461 could stabilize G-quadruplex DNA and cause DNA damage [40], and that BMH-21, which is an inhibitor of RNA polymerase I, suppresses the nucleolar translocation of both TopoIIα and TopoIIβ in ATP-depleted cells [41]. Similarly, in United Kingdom, Cowell et al. [42] explored PR-619 [*2,6-Diaminopyridine-3,5-bis(thiocyanate)*] **31**, a broad-spectrum deubiquitinating enzyme (DUB) inhibitor [43], and they found it to be a potent TopoII poison, inducing both TopoIIα and TopoIIβ covalent DNA complexes with an efficiency equal to that of etoposide.

On prostate cancer (PCa)-targeted TopoII inhibition, Jeon et al. [44] investigated the mechanism by which AK-I-190 [*2-(3-trifluorophenyl)-4-(3-hydroxyphenyl)-5H-indeno[1,2-b]pyridin-6-ol*] inhibited TopoII by using various types of biological and spectroscopic evaluations, and they found that its inhibitory activity was through intercalating into DNA without stabilizing the DNA–enzyme cleavage complex, which resulted in significantly less DNA toxicity than etoposide and inhibited the growth of AR-negative PCa cells. This work served as a therapeutic strategy against castration-resistant prostate cancer (CRPC), resulting from androgen independence in cellular growth [45]. The lead compound, T60 (PubChem CID: 36589274) **13**, was reported as a potent inhibitor of both TopoIIα and TopoIIβ enzymatic activities, as well as having dual inhibitory activity on the androgen receptor (AR) (an oncogenic transcriptional factor that requires TopoIIβ) and AR-positive PCa cell growth [46]. Matias-Barrios et al. [47] in Canada investigated the derivatives of T60 in order to improve its pharmacokinetic properties and further enhance its efficacy to inhibit TopoII proteins, and they found that T638, an amino derivative on the central benzene ring of T60, retained TopoII inhibitory activities and showed improved solubility and better metabolic stability, with the possibility of high dosage administration due to low cytotoxicity [47]. In a similar manner, heteronemin (a marine sponge *Hyrtios* sp. Sesterterpene) promoted apoptosis and autophagy through the inhibition of TopoIIα and HSP90 as well as through protein tyrosine phosphatase (PTP) activation in PCa cells [48], and inhibited TNFα-induced NF-κB activation through proteasome and induced apoptotic cell death [49].

Ortega et al. [50] in Italy studied a novel class of 6-amino-tetrahydroquinazoline derivatives, and they pinpointed *N^4-[4-(Dimethylamino)phenyl]-2-(4-pyridyl)-5,6,7,8-tetrahydroquinazoline-4,6-diamine* **15** as the main lead compound for the inhibition of DNA relaxation, which possessed excellent metabolic stability and solubility than etoposide; this compound showed about 100-fold selectivity for TopoIIα over TopoIIβ, with a broad antiproliferative activity against cultured human cancer cells, a satisfactory in vivo pharmacokinetic profile, and penetrability of the blood−brain barrier. These excellent properties indicated this compound 36 as a highly promising lead for the development of novel and potentially safer TopoII-targeted anticancer drugs. Additionally, in China, Chen et al. [51] evaluated the derivatives of acridine hydroxamic acid, and they found that *7-(4-(4-(Acridin-9-ylamino)-phenyl)-1H-1,2,3-triazol-1-yl)-N-hydroxyheptanamide* **27** showed the best inhibitory activities against TopoII and histone deacetylase (HDAC), and it could intercalate into DNA and induce U937 apoptosis. A combination of Topo and HDAC inhibitors has been found to show synergistic anticancer effects with enhanced cytotoxicity [52,53]. These dual inhibitory compounds are promising drug candidates that could serve as a double-edge sword to effectively inhibit tumor growth and progression.

In USA, Oyedele et al. [54] worked on a novel series of acridone derivatives, of which five derivatives, including 7-chloro-3-phenyl-3,4-dihydroacridin-1(2H)-one], 7-bromo-3-phenyl-3,4-dihydroacridin-1(2H)-one, 7-methoxy-3-(trifluoromethyl)-3,4-dihydroacridin-1(2H)-one, 7-methoxy-3-phenyl-3,4-dihydroacridin-1(2H)-one, and 5,7-dibromo-3-phenyl-3,4-dihydroacridin-1(2H)-one, showed excellent in vitro antiproliferative activities against 60 human cancer cell lines. Overall, 5,7-dibromo-3-phenyl-3,4-dihydroacridin-1 (2H)-one was found to be the most active and sensitive agent in all the nine cancer panels in the order of prostate > leukemia > non-small cell lung cancer > colon cancer > CNS cancer > melanoma > renal cancer > ovarian cancer > breast cancer, and the authors suggested possible inhibition of TopoIIα based on molecular binding interaction with the active site of the ATPase domain [54]. A limitation of this work was that no standard drug was used to compared against the activity of the synthesized compounds on the nine cancer panels, and an actual assay for TopoIIα was not conducted. Moreover, in China, Li et al. [55] reported that a newly developed acridone derivative, 1-((3-(dimethylamino)propyl)amino)-7-hydroxy-4-nitroacridin-9(10H)-one, could inhibit TopoIIα, intercalated with DNA, and showed significant and long-term antiproliferative activity at relatively high concentrations. Previous studies have identified several acridone derivatives as TopoII inhibitors and DNA intercalators with cell cycle arrest and apoptosis [56,57]. Furthermore, in Egypt, Nemr et al. [58] and Nemr and AboulMagd [59] worked on a novel series of thiazolopyrimidines and fused thiazolopyrimidines, and screened for anticancer activity against 60 human cancer cell lines. They found Ethyl 4-(4-bromophenyl)-2-imino-9-(3,4,5-trimethoxyphenyl)-7-phenyl-1,2-dihydro-9H-pyrimido[4′,5′:4,5]thiazolo[3,2-a]pyrimidine-8-carboxylate and Ethyl 3-(4-chlorophenyl)-5-(4-chlorophenyl)-7-phenyl-5H-thiazolo-[3,2-a] pyrimidine-6-carboxylate hydrobromide to be potent inhibitors against a renal cell line (A-498) and induce cell cycle arrest at the G2/M phase, leading to cell proliferation inhibition and apoptosis, and their fused derivative both showed potent TopoII inhibitory activity, with IC_{50} slightly higher than that of the standard drug, doxorubicin.

Moreover, in Egypt, *2-(1-Ethyl-7-methyl-4-oxo-1,4-dihydro-1,8-naphthyridine-3-carbonyl)-N-(m-tolyl)-hydrazinecarbothioamide,* a derivative of nalidixic acid (**14**), has been shown to be a potent inhibitor of TopoIIα and TopoIIβ and induce cell cycle arrest at the G2-M phase, leading to inhibition of cell proliferation and apoptosis [60]. According to Jiang et al. [61], four compounds from a series of carbazole-rhodanine conjugates were found to possess topoisomerase II inhibitory activity, with potency at 20 μM. However, this study did not use any human cell lines but used plasmid (pBR322 DNA), did not extensively explore biological activities, and failed to differentiate between TopoIIα and TopoIIβ. Shrestha et al. [62] in Korea investigated a series of new benzofuro[3,2-b]pyridin-7-ols derivatives, and their results showed a chemical structure named compound 11, that has *meta*-OH positions in the 2,4-diphenol moieties of benzofuro[3,2-b]pyridin-7-ol ring, as having the most selective and potent TopoII inhibition, with the sturdiest antiproliferative activity in HeLa cell line. Although this work could differentiate between Topo I and TopoII activities, it failed to classify the inhibition of TopoII as whether coming from IIα or IIβ. A collaborative work by Oviatt et al. [63] in Italy and USA reported that etoposide derivatives, in which the C4 sugar moiety was replaced with a variety of polyamine tails, induced higher levels of DNA cleavage with human topoisomerases IIα and IIβ than did the parent drug. Although some of the hybrid compounds showed better cleavage on TopoIIα than etoposide, the interaction of all these derivatives on TopoIIβ showed a greater fold cleavage than etoposide, implicated Gln778, and limited their further clinical usefulness.

In China, Song et al. [64] worked on a novel series of pyrazoline **22**, **23** derivatives, and they reported 8-chloro-3-(1H-indol-3-yl)-2-phenyl-2,3,3a,4-tetrahydrothiochromeno[4,3-c] pyrazole and 6,8-dichloro-3-(1H-indol-3-yl)-2-phenyl-2,3,3a,4-tetrahydrothiochromeno-[4,3-c]pyrazole as having antiproliferative activity in four human cancer cell lines (MGC-803, Hela, MCF-7, and Bel-7404) and a low cytotoxicity in the normal cell line L929 in vitro. Both were non-intercalative Topo II catalytic inhibitors and were able to induce G2/M cell cycle arrest and apoptosis in MGC-803 cells. Moreover, the derivatives of 4,5′-bithiazoles were reported to be acting as catalytic inhibitors of TopoIIα via the competitive inhibition of ATP hydrolysis, and they were able to reduce cell proliferation and stop the cell cycle mainly in the G1 phase [65]. Li et al. [66] in China reported that N-(3-(4-Methylpiperazin-1-yl)propyl)-50-methyl-10H-ursa-2,12-dieno[3,2-b]indol-28-carboxamide, a new indole derivative of ursolic acid, exhibited the most effective activity against two human hepatocarcinoma cell lines (SMMC-7721 and HepG2) and normal hepatocyte cell line (LO2) via a MTT assay. The results showed that the compound significantly inhibited TopoII activity, elevated intracellular ROS levels, decreased mitochondrial membrane potential, and caused apoptosis of SMMC-7721 cells.

Moreover, Legina et al. [67] in Austria used biological assays and molecular dynamic simulations to show that thiomaltol-containing ruthenium (Ru^{II})-, osmium (Os^{II})-, rhodium (Rh^{III})-, and iridium (Ir^{III})-based organometallic complexes bearing 1-methylimidazole or chloride as the leaving group possessed cytotoxic and DNA-damaging activity in human mammary carcinoma cell lines. A study on the anticancer properties of novel Ru^{II}, Os^{II}, Rh^{III}, and Ir^{III} thiomaltol complexes showed that they acted as inhibitors of TopoII catalytic activity and had a significantly higher enzyme inhibitory capacity than the free ligand [68]. In 2014, Bau, Kang and their group [69] reported that salicylate **29** showed selectivity for topo IIα-isoform in DNA cleavage assay, thus acting as a catalytic inhibitor. However, further studies are needed to confirm the basis for its isoform selectivity. In their studies, they reported that salicylate did not intercalate DNA and did not prevent the enzyme from interacting with DNA. Furthermore, salicylate did not stabilize the cleavable "complex".

Ziga et al. [14] reported **30** as a novel ATP-competitive inhibitor of hDNA topo IIα containing pyrrolamide pharmacophore. The compound showed high kinetic ATPase activity (IC_{50} 0.43 μM). In another study, Kamila et al. [70] reported that **32**, a quinolone derivative, is being used in phase I and II clinical trials in combination with azacitidine and infusional cytabarin. Recently, Sisodiya et al. [71] reported the synthesis of benzo-fused carbazolequinone derivatives that contain both indole and quinone moieties that are found in numerous drugs, including natural products. Compound **28** displayed significant apoptotic antiproliferation in cancer cells with cell cycle arrest at the S phase. It also inhibited topoIIα with more efficiency compared to etoposide. The structures of topoisomerase II inhibitors (**4–17** and **18–32**) are shown in Figures 2 and 3.

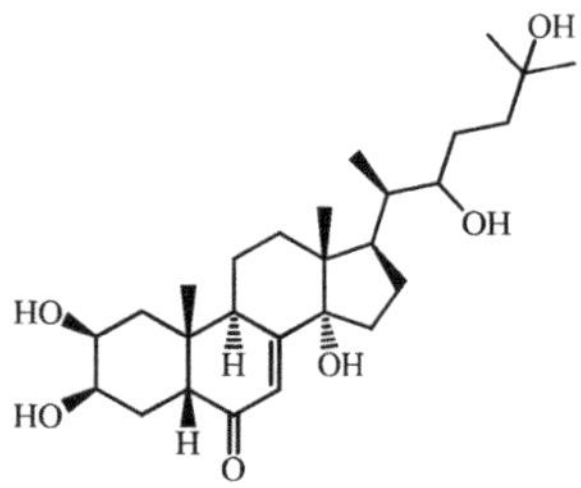

20-beta-ecdysone, **8**

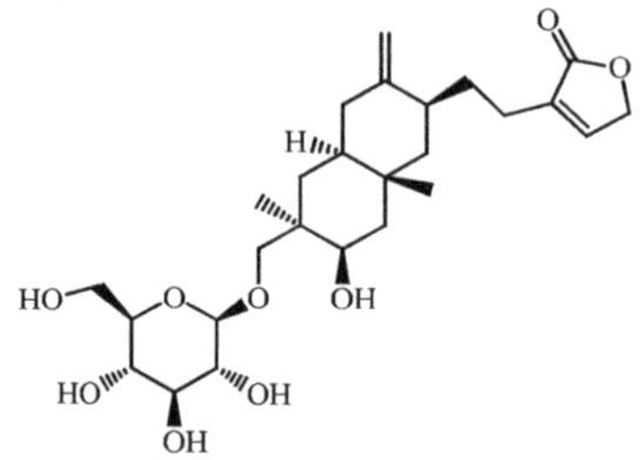

Adropanoside, 9

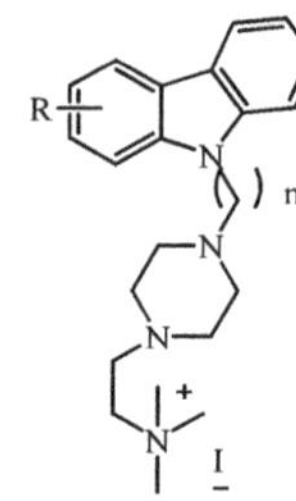

R = H: n = 3,4,5,6,7

R = Cl: n = 5,6

N-Alkylcarbazole, **10**

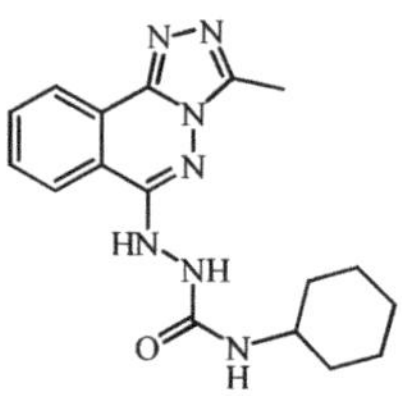

Triazolo-phthalazine derivative, IX_b **11**,

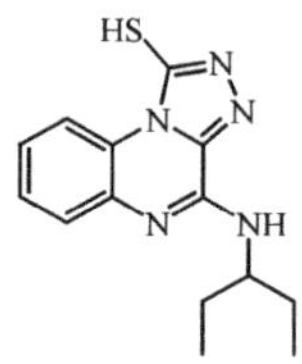

11, topo II = 0.384 μM

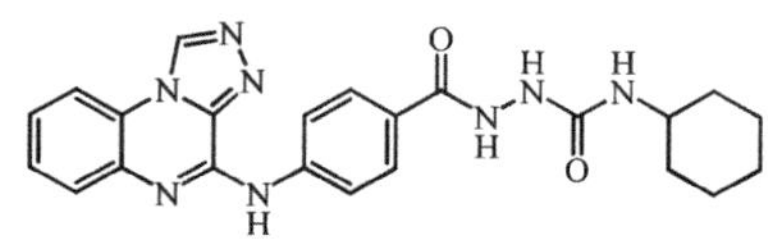

12, topo II = 0.379 μM

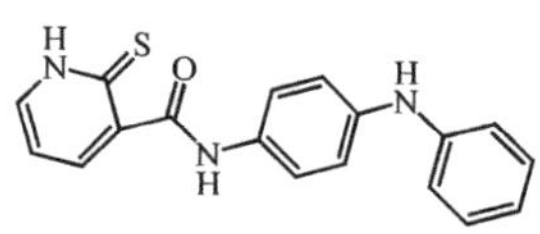

T60, Catalytic Topo 2 inhibitor, **13**

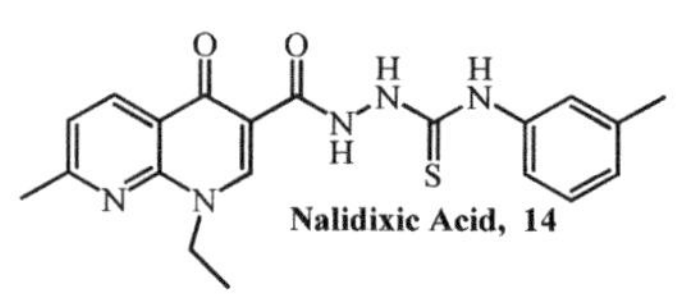

Nalidixic Acid, 14

topo II-alpha IC50 = 1.30 μM

topo II-beta IC50 = 0.017 μM

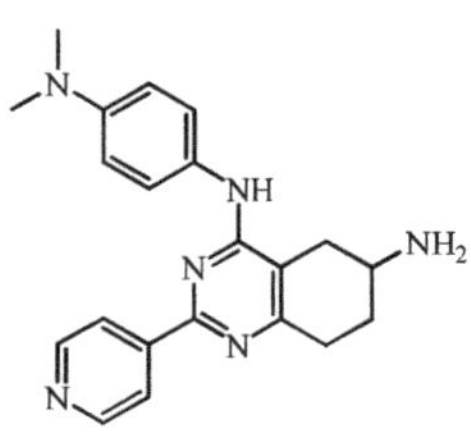

Selective catalytic inhibitor of Topo II, **15**

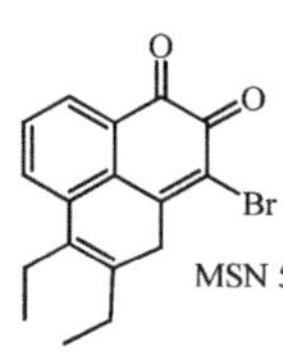

MSN 54

16

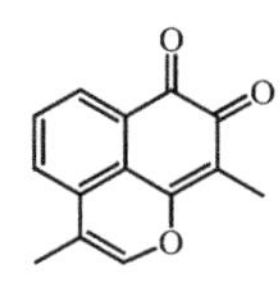

Mansanone F, **17**

Figure 2. New Topoisomerase II Inhibitors: Part 1.

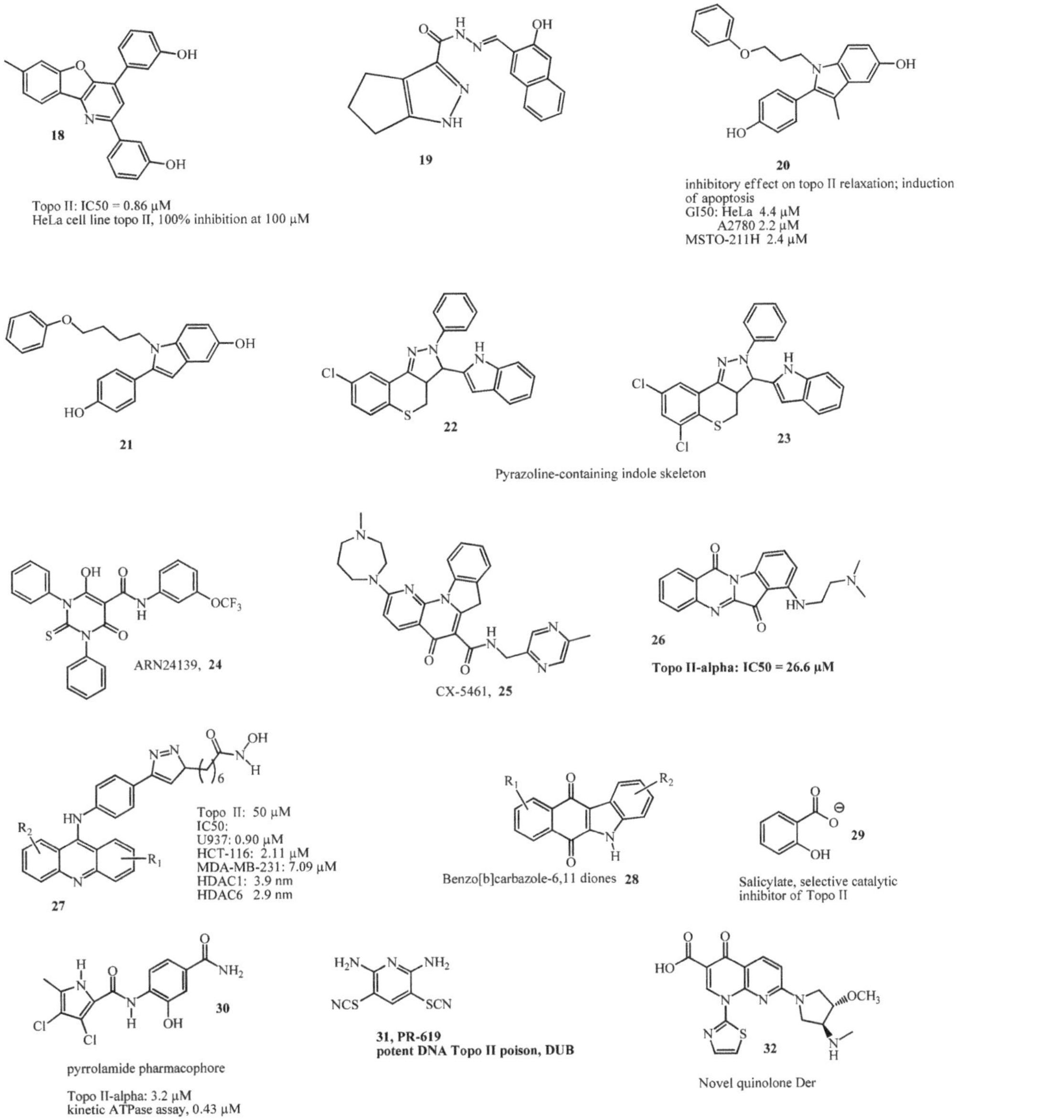

Figure 3. Structures of Topoisomerase Inhibitors: Part 2.

2. Conclusions

In this review we see a significant number of reports of small molecule inhibitors of topoisomerase II. Topoisomerase II poisons continue to dominate the literature despite reports of cardiotoxicity and multi-drug resistance, including secondary malignancy. In the past three years, we have also seen a gradual increase in the number of catalytic inhibitors, which appear more attractive from a safety standpoint. However, no catalytic inhibitor has received FDA approval. The large number of reports of topoisomerase II inhibitors in the recent literature reflect the high level of interest in topoisomerase II inhibitors as therapeutic targets. Several reports have confirmed the structural similarity between topo

IIα and topo IIβ. The two isoforms are similar except in the C-terminus. The above calls for structure-based drug design beyond molecular docking. Docking simulations are prone to inaccuracy because the scoring functions used make estimates of the binding energy. In addition, docking often excludes hydrogens and solvents. On the other hand, molecular dynamic simulation of the drug–protein complex gives a more accurate binding energy since it considers protonation and solvation. Thus, docking and MD simulation would undoubtedly reveal the structural differences that exist at the C-terminus. In addition, structural modification of natural products and hit molecules will continue to be an integral part of the drug discovery of novel topoisomerase II inhibitors.

Funding: Research funding by Title III US department of Education, Tennessee State University.

Informed Consent Statement: Not applicable.

Data Availability Statement: Not applicable.

Conflicts of Interest: The authors declare no conflict of interest.

References

1. Wang, J.C. Cellular roles of DNA topoisomerases: A molecular perspective. *Nat. Rev. Mol. Cell Biol.* **2002**, *3*, 430–440. [CrossRef]
2. Leppard, J.B.; Champoux, J.J. Human DNA topoisomerase I: Relaxation, roles, and damage control. *Chromosoma* **2005**, *114*, 75–85. [CrossRef] [PubMed]
3. Pogorelcnik, B.; Perdih, A.; Solmajer, T. Recent advances in the development of catalytic inhibitors of human DNA topoiso-merase IIα as novel anticancer agents. *Curr. Med. Chem.* **2013**, *20*, 694–709. [CrossRef] [PubMed]
4. Pogorelcnik, B.; Perdih, A.; Solmajer, T. Recent Developments of DNA Poisons-Human DNA Topoisomerase IIα Inhibitors-as Anticancer Agents. *Curr. Pharm. Des.* **2013**, *19*, 2474–2488. [CrossRef] [PubMed]
5. Baldwin, E.L.; Osheroff, N. Etoposide, topoisomerase II and cancer. *Curr. Med. Chem. Anti Cancer Agents* **2005**, *5*, 363–372. [CrossRef] [PubMed]
6. Lee, C.J.; Kang, J.S.; Kim, M.S.; Lee, K.P.; Lee, M.S. The study of doxorubicin and its complex with DNA by SERS and UV-resonance Raman spectroscopy. *Bull Korean Chem. Soc.* **2004**, *25*, 1211–1216.
7. Bodley, A.; Liu, L.F.; Israel, M.; Seshadri, R.; Koseki, Y.; Giuliani, F.C.; Kirschenbaum, S.; Silber, R.; Potmesil, M. DNA topoisomerase II-mediated interaction of doxorubicin and daunorubicin congeners with DNA. *Cancer Res.* **1989**, *49*, 5969–5987.
8. Elmore, R.H.; Wadkins, R.M.; Graves, D.E. Cooperative binding of m-AMSA to nucleic acids. *Nucleic Acids Res.* **1988**, *16*, 9707–9719. [CrossRef]
9. Ketron, A.C.; Dennyl, W.A.; Graves, D.E.; Osheroff, N. Amsacrine as a Topo II Poison: Importance of Drug-DNA Interaction. *Biochemistry* **2012**, *51*, 1730–1739. [CrossRef]
10. Wendorff, T.J.; Schmidt, B.H.; Heslop, P.; Austin, C.A.; Berger, J.M. The structure of DNA-bound human topoisomerase IIα: Conformational mechanisms for coordinating inter-subunit interactions with DNA cleavage. *J. Mol. Biol.* **2012**, *424*, 109–124. [CrossRef]
11. Bollimpelli, V.S.; Dholaniya, P.S.; Kondapi, A.K. Topoisomerase IIβ and its role in different biological contexts. *Arch. Biochem. Biophys.* **2017**, *633*, 78–84. [CrossRef] [PubMed]
12. Ertan-Bolelli, T.; Bolelli, K. Discovery of New DNA Topoisomerase II Inhibitors using Structure Based Virtual Screening Method. *J. Turk. Chem. Soc.* **2019**, *6*, 71–78. [CrossRef]
13. Adeniran, O.Y.; Metibemu, A.O.; Boboye, S.O. Virtual high-throughput screening (VHTS), three-dimensional quantitative structureactivity and relationship (3D-QSAR) and molecular docking studies of novel phytoinhibtors of topoisomerase II alpha. *GSC Biol. Pharm. Sci.* **2021**, *15*, 072–082. [CrossRef]
14. Skok, Ž.; Durcik, M.; Skledar, D.G.; Barančoková, M.; Mašič, L.P.; Tomašič, T.; Zega, A.; Kikelj, D.; Zidar, N.; Ilaš, J. Discovery of new ATP-competitive inhibitors of human DNA topoisomerase IIα through screening of bacterial topoisomerase inhibitors. *Bioorg. Chem.* **2020**, *102*, 104049. [CrossRef] [PubMed]
15. Knölker, H.-J.; Reddy, K.R. Isolation and Synthesis of Biologically Active Carbazole Alkaloids. *Chem. Rev.* **2002**, *102*, 4303–4428. [CrossRef]
16. Kizek, R.; Adam, V.; Hrabeta, J.; Eckschlager, T.; Smutny, S.; Burda, J.V.; Frei, E.; Stiborova, M. Anthracyclines and ellipticines as DNA-damaging anticancer drugs: Recent advances. *Pharmacol. Ther.* **2012**, *133*, 26–39. [CrossRef]
17. Stefania Sinicropi, M.; Lappano, R.; Caruso, A.; Francesca Santolla, M.; Pisano, A.; Rosano, C.; Capasso, A.; Panno, A.; Charles Lancelot, J.; Rault, S.; et al. (6-bromo-1, 4-dimethyl-9H-carbazol-3-yl-methylene)-hydrazine (carbhydraz) acts as a GPER agonist in breast cancer cells. *Curr. Top. Med. Chem.* **2015**, *15*, 1035–1042. [CrossRef]
18. Saturnino, C.; Caruso, A.; Iacopetta, D.; Rosano, C.; Ceramella, J.; Muià, N.; Mariconda, A.; Bonomo, M.G.; Ponassi, M.; Rosace, G.; et al. Inhibition of Human Topoisomerase II by N,N,N-Trimethylethanammonium Iodide Alkylcarbazole Derivatives. *Chemmedchem* **2018**, *13*, 2635–2643. [CrossRef]

19. Catanzaro, E.; Betari, N.; Arencibia, J.M.; Montanari, S.; Sissi, C.; De Simone, A.; Vassura, I.; Santini, A.; Andrisano, V.; Tumiatti, V.; et al. Targeting topoisomerase II with trypthantrin derivatives: Discovery of 7-((2-(dimethylamino)ethyl)amino)indolo [2,1-b]quinazoline-6,12-dione as an antiproliferative agent and to treat cancer. *Eur. J. Med. Chem.* **2020**, *202*, 112504. [CrossRef]
20. Tosa, H.; Iinuma, M.; Tanaka, T.; Nozaki, H.; Ikeda, S.; Tsutsui, K.; Tsutsui, K.; Yamada, M.; Fujimori, S. Inhibitory activity of xanthone derivatives isolated from some guttiferaeous plants against DNA toposiomerases I and II. *Chem. Pharm. Bull.* **1997**, *45*, 418–420. [CrossRef]
21. Mai, Y.-W.; Liang, C.-C.; Ou, J.-B.; Xie, H.-T.; Chen, S.-B.; Zhou, D.-C.; Yao, P.-F.; Huang, Z.-S.; Wang, H.; Huang, S.-L. 9-Bromo-2,3-diethylbenzo[de]chromene-7,8-dione (MSN54): A novel non-intercalative topoisomerase II catalytic inhibitor. *Bioorg. Chem.* **2021**, *114*, 105097. [CrossRef] [PubMed]
22. Xie, H.T.; Zhou, D.C.; Mai, Y.W.; Huo, L.; Yao, P.F.; Huang, S.L.; Wang, H.G.; Huang, Z.S.; Gu, L.Q. Construction of the oxaphenalene skeletons of mansonone F derivatives through C–H bond functionalization and their evaluation for anti-proliferative activities. *RSC Adv.* **2017**, *7*, 20919–20928. [CrossRef]
23. Wu, W.-B.; Ou, J.-B.; Huang, Z.-H.; Chen, S.-B.; Ou, T.-M.; Tan, J.-H.; Li, D.; Shen, L.-L.; Huang, S.-L.; Gu, L.-Q.; et al. Synthesis and evaluation of mansonone F derivatives as topoisomerase inhibitors. *Eur. J. Med. Chem.* **2011**, *46*, 3339–3347. [CrossRef]
24. Di Micco, S.; Masullo, M.; Bandak, A.F.; Berger, J.M.; Riccio, R.; Piacente, S.; Bifulco, G. Garcinol and Related Polyisoprenylated Benzophenones as Topoisomerase II Inhibitors: Biochemical and Molecular Modeling Studies. *J. Nat. Prod.* **2019**, *82*, 2768–2779. [CrossRef]
25. Zidar, N.; Secci, D.; Tomašič, T.; Mašič, L.P.; Kikelj, D.; Passarella, D.; Argaez, A.N.G.; Hyeraci, M.; Via, L.D. Synthesis, Antiproliferative Effect, and Topoisomerase II Inhibitory Activity of 3-Methyl-2-phenyl-1H-indoles. *ACS Med. Chem. Lett.* **2020**, *11*, 691–697. [CrossRef]
26. Zhou, D.-C.; Lu, Y.-T.; Mai, Y.-W.; Zhang, C.; Xia, J.; Yao, P.-F.; Wang, H.-G.; Huang, S.-L.; Huang, Z.-S. Design, synthesis and biological evaluation of novel perimidine o-quinone derivatives as non-intercalative topoisomerase II catalytic inhibitors. *Bioorg. Chem.* **2019**, *91*, 103131. [CrossRef]
27. Sakr, H.; Ayyad, R.R.; El-Helby, A.A.; Khalifa, M.M.; Mahdy, H.A. Discovery of novel triazolophthalazine derivatives as DNA intercalators and topoisomerase II inhibitors. *Arch. Pharm.* **2021**, *354*, 2000456. [CrossRef]
28. Pons, M.; Campayo, L.; Martinez-Balbás, M.; Azorín, F.; Navarro, P.; Giralt, E. A new ionizable chromophore of 1,4-bis (alkylamino) benzo [g] phthalazine which interacts with DNA by intercalation. *J. Med. Chem.* **1991**, *34*, 82–86. [CrossRef]
29. Arencibia, J.M.; Brindani, N.; Franco-Ulloa, S.; Nigro, M.; Kuriappan, J.A.; Ottonello, G.; Bertozzi, S.M.; Summa, M.; Girotto, S.; Bertorelli, R.; et al. Design, Synthesis, Dynamic Docking, Biochemical Characterization, and *In Vivo* Pharmacokinetics Studies of Novel Topoisomerase II Poisons with Promising Antiproliferative Activity. *J. Med. Chem.* **2020**, *63*, 3508–3521. [CrossRef]
30. Ibrahim, M.K.; Taghour, M.S.; Metwaly, A.M.; Belal, A.; Mehany, A.B.M.; Elhendawy, M.A.; Radwan, M.M.; Yassin, A.M.; El-Deeb, N.M.; Hafez, E.E.; et al. Design, synthesis, molecular modeling and anti-proliferative evaluation of novel quinoxaline derivatives as potential DNA intercalators and topoisomerase II inhibitors. *Eur. J. Med. Chem.* **2018**, *155*, 117–134. [CrossRef]
31. El-Adl, K.; El-Helby, A.A.; Sakr, H.; Elwan, A. Design, synthesis, molecular docking and anti-proliferative evaluations of [1,2,4] triazolo [4,3-a] quinoline derivatives as DNA intercalators and Topoisomerase II inhibitors. *Bioorg. Chem.* **2020**, *105*, 104399. [CrossRef] [PubMed]
32. Oyallon, B.; Brachet-Botineau, M.; Logé, C.; Bonnet, P.; Souab, M.; Robert, T.; Ruchaud, S.; Bach, S.; Berthelot, P.; Gouilleux, F.; et al. Structure-based design of novel quinoxaline-2-carboxylic acids and analogues as Pim-1 inhibitors. *Eur. J. Med. Chem.* **2018**, *154*, 101–109. [CrossRef] [PubMed]
33. Park, Y.-S.; Shin, W.-S.; Kim, C.-S.; Ahn, C.M.; Qi, X.-F.; Kim, S.-K. Molecular and cellular toxicological profiling of DNA bis-intercalator, quinoxaline compounds: Echinomycin as the versatile lead. *Mol. Cell. Toxicol.* **2018**, *14*, 9–18. [CrossRef]
34. Eissa, I.H.; El-Naggar, A.M.; Abd El-Sattar, N.E.; Youssef, A.S. Design and Discovery of Novel Quinoxaline Derivatives as Dual DNA Intercalators and Topoisomerase II Inhibitors. *Anti Cancer Agents Med. Chem.* **2018**, *18*, 195–209. [CrossRef]
35. Abbass, E.M.; Khalil, A.K.; Mohamed, M.M.; Eissa, I.H.; El-Naggar, A.M. Design, efficient synthesis, docking studies, and anticancer evaluation of new quinoxalines as potential intercalative Topo II inhibitors and apoptosis inducers. *Bioorg. Chem.* **2020**, *104*, 104255. [CrossRef] [PubMed]
36. Bruno, P.M.; Lu, M.; Dennis, K.A.; Inam, H.; Moore, C.J.; Sheehe, J.; Elledge, S.J.; Hemann, M.T.; Pritchard, J.R. The primary mechanism of cytotoxicity of the chemotherapeutic agent CX-5461 is topoisomerase II poisoning. *Proc. Natl. Acad. Sci. USA* **2020**, *117*, 4053–4060. [CrossRef]
37. Lin, A.; Giuliano, C.J.; Palladino, A.; John, K.M.; Abramowicz, C.; Yuan, M.L.; Sausville, E.L.; Lukow, D.A.; Liu, L.; Chait, A.R.; et al. Off-target toxicity is a common mechanism of action of cancer drugs undergoing clinical trials. *Sci. Transl. Med.* **2019**, *11*, eaaw8412. [CrossRef] [PubMed]
38. Haddach, M.; Schwaebe, M.K.; Michaux, J.; Nagasawa, J.; O'Brien, S.E.; Whitten, J.P.; Pierre, F.; Kerdoncuff, P.; Darjania, L.; Stansfield, R.; et al. Discovery of CX-5461, the First Direct and Selective Inhibitor of RNA Polymerase I, for Cancer Therapeutics. *ACS Med. Chem. Lett.* **2012**, *3*, 602–606. [CrossRef]
39. Ray, S.; Panova, T.; Miller, G.; Volkov, A.; Porter, A.C.; Russell, J.; Panov, K.I.; Zomerdijk, J.C. Topoisomerase IIα promotes activation of RNA polymerase I transcription by facilitating pre-initiation complex formation. *Nat. Commun.* **2013**, *4*, 1598. [CrossRef]

40. Xu, H.; Di Antonio, M.; McKinney, S.; Mathew, V.; Ho, B.; O'Neil, N.J.; Dos Santos, N.; Silvester, J.; Wei, V.; Garcia, J.; et al. CX-5461 is a DNA G-quadruplex stabilizer with selective lethality in BRCA1/2 deficient tumours. *Nat. Commun.* **2017**, *8*, 14432. [CrossRef]
41. Morotomi-Yano, K.; Yano, K.I. Nucleolar translocation of human DNA topoisomerase II by ATP depletion and its disruption by the RNA polymerase I inhibitor BMH-21. *Sci. Rep.* **2021**, *11*, 21533. [CrossRef] [PubMed]
42. Cowell, I.G.; Ling, E.M.; Swan, R.L.; Brooks, M.L.; Austin, C.A. The Deubiquitinating Enzyme Inhibitor PR-619 is a Potent DNA Topoisomerase II Poison. *Mol. Pharmacol.* **2019**, *96*, 562–572. [CrossRef] [PubMed]
43. Altun, M.; Kramer, H.B.; Willems, L.I.; McDermott, J.L.; Leach, C.A.; Goldenberg, S.J.; Kumar, K.G.S.; Konietzny, R.; Fischer, R.; Kogan, E.; et al. Activity-Based Chemical Proteomics Accelerates Inhibitor Development for Deubiquitylating Enzymes. *Chem. Biol.* **2011**, *18*, 1401–1412. [CrossRef] [PubMed]
44. Jeon, K.-H.; Park, S.; Jang, H.J.; Hwang, S.-Y.; Shrestha, A.; Lee, E.-S.; Kwon, Y. AK-I-190, a New Catalytic Inhibitor of Topoisomerase II with Anti-Proliferative and Pro-Apoptotic Activity on Androgen-Negative Prostate Cancer Cells. *Int. J. Mol. Sci.* **2021**, *22*, 11246. [CrossRef] [PubMed]
45. Kirby, M.; Hirst, C.; Crawford, E.D. Characterising the castration-resistant prostate cancer population: A systematic review. *Int. J. Clin. Pract.* **2011**, *65*, 1180–1192. [CrossRef] [PubMed]
46. Matias-Barrios, V.M.; Radaeva, M.; Song, Y.; Alperstein, Z.; Lee, A.R.; Schmitt, V.; Lee, J.; Ban, F.; Xie, N.; Qi, J.; et al. Discovery of New Catalytic Topoisomerase II Inhibitors for Anticancer Therapeutics. *Front. Oncol.* **2021**, *10*, 633142. [CrossRef]
47. Matias-Barrios, V.M.; Radaeva, M.; Ho, C.-H.; Lee, J.; Adomat, H.; Lallous, N.; Cherkasov, A.; Dong, X. Optimization of New Catalytic Topoisomerase II Inhibitors as an Anti-Cancer Therapy. *Cancers* **2021**, *13*, 3675. [CrossRef]
48. Lee, M.-G.; Liu, Y.-C.; Lee, Y.-L.; El-Shazly, M.; Lai, K.-H.; Shih, S.-P.; Ke, S.-C.; Hong, M.-C.; Du, Y.-C.; Yang, J.-C.; et al. Heteronemin, a Marine Sesterterpenoid-Type Metabolite, Induces Apoptosis in Prostate LNcap Cells via Oxidative and ER Stress Combined with the Inhibition of Topoisomerase II and Hsp90. *Mar. Drugs* **2018**, *16*, 204. [CrossRef]
49. Schumacher, M.; Cerella, C.; Eifes, S.; Chateauvieux, S.; Morceau, F.; Jaspars, M.; Dicato, M.; Diederich, M. Heteronemin, a spongean sesterterpene, inhibits TNF α-induced NF-κB activation through proteasome inhibition and induces apoptotic cell death. *Biochem. Pharmacol.* **2010**, *79*, 610–622. [CrossRef]
50. Ortega, J.A.; Arencibia, J.M.; Minniti, E.; Byl, J.A.W.; Franco-Ulloa, S.; Borgogno, M.; Genna, V.; Summa, M.; Bertozzi, S.M.; Bertorelli, R.; et al. Novel, Potent, and Druglike Tetrahydroquinazoline Inhibitor That Is Highly Selective for Human Topoisomerase II α over β. *J. Med. Chem.* **2020**, *63*, 12873–12886. [CrossRef]
51. Chen, J.; Li, D.; Li, W.; Yin, J.; Zhang, Y.; Yuan, Z.; Gao, C.; Liu, F.; Jiang, Y. Design, synthesis and anticancer evaluation of acridine hydroxamic acid derivatives as dual Topo and HDAC inhibitors. *Bioorg. Med. Chem.* **2018**, *26*, 3958–3966. [CrossRef] [PubMed]
52. Sarcar, B.; Kahali, S.; Chinnaiyan, P. Vorinostat enhances the cytotoxic effects of the topoisomerase I inhibitor SN38 in glioblastoma cell lines. *J. Neuro Oncol.* **2010**, *99*, 201–207. [CrossRef] [PubMed]
53. Gray, J.; Cubitt, C.L.; Zhang, S.; Chiappori, A. Combination of HDAC and topoisomerase inhibitors in small cell lung cancer. *Cancer Biol. Ther.* **2012**, *13*, 614–622. [CrossRef] [PubMed]
54. Oyedele, A.S.; Bogan, D.N.; Okoro, C.O. Synthesis, biological evaluation and virtual screening of some acridone derivatives as potential anticancer agents. *Bioorg. Med. Chem.* **2020**, *28*, 115426. [CrossRef] [PubMed]
55. Li, Z.-Y.; Xu, G.-S.; Li, X. A Unique Topoisomerase II Inhibitor with Dose-Affected Anticancer Mechanisms and Less Cardiotoxicity. *Cells* **2021**, *10*, 3138. [CrossRef] [PubMed]
56. Zhang, W.; Zhang, B.; Yang, T.; Wang, N.; Gao, C.; Tan, C.; Liu, H.; Jiang, Y. Synthesis and antiproliferative activity of 9-benzylamino-6-chloro-2-methoxy-acridine derivatives as potent DNA-binding ligands and topoisomerase II inhibitors. *Eur. J. Med. Chem.* **2016**, *116*, 59–70. [CrossRef] [PubMed]
57. Prasher, P.; Sharma, M. Medicinal chemistry of acridine and its analogues. *MedChemComm* **2018**, *9*, 1589–1618. [CrossRef]
58. Nemr, M.T.M.; Sonousi, A.; Marzouk, A.A. Design, synthesis and antiproliferative evaluation of new tricyclic fused thiazolopyrimidines targeting topoisomerase II: Molecular docking and apoptosis inducing activity. *Bioorg. Chem.* **2020**, *105*, 104446. [CrossRef]
59. Nemr, M.T.M.; AboulMagd, A.M. New fused pyrimidine derivatives with anticancer activity: Synthesis, topoisomerase II inhibition, apoptotic inducing activity and molecular modeling study. *Bioorg. Chem.* **2020**, *103*, 104134. [CrossRef]
60. Khalila, O.M.; Gedawy, E.M.; El-Malaha, A.A.; Adly, M.E. Novel nalidixic acid derivatives targeting topoisomerase II enzyme; Design, synthesis, anticancer activity and effect on cell cycle profile. *Bioorg. Chem.* **2019**, *83*, 262–276. [CrossRef]
61. Jiang, H.; Zhang, W.-J.; Li, P.-H.; Wang, J.; Dong, C.-Z.; Zhang, K.; Chen, H.-X.; Du, Z.-Y. Synthesis and biological evaluation of novel carbazole-rhodanine conjugates as topoisomerase II inhibitors. *Bioorg. Med. Chem. Lett.* **2018**, *28*, 1320–1323. [CrossRef] [PubMed]
62. Shrestha, A.; Jo, H.; Kwon, Y.; Lee, E.-S. Design, synthesis, and structure-activity relationships of new benzofuro [3,2-b] pyridine-7-ols as DNA topoisomerase II inhibitors. *Bioorg. Med. Chem. Lett.* **2018**, *28*, 566–571. [CrossRef] [PubMed]
63. Oviatt, A.A.; Kuriappan, J.A.; Minniti, E.; Vann, K.R.; Onuorah, P.; Minarini, A.; De Vivo, M.; Osheroff, N. Polyamine-containing etoposide derivatives as poisons of human type II toposiomerases: Differential effects on topoisomerase IIα and IIβ. *Bioorg. Med. Chem. Lett.* **2018**, *28*, 2961–2968. [CrossRef] [PubMed]
64. Song, Y.; Feng, S.; Feng, J.; Dong, J.; Yang, K.; Liu, Z.; Qiao, X. Synthesis and biological evaluation of novel pyrazoline derivatives containing indole skeleton as anti-cancer agents targeting topoisomerase II. *Eur. J. Med. Chem.* **2020**, *200*, 112459. [CrossRef] [PubMed]

65. Bergant Loboda, K.; Janežič, M.; Štampar, M.; Žegura, B.; Filipic, M.; Perdih, A. Substituted 4,5′-Bithiazoles as Catalytic Inhibitors of Human DNA Topoisomerase IIα. *J. Chem. Inf. Model.* **2020**, *60*, 3662–3678. [CrossRef]
66. Li, A.-L.; Hao, Y.; Wang, W.-Y.; Liu, Q.-S.; Sun, Y.; Gu, W. Design, Synthesis, and Anticancer Evaluation of Novel Indole Derivatives of Ursolic Acid as Potential Topoisomerase II Inhibitors. *Int. J. Mol. Sci.* **2020**, *21*, 2876. [CrossRef]
67. Legina, M.S.; Nogueira, J.J.; Kandioller, W.; Jakupec, M.A.; González, L.; Keppler, B.K. Biological evaluation of novel thiomaltol-based organometallic complexes as topoisomerase IIα inhibitors. *J. Biol. Inorg. Chem.* **2020**, *25*, 451–465. [CrossRef]
68. Hackl, C.M.; Legina, M.S.; Pichler, V.; Schmidlehner, M.; Roller, A.; Dömötör, O.; Enyedi, E.A.; Jakupec, M.A.; Kandioller, W.; Keppler, B.K. Thiomaltol-based organometallic complexes with 1 methylimidazole as leaving group: Synthesis, stability, and biological behavior. *Chem. Eur. J.* **2016**, *22*, 17269–17281. [CrossRef]
69. Bau, J.T.; Kang, Z.; Austin, C.A.; Kurz, E.U. Salicylate, a catalytic Inhibitor of Topoisomerase II, Inhibits DNA Cleavage and Is Selective for the α-isoform. *Mol. Pharmacol.* **2014**, *85*, 198–207. [CrossRef]
70. Kamila, B.; Anna, B.; Krzysztof, B.; Agnieszka, G. DNA topoisomerases as molecular targets for anticancer drugs. *J. Enzym. Inhib. Med. Chem.* **2020**, *35*, 1781–1799.
71. Sisodiya, S.; Paul, S.; Chaudhary, H.; Grewal, P.; Kumar, G.; Daniel, D.P.; Das, B.; Nayak, D.; Guchhait, S.K.; Kundu, C.N.; et al. Exploration of Benzo [b] carbazole-6,11-diones as anticancer agents: Synthesis and studies of hTopoIIα inhibition and apoptotic effects. *Bioorg. Med. Chem. Lett.* **2021**, *49*, 128274. [CrossRef] [PubMed]

International Journal of *Molecular Sciences*

MDPI

Article

A 2.8 Å Structure of Zoliflodacin in a DNA Cleavage Complex with *Staphylococcus aureus* DNA Gyrase

Harry Morgan [1], Magdalena Lipka-Lloyd [1], Anna J. Warren [2], Naomi Hughes [3], John Holmes [4], Nicolas P. Burton [4], Eshwar Mahenthiralingam [3] and Ben D. Bax [1,*]

[1] Medicines Discovery Institute, Cardiff University, Cardiff CF10 3AT, UK
[2] Diamond Light Source, Harwell Science and Innovation Campus, Didcot OX11 0DE, UK
[3] Cardiff School of Biosciences, Cardiff University, Cardiff CF10 3AX, UK
[4] Inspiralis Limited, Norwich Research Park Innovation Centre, Norwich NR4 7GJ, UK
* Correspondence: baxb@cardiff.ac.uk

Abstract: Since 2000, some thirteen quinolones and fluoroquinolones have been developed and have come to market. The quinolones, one of the most successful classes of antibacterial drugs, stabilize DNA cleavage complexes with DNA gyrase and topoisomerase IV (topo IV), the two bacterial type IIA topoisomerases. The dual targeting of gyrase and topo IV helps decrease the likelihood of resistance developing. Here, we report on a 2.8 Å X-ray crystal structure, which shows that zoliflodacin, a spiropyrimidinetrione antibiotic, binds in the same DNA cleavage site(s) as quinolones, sterically blocking DNA religation. The structure shows that zoliflodacin interacts with highly conserved residues on GyrB (and does not use the quinolone water–metal ion bridge to GyrA), suggesting it may be more difficult for bacteria to develop target mediated resistance. We show that zoliflodacin has an MIC of 4 μg/mL against *Acinetobacter baumannii* (*A. baumannii*), an improvement of four-fold over its progenitor QPT-1. The current phase III clinical trial of zoliflodacin for gonorrhea is due to be read out in 2023. Zoliflodacin, together with the unrelated novel bacterial topoisomerase inhibitor gepotidacin, is likely to become the first entirely novel chemical entities approved against Gram-negative bacteria in the 21st century. Zoliflodacin may also become the progenitor of a new safer class of antibacterial drugs against other problematic Gram-negative bacteria.

Keywords: zoliflodacin; quinolones; DNA gyrase; topoisomerase IV; ESKAPE; antibiotic; spiropyrimidinetrione; NBTI; gepotidacin; structure

Citation: Morgan, H.; Lipka-Lloyd, M.; Warren, A.J.; Hughes, N.; Holmes, J.; Burton, N.P.; Mahenthiralingam, E.; Bax, B.D. A 2.8 Å Structure of Zoliflodacin in a DNA Cleavage Complex with *Staphylococcus aureus* DNA Gyrase. *Int. J. Mol. Sci.* **2023**, *24*, 1634. https://doi.org/10.3390/ijms24021634

Academic Editor: Joseph E. Deweese

Received: 22 November 2022
Revised: 4 January 2023
Accepted: 7 January 2023
Published: 13 January 2023

1. Introduction

Zoliflodacin is an oral spiropyrimidinetrione antibiotic currently in a phase III clinical trial for the treatment of gonorrhea, a sexually transmitted infection (STI) caused by the Gram-negative bacteria *Neisseria gonorrhoeae* (*N. gonorrhoeae*) [1–4]. Zoliflodacin (Figure 1a) was developed from QPT-1 (or PNU-286607), a compound discovered in Pharmacia via whole-cell screening against Gram-negative (and Gram-positive) bacteria [5]. QPT-1 (Figure 1b), discovered for its antibacterial whole-cell activity, was found to inhibit the bacterial type IIA topoisomerases, *Escherichia coli* (*E. coli*) DNA gyrase (IC_{50} 9 μM) and *E. coli* topo IV (IC_{50} 30 μM) [5]. This method of discovery is reminiscent of the discovery of quinolone and fluoroquinolone antibiotics, which were also initially discovered for whole-cell activity and then found to be inhibitors of the bacterial type IIA topoisomerases [6].

Topoisomerases are essential enzymes needed to relieve topological problems when the DNA double helix is unwound for both DNA replication and transcription [7]. Topoisomerases are divided into type I topoisomerases, which introduce single-stranded DNA breaks to modify the DNA topology, and type II topoisomerases, which modify the topology by introducing double-stranded DNA breaks [7–9]. Most bacteria possess two type IIA topoisomerases, DNA gyrase and topo IV. While DNA gyrase can uniquely introduce negative supercoils into DNA, topo IV has good decatenase activity [9,10]. A mechanism for topological changes introduced by DNA gyrase is shown in Figure 1.

The introduction of double-stranded breaks into DNA is potentially hazardous for the cell, and the stabilization of DNA cleavage complexes by quinolones is often bactericidal [11,12].

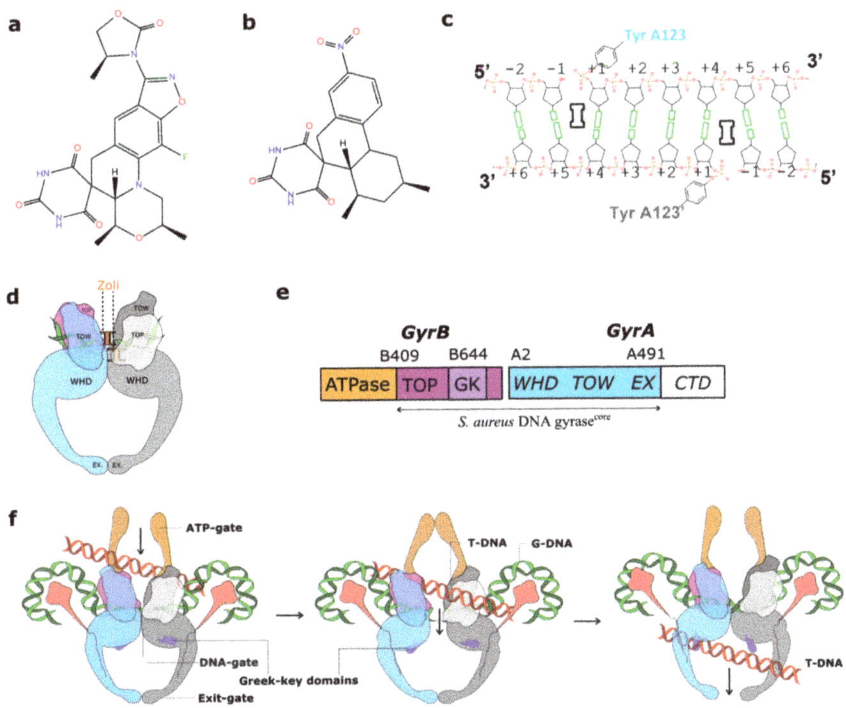

Figure 1. Zoliflodacin, QPT-1, and *Staphylococcus aureus (S. aureus)* DNA gyrase DNA cleavage complexes. (**a**) Chemical structure of zoliflodacin (oxygens shown in red, nitrogens in blue, fluorine in green) and (**b**) chemical structure of QPT-1. (**c**) A schematic of the central eight base-pairs of DNA, with two inhibitors (I) binding in the cleaved DNA and inhibiting DNA religation. Note that DNA cleavage takes place between the −1 and +1 nucleotides on both the Watson and Crick strands. (**d**) A schematic of the DNA cleavage complex with two zoliflodacins of *S. aureus* DNA gyrase and DNA presented in this paper. (**e**) The *S. aureus* DNA gyrase^{CORE} construct consists of residues B409 to B644 from GyrB, fused to A2 to A491 from GyrA. The small Greek key (GK) domain has been deleted from GyrB [13]. (**f**) A simplified schematic of DNA gyrase, in which a G-DNA duplex (green) is cleaved by the enzyme, and another DNA duplex (known as the T or transported DNA, red) is moved through the enzyme. The Greek key domains are not involved in cleaving the gate (or G-) DNA segment [10,13]. The C-terminal domains (CTD) are shown in pink (approximate positions as in full-length *E. coli* structures [14]).

The proposal that Gram-negative bacteria evolved a second cell wall to protect them from antibiotics produced by other micro-organisms [15] may partly explain the failure of new classes of antibiotics targeting Gram-negative bacteria to date in the 21st century [16–18]. Perhaps for Gram-negative bacteria the hardest task is to get antibiotics into the cells, and a whole-cell screening approach followed by the target identification of proven targets is more likely to be successful [19,20]. Indeed, GlaxoSmithKline discovered and developed the NBTI gepotidacin, another new class of DNA-gyrase-targeting antibiotics currently in phase III clinical trials [21], from a hit compound active in a screen for whole-cell antibacterial activity [13]. The chemical diversity of NBTIs such as gepotidacin, which stabilize single-stranded DNA cleavage complexes with bacterial type IIA topoisomerases, suggested that this class of compounds could not have a chemistry-based name [13,22–27]. The name NBTI, although originally a pneumonic for novel bacterial topoisomerase inhibitor [13], could also be taken to stand for non-DNA cleavage pocket binding on the two-fold axis inhibitor (as this describes the binding mode of the chemically diverse NBTIs [13,22–27]).

The occurrence of antimicrobial resistance in hospital-acquired ESKAPE pathogens (*Enterococcus faecium*, *Staphylococcus aureus*, *Klebsiella pneumoniae*, *Acinetobacter baumannii*, *Pseudomonas aeruginosa*, and *Enterobacter* species) was a major cause for concern in 2009 [28]. New classes of antibiotics have now been developed for Gram-positive bacteria, such as the tiacumicin Fidaxomicin for *Clostridioides difficile* [29]. However, Gram-negative bacteria (the KAPE in ESKAPE) remain a major cause for concern. The popular quinolone and fluoroquinolone antibacterial agents were discovered over sixty years ago from a whole-cell screening approach against Gram-negative bacteria [6]. Since then, the field of chemistry has expanded the quinolone activity to include such agents as delafloxacin, approved in 2017 for treating acute bacterial skin infections caused by the Gram-positive *S. aureus*. Some thirteen out of thirty-eight new antibiotics introduced between 2000 and 2019 were quinolones [16–18,30]. However, safety concerns about quinolone side effects have prompted regulatory recommendations to limit the use of quinolones to patients who do not have other treatment options in both Europe (https://www.ema.europa.eu/en/medicines/human/referrals/quinolone-fluoroquinolone-containing-medicinal-products, accessed on 4 January 2023) and the USA (https://www.fda.gov/drugs/drug-safety-and-availability/fda-drug-safety-communication-fda-advises-restricting-fluoroquinolone-antibiotic-use-certain, accessed on 4 January 2023).

The determination of the specific DNA sequences cleaved by DNA gyrase or topo IV [31,32] was important in determining the structures of quinolones in DNA cleavage complexes. In particular, structural studies showed that quinolone antibiotics stabilize double-stranded DNA cleavage complexes with the two bacterial topoisomerases, topo IV [33] and DNA gyrase [34,35], by interacting with ParC or GyrA via a water–metal ion bridge [12,33]. Although the original DNA sequences used in two papers describing structures showing the water–metal ion bridge [33,35] were defined in 2005 [32] and were initially used in structures with *S. pneumoniae* topo IV [36,37], they are asymmetric, and in high enough resolution structures the DNA was clearly averaged around the two-fold axis of the complex [35]. In this paper we used a two-fold symmetric 20-mer DNA duplex to avoid such problems [34,38]. This 20-mer homoduplex DNA was previously used in determining structures with the progenitor of zoliflodacin, QPT-1 [34].

Herein, we describe a 2.8 Å X-ray crystal structure of zoliflodacin in a DNA cleavage complex with *S. aureus* DNA gyrase. The structure is compared with a structure with the quinolone moxifloxacin, also in a DNA cleavage complex with *S. aureus* DNA gyrase. We also show that zoliflodacin has reasonable activity against *A. baumannii* (MICs of 4 μg/mL; the A in ESKAPE). In 2018, a World Health Organization (WHO) priority list [39] proposed developing new drugs active against multidrug-resistant tuberculosis and Gram-negative bacteria. While spiropyrimidinetriones related to zoliflodacin are being developed against *M. tuberculosis* [40,41], the WHO critical priority, carbapenem-resistant *A. baumannii* [39], still urgently requires the development of new antibiotics [42].

The DNA cleavage gate of bacterial type IIA topoisomerases, when complexed with DNA and compounds, seems inherently flexible and usually gives low- or medium-low-resolution data [13,14,37] (we define low (>3 Å), medium-low (2.5–2.99 Å), medium-high (2.01–2.49 Å), and high (<2 Å) based on the confidence in determining the water structures around ligands and metal ions). Deleting the Greek key domain from a *S. aureus* GyrBA fusion truncate allowed the resolution of a complex with GSK299423 and DNA to be improved from 3.5Å to 2.1 Å [13], and this *S. aureus* DNA gyrase fusion truncate (Figure 1e) construct (used in this paper) has given the only high-resolution structures of DNA complexes of bacterial type IIA topoisomerases obtained to date [10].

2. Results

2.1. A 2.8 Å Zoliflodacin DNA Cleavage Complex with S. aureus DNA Gyrase

Crystals of zoliflodacin in a complex with a 20-mer DNA homoduplex (20-447T) and *S. aureus* DNA gyrase were grown by a microbatch crystallization method, and a 2.8 Å dataset was collected on beamline I24 at Diamond Light Source (see Materials and Methods for details). The data were phased using the 2.5 Å QPT-1 complex with the same 20-447T DNA and *S. aureus* DNA gyrase in the same $P6_1$ space group (PDB code: 5CDM; a = b = 93.9 Å, c = 412.5 Å) and then refined (see Materials and Methods for details and Figure 2 below for electron density). The 20-447T DNA homoduplex contains 18 base-pairs and a G-T mismatch at either end of the DNA.

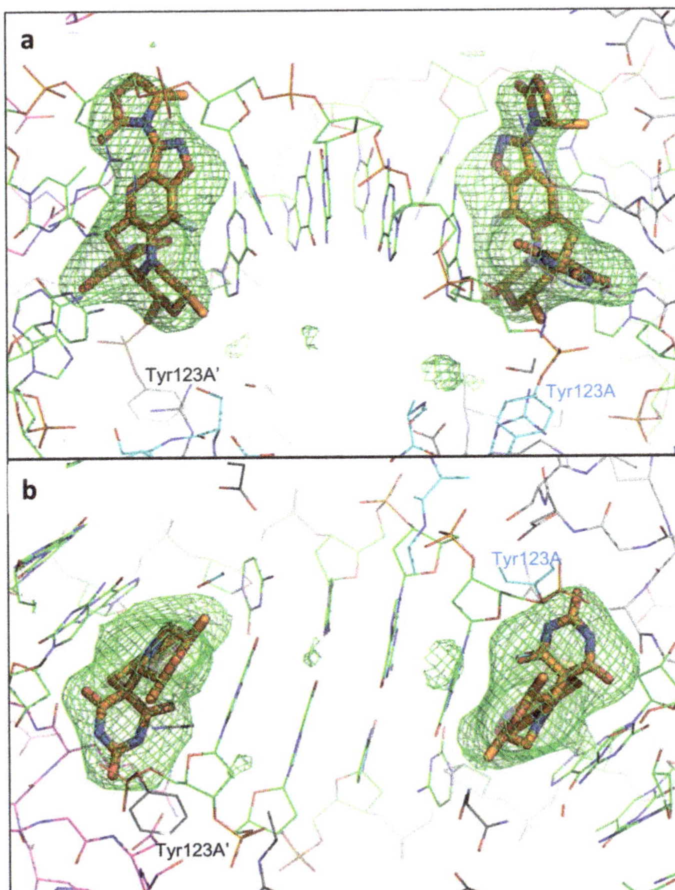

Figure 2. Final Fo-Fc omit map (+3 sigma) for the two zoliflodacins: orthogonal (90°) views. A final Fo-Fc omit map was calculated by omitting the two zoliflodacins from the coordinates, which was refined with refmac [43]. The initial omit maps calculated from 5cdm coordinates were of similar quality but showed some additional density around the novel five-membered ring (top (**a**)). The zoliflodacins are shown as sticks with orange carbon atoms; gyrA has cyan or grey carbons, gyrB has magenta or grey carbons, and the DNA has green carbons (water molecules are not shown for clarity). The pyrimidinetrione rings are clearly seen in the view in panel (**b**).

The structure shows two zoliflodacins binding in the cleaved DNA, physically blocking religation (Figure 3). The DNA has been cleaved by and is covalently attached to tyrosine 123A from the GyrA subunit (and to the symmetry related tyrosine 123A′, from the second GyrA subunit in the complex). Catalytic metal ions (normally Mg^{2+} in bacteria) are required for DNA cleavage, and in our structure we can see two Mn^{2+} ions occupying the 'B-site' in the GyrB and GyrB′ subunits [10].

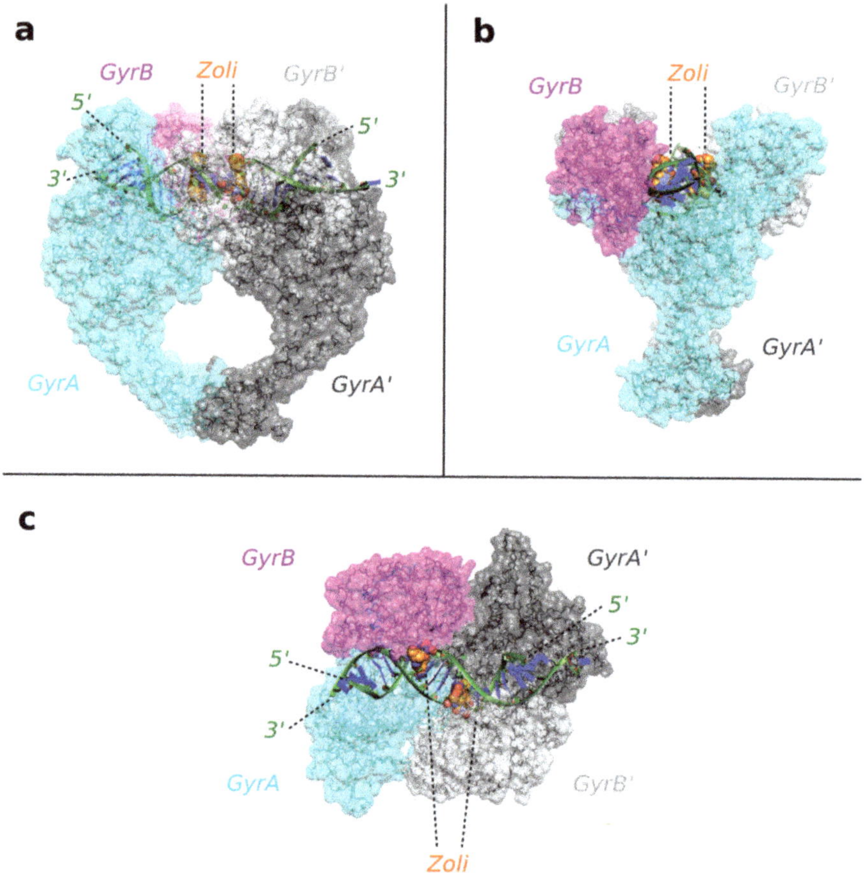

Figure 3. The 2.8 Å zoliflodacin crystal structure with *S. aureus* DNA gyrase. (**a**) View of the 2.8 Å zoliflodacin crystal structure. The DNA (cartoon; green backbone and blue bases) has been cleaved by *S. aureus* gyrase (shown as backbone trace with semi-transparent surface). Tyr 123 (and Tyr 123′) have cleaved the DNA and are covalently attached. The compounds (Zoli: zoliflodacin) are shown as solid spheres (carbons as orange, oxygens as red, nitrogens as blue). (**b**) An orthogonal (90°) view of the same complex. (**c**) An orthogonal (90°) view looking down the two-fold axis of the complex. The two ends of the DNA duplex adopt different conformations due to crystal packing. Figure produced using ChimeraX [44,45].

2.2. Zoliflodacin Interacts with GyrB, Whereas Moxifloxacin Interacts with GyrA

Figure 4 compares the binding sites of compounds in our 2.8 Å zoliflodacin structure with a 2.95 Å *S. aureus* DNA gyrase DNA cleavage complex with the widely used quinolone antibiotic moxifloxacin (PDB code: 5CDQ [34]). Figure 4a shows the binding mode of zoliflodacin with the pyrimidinetrione (or barbituric acid moiety) of the compound, making direct interactions with GyrB. In particular, the terminal oxygen of the pyrimidinetrione makes a hydrogen bond with the main-chain NH of aspartic acid B437.

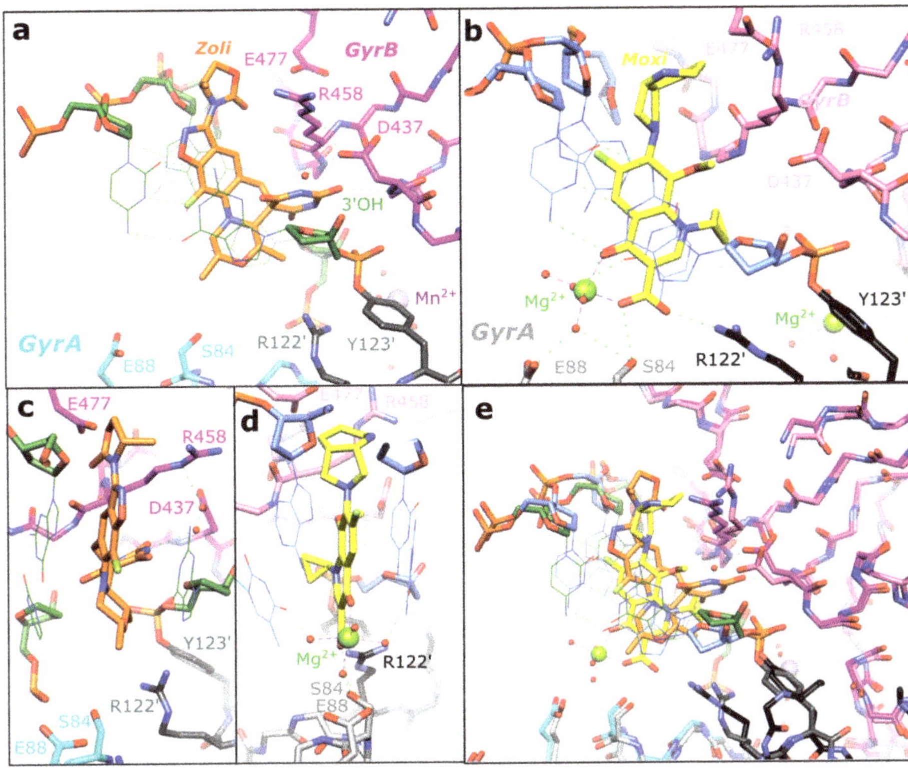

Figure 4. Equivalent views of zoliflodacin and moxifloxacin in *S. aureus* DNA gyrase DNA cleavage complexes. (**a**) A close-up view of a zoliflodacin (Zoli: orange carbons) binding site in the 2.8 Å structure. The pyrimidinetrione moiety of zoliflodacin interacts directly and indirectly (via a water) with Asp437 of GyrB (dotted green lines), in a similar manner to that described in the 2.5 Å QPT-1 structure. Y123′ has cleaved the DNA and formed a 'phosphotyrosine' type linkage with the cleaved DNA. The DNA-backbone is shown with a fatter 'stick' representation, with the bases drawn in thinner 'line' (base-pair H-bonds only shown for the +1, +4 base-pair). (**b**) In the 2.95 Å moxifloxacin (Moxi: yellow carbons) structure, the quinolone-bound Mg^{2+} ion (green sphere) and coordinating water molecules (red spheres) make hydrogen bonds (dotted red lines) to S84, E88, and the bases either side of the DNA cleavage site (at the +1 and -1 positions; see panel d). (**c**,**d**) Orthogonal (90°) views of the compound binding sites in the zoliflodacin structure (**c**) and moxifloxacin structure (**d**). (**e**) Superposition of (**a**,**b**). Figure produced using ChimeraX [44,45].

This contrasts with moxifloxacin, where the compound (Figure 4b,e) interacts with S84A and E88A from the GyrA subunit via the now well-characterized water–metal ion (Mg^{2+}) bridge [12,33–35,46,47]. The lack of interactions with GyrA and the interactions with GyrB account for the much of the activity of zoliflodacin against quinolone-resistant strains of bacteria (e.g., Table 4 in [3]; target-mediated resistance is common in quinolone resistant bacteria [11]). The interactions of the quinolones with the GyrA (or ParC) subunit via the flexible water–metal ion bridge may account for some of the specificity of the quinolones for DNA gyrase and topo IV (see sequence alignment in Figure 5) over the two human type IIA topoisomerases, Top2α and Top2β.

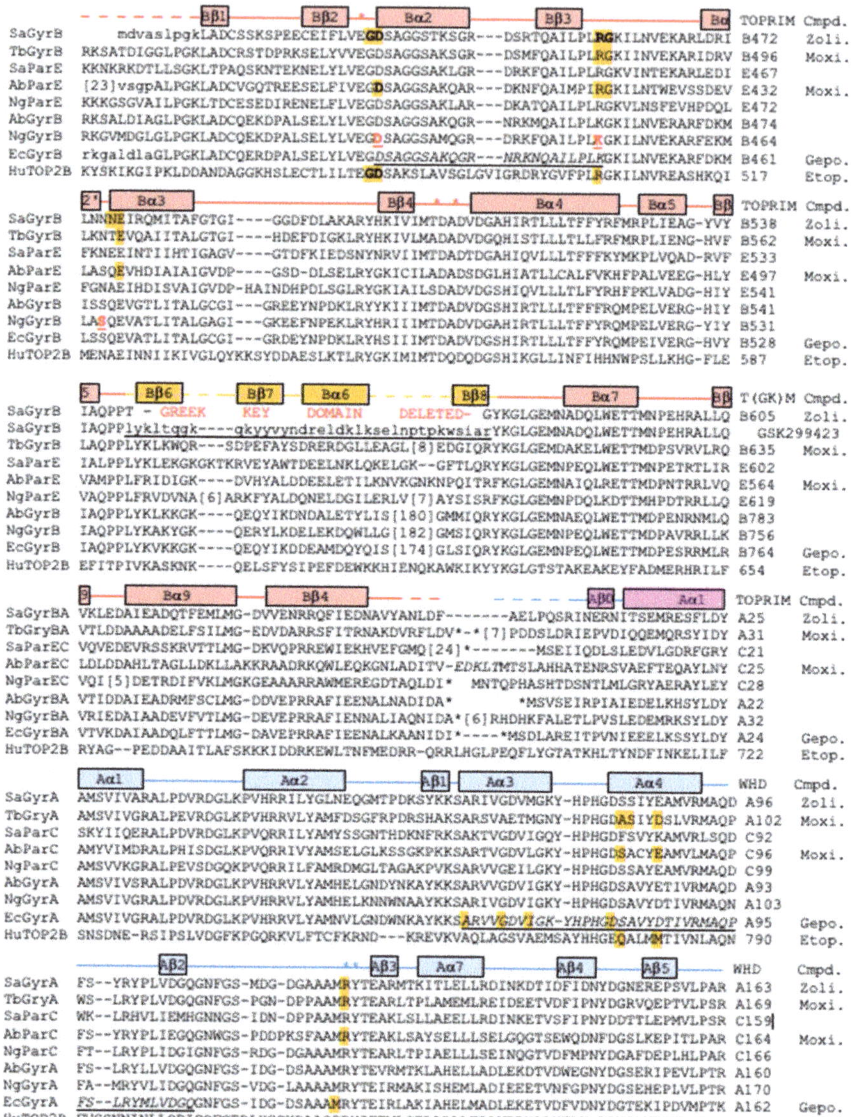

Figure 5. Sequence alignment highlighting residues contacting zoliflodacin, moxifloxacin, gepotidacin, or etoposide. An alignment of the residues in the TOPRIM, Greek key, and WHD domains of *S. aureus* GyrB/GyrA (SaGyrBA), *M. tuberculosis* GyrB/GyrA (TbGYRba), *S. aureus* ParE/ParC (SaParEC), *A. baumannii* ParE/ParC (AbParEC), *N. gonorrhoeae* ParE/ParC (NgParEC), *A. baumannii* GyrB/GyrA (AbGyrBA), *N. gonorrhoeae* GyrB/GyrA (NgGyrBA), *E. coli* GyrB/GyrA (EcGyrBA), and human Top2B (HuTOP2B). Numbers in brackets (e.g., [180]) show numbers of residues not included in the alignment; note the large insertion within the Greek key domain in Gram-negative DNA gyrase sequences. The three *N. gonorrhoeae* GyrB residues whose mutations give low levels of resistance are highlighted in red [48]. Amino acids are **highlighted** on sequences if they contact (<3.8 Å) compounds in *S. aureus* DNA gyrase structures with compounds. Zoli. = contacts in the 2.8 Å

zoliflodacin structure (pdb code: 8BP2). Moxi. = contacts from the 2.4 Å *M. tuberculosis* complex with moxifloxacin (pdb code: 5bs8) or contacts from the 2.95 Å moxifloxacin complex (pdb code: 5cdq, note for moxifloxacin, the Mg^{2+} ion and water molecules of the water–metal ion bridge are taken as part of the compound; contacts in the 3.25 Å *A. Baumannii* topoIV moxifloxacin complex structure, 2XKK, are nearly identical to those shown). Gepo. = from the 2.37 Å gepotidacin with uncleaved DNA (pdb code: 6qtp; note contacts in the 2.31 Å structure 6qtk and in the full-length *E.coli* cryoEM gepotidacin structures, e.g., pdb code:6rks, are very similar). The contacts mapped onto the HuTOP2B structure are from the 2.16 Å etoposide complex with human Top2β (pdb code: 3QX3; but are similar in *S. aureus* crystal structures: 5cdp and 5cdn). The secondary structural elements in the 3.5 Å *S. aureus* gyrase complex with DNA and GSK299423 (2xcr) are shown above the alignment; note deletion of the Greek key domain gave a GSK299423 structure at 2.1 Å (2xcs). Note: * above the sequence alignment indicates positions of catalytic residues (Glu B435, AspB508, Asp B510, Arg A122, and Tyr A123 in S. aureus DNA gyrase). The quinolone-resistance-determining region (QRDR), defined as 426-447 in *E.coli* GyrB 67-106 in *E.coli* GyrA, is underlined in italics on the EcGyrB/A sequences [11].

However, the residues on GyrB, which partly form the DNA gyrase–zoliflodacin binding interactions, are conserved not only in the bacterial type IIA topoisomerases but also in the human enzymes. As shown in the sequence alignment in Figure 5 and in Figure 4a, zoliflodacin recognizes and interacts with the GD from the conserved EGDSA motif and the RG from the PLRGK (or PLKGK in Gram-negative DNA gyrases). While E435 at the start of the EGDSA motif is a catalytic residue, the other residues from the EGDSA are not catalytic, neither are the PLRGK motif residues. In QPT-1, only the GD and RG residues from GyrB contact the compound.

The specificity of spiropyrimidinetriones, such as zoliflodacin, towards bacterial type IIA topoisomerases such as human topoisomerases was proposed to be because such compounds can be squeezed out of the pocket when the DNA-gate closes in human topoisomerases [34]. An alternative explanation could be because the DNA gate of DNA gyrase acts like a pair of swing doors, closing automatically once the transport segment has been pushed through [34]. This alternative explanation might account for the lower activity seen against both human topo 2s (liabilities) and bacterial topo IVs, which tend to only act on supercoiled DNA (DNA cross-overs).

Conformational flexibility in spiropyrimidinetrione ligands, such as QPT1 and zoliflodacin, may be important in allowing ligands to maintain favorable interactions within the binding sites as the DNA wriggles the protein [34,49,50]. In addition to the multiple tautomeric forms that the pyrimidinetrione moiety can adopt (only one of which is chemically called a 'pyrimidinetrione') and the conformational flexibility of the anilino-nitrogen [34], methyl-oxazolidine-2-one may also be able to adopt more than one conformation. Multiple high-resolution structures will be required to fully discern how the compound wriggles (when its binding pocket changes shape) as the enzyme is moved around by its substrate DNA [50]. However, from this initial 2.8 Å zoliflodacin structure, it is clear that the major protein interactions made by zoliflodacin are clearly with the GD and the RG from the highly conserved EGDSA and PLRGK motifs (Figures 4 and 5).

In the 2.1 Å crystal structure of the NBTI GSK299423 with the *S. aureus* $gyrase^{CORE}$ and DNA (PDB code: 2XCS; [13]), a Y123F mutant was used so that the DNA could not be cleaved. In this 2.1 Å GSK299423 structure, the +1:+4 base-pair (Figure 1c) occupies a similar space to the inhibitors in the zoliflodacin and moxifloxacin structures. Some reasons for the conservation of the EGDSA and PLRGK motifs (Figure 5) may be discerned from this 2.1 Å structure. While the side-chain of E435 (the first residue of the EGDSA motif) coordinates the catalytic metal (at the 'A' position, poised to cleave the DNA), both the main-chain NH and side-chain hydroxyl of serine 438 are within the hydrogen-bonding distance of the phosphate between nucleotides 1 and 2. The main-chain C = O of Arg 458 and the main-chain NH of Lys 460 (from the PLRGK motif) accept and donate hydrogen bonds to the -1 guanine base, helping to hold it firmly in place. NBTIs can stabilize complexes with one strand cleaved or with no DNA cleavage [13,24,51]; however, experimental nucleotide preferences for NBTI cleavage have not yet, to the best of our knowledge, been determined [52].

2.3. Target Mediated Resistance to Zoliflodacin in N. Gonorrhoeae

The binding of zoliflodacin to the conserved motifs on GyrB correlates well with the low prevalence of target-mediated resistance; only one of some 12,493 *N. gonorrhoeae* genomes from the PathogenWatch database has a predicted first-level resistance mutation [53]. Assessing the probability of developing resistance is an important step in the development of any new antibiotic. The development of zoliflodacin (AZD0914) for gonorrhea followed from a 2015 paper assessing the likelihood of developing resistance in *N. gonorrhoeae* [48]. This paper showed that higher MIC resistance was associated with target mutations in three amino acids in *N. gonorrhoeae* DNA gyrase, namely GyrB:D429N, K450T, and S467N [48]. These mutations were identified via the in vitro selection of resistance and can give a four-fold to sixteen-fold increase in the MIC of zoliflodacin [48,54]. Interestingly, these *N. gonorrhoeae* GyrB mutations correspond to D437, R458, and N475 in *S. aureus* DNA gyrase. The D429N mutation is associated with the slower growth of bacteria [55]. All three regions are close to the compound (see Figure 4a). In the D437N mutant (*S. aureus* DNA gyrase), the asparagine side chain may have its NH_2 group pointing towards the compound (because if the sidechain was in the opposite orientation, the hydrogens on the NH_2 would clash with hydrogens on proline 56, i.e., the P in PLRGK).

Zoliflodacin has an extra methyl-oxazolidine-2-one ring, which QPT-1 does not possess (Figure 1a,b), and this extra ring makes van der Waals contacts with residues N476 and E477. While there is clear electron density for the both the additional fluorine and the extra ring (which are not in the QPT-1 structure; Figure 6), the 2.8 Å electron density map is not able to clearly define all water structures or totally unambiguously define the orientation of the extra five-membered ring (see Figures 2 and 6). N475 is equivalent to the third mutated residue in *N. gonorrhoeae* GyrB, Ser 467 [48]. The mutation of this residue, which is adjacent to residues contacting the compound, presumably affects their conformations. A similar effect is perhaps seen in the *S. aureus* ParC V67A, found in a strain of *S. aureus* resistant to gepotidacin [56]. In high-resolution *S. aureus* DNA gyrase NBTI crystal structures, three residues (A, G, and M) from the GyrA motif 68-ARIVGDVM-75 are within the van der Waals distance of the compounds [13,22,24]. ParC V67A is the first V in the equivalent *S. aureus* ParC sequence 64-AKTVGDVI-71, i.e., Val 67 is adjacent to an amino acid making direct van der Waals contacts with the compounds.

Most bacteria (including *N. gonorrhoeae*) have two type IIA topoisomerases—DNA gyrase and topo IV. The target-mediated resistance to dual targeting quinolones, which form bactericidal DNA cleavage complexes with both DNA gyrase and topo IV, is only significant after mutations have occurred in both DNA gyrase and topo IV [57]. The observation by Alm et al. [48] of mutations only occurring in *N. gonorrhoeae* DNA gyrase when bacteria were challenged with zoliflodacin (AZD0914) suggests the compound has limited activity against *N. gonorrhoeae* topo IV. While third-generation cephalosporin-resistant, fluoroquinolone-resistant *N. gonorrhoeae* was listed as a high-piority target in 2018 [39], *A. baumanii* was a higher priority target.

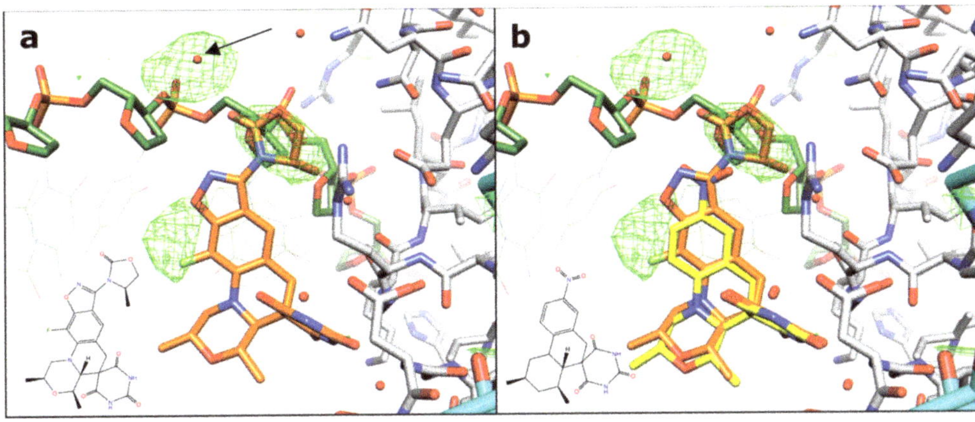

Figure 6. Difference in map densities from refined QPT-1 structure against zoliflodacin data. (**a**) The F_o-F_c map from refined QPT-1 coordinates contoured at 3σ shows extra features in the zoliflodacin structure not in the QPT-1 starting coordinates. The arrow points to extra density modelled as water, which could be a metal ion (it is close to the oxygen of a phosphate from the DNA backbone and also an oxygen from methyl-oxazolidine-2-one). (**b**) Refined QPT-1 coordinates are also shown with yellow carbons.

2.4. Improved Activity of Zoliflodacin against A. baumannii Compared to QPT-1

A. baumannii was selected as a WHO critical priority AMR pathogen [39] to test for susceptibility to zoliflodacin. The MIC of zoliflodacin against two carbapenem-resistant outbreak strains of *Acinetobacter baumannii* [58] was determined (see Section 4 for details) as 4 μg/mL (Table 1). The tested outbreak strains of *A. baumannii* (Table 1) possessed imipenem and meropenem MICs in excess of 4 μg/mL, precluding their treatment with these carbapenems [58]. The zoliflodacin MIC of 4 μg/mL suggested that although its activity has been optimized against other Gram-negative bacteria, the potency of zoliflodacin against *A. baumannii* is better than that of QPT-1, from which it was developed (the activity of QPT-1 against *A. baumannii* is from Supplementary Table S4 in the paper by Chan et al. [34]; note that QPT-1 is considerably more active against an efflux knock-out strain, *A. baumannii* BM4454 (ΔadeABC ΔadeIJK) [59]). As expected from previous testing, the analysis of the *S. aureus* reference strains NCTC 12981 showed good zoliflodacin susceptibility (<0.313 μg/mL).

Table 1. MICs of zoliflodacin for *A. baumannii*.

Compound	Species	MIC (μg/mL)
Zoliflodacin	*A. baumannii* BCC 807 (UK OXA-23 clone)	4
Zoliflodacin	*A. baumannii* BCC 810 (South East OXA-23 clone)	4
QPT-1	*A. baumannii* BM4454*A. baumannii*	16
QPT-1	BM4454 (ΔadeABC ΔadeIJK) *	0.125

* *A. baumannii* BM4454 (adeABC and adeIJK) is equivalent to BM4652 [59].

3. Discussion

Some Gram-negative bacteria are difficult to kill with antibiotics. Not only do they have two cell walls but they also have export pumps that can rapidly pump antibiotics out of the bacteria [59]. Such bacterial export pumps can play a role in antimicrobial resistance [60]. There is much interest in compounds that can inhibit antibiotic efflux pumps [61], as there is clearly a potential for combination therapies. If the MICs of a compound such as zoliflodacin could be lowered, the dose might be lowered and the therapeutic window would be increased.

However, the success of whole-cell screening, including early counterscreening of human cells for safety, seems to have been effective in discovering two new classes of Gram-negative targeting antibiotics [5,13]. A similar approach, although starting with a natural product, recently lead to the discovery of evybactin, a new *M. tuberculosis* DNA gyrase-targeting compound [62]; this compound appears to work in a similar manner to the thiophene inhibitors [63] that allosterically stabilize DNA cleavage complexes [64] by binding to a 'third site'—a hinge pocket [10].

Interestingly, two of the mutations in *N. gonorrhoeae* GyrB that give rise to resistance to zoliflodacin (AZD0914) are Asp429Asn and Lys450Thr, which correspond in *S. aureus* crystal structures to residues involved in making up the binding pocket of the compound (Figure 4). Namely, Asp 437 (=Asp429) is the D from the conserved EGDSA motif and Arg 458 (=Lys450) is from the conserved PLRGK motif.

In a previous paper describing the crystals structures of QPT-1, moxifloxacin, and etoposide in DNA cleavage complexes with *S. aureus* DNA gyrase [34], the DNA gate of DNA gyrase was proposed to act like a pair of swing doors, through which the T-segment could be pushed (Figure 1f) but that would then swing close. Such a model might partly account for why in *N. gonorrhoeae* mutations are only seen in GyrB and not in ParE [48]. The swinging close of the DNA gate in DNA gyrase might be predicted to give slower 'off' rates for zoliflodacin compared to topo IV; it was also proposed that zoliflodacin would be squeezed out of a slightly larger equivalent pocket in human topo2s [34]. Much work remains to be done; for example, one current model suggests that before the C-gate (or exit gate) can be opened, the small Greek key domain senses the presence of the T-DNA segment (once it has passed through the G-gate) and then moves the catalytic metal away from the active site (see the Supplementary Discussion and Supplementary Figures S12 and S13 in [34]). This model allows the DNA to be religated by the lysine residue from the highly conserved Y<u>K</u>GLG motif at the C-terminus of the Greek key domain (see Figure 5), while not allowing DNA cleavage by the catalytic metal when the exit gate is opened and not allowing exit gate opening while the gate DNA is cleaved. In this model, this is a 'safety feature' of type IIA topoisomerases, allowing DNA religation by the Y<u>K</u>GLG lysine but inhibiting DNA cleavage by the catalytic metal. Interestingly, it has also been shown that DNA gyrase can catalyze supercoiling by introducing a single nick in the DNA [65]; perhaps this mechanism is also a safe way of introducing negative supercoils into DNA without opening the C-gate.

The safety and size of the therapeutic window are clearly important in antibacterial drug discovery. It will be interesting to see if the new spiropyrimidinetrione class of compounds, such as zoliflodacin, can be developed to be safer and more efficacious medicines with less of a tendency for target-mediated antibiotic resistance than the quinolones.

4. Materials and Methods

4.1. Protein Purification and Crystallization of a Zoliflodacin DNA Cleavage Complex

The *S. aureus* DNA gyrase fusion truncate GyrB27:A56 (GKdel) (M_w 78,020) was expressed in *E. coli* and purified based on the procedure used by Bax et al. [13], modified as described [25]. The protein was purified (at 10 mg/mL = 0.128 mM) in 20 mM HEPES pH 7.0, 5 mM $MnCl_2$, and 100 mM $NaSO_4$. The DNA oligonucleotide used in crystallizations, 20-447T, was custom-ordered from Eurogentec (Seraing, Belgium). Received in lyophilized form, the DNA was resuspended in nuclease-free water and annealed from 86 to 21°C over 45 min to give the duplex DNA at a concentration of 2 mM. The zoliflodacin was purchased from MedChemExpress (South Brunswick Township, NJ, USA) as a solid and was dissolved in 100% DMSO, forming a 100 mM stock solution.

Crystallization complexes were formed by mixing a protein, HEPES buffer, DNA, and compound and incubating the mixture on ice for 1 h 15 min. Crystals of *S. aureus* GyrB27:A56 (GKdel)-zoliflodacin-20-447T were grown using the microbatch under oil method [38], with streak seeding being implemented for subsequent plates after the first plate gave crystals. Following established protocols, a crystallization screen consisting of Bis-Tris buffer pH 6.3 to 6.0 (90, 150 mM) and PEG 5*k*MME (13–7%) was used. For a single drop, 1 μL of complex mixture was mixed with 1 μL of crystallization buffer in a 72-well Terasaki microbatch plate, prior to covering with paraffin oil. The plates were incubated at 20 °C and crystal growth was observed between 5 and 30 days. A seed solution was prepared by crushing several previously grown hexagonal rod-shaped *S. aureus* GyrB27:A56 (GKdel)-zoliflodacin-20-447T crystals in 20 μL of crystallization buffer. The crystal, which gave a 2.8 Å dataset, was grown in a crystallization plate, where 1 μL of complex mixture (0.066 mM GyrB27:A56 dimer, 0.171 mM 20-447T DNA duplex, 5.714 mM zoliflodacin, 2.571 mM $MnCl_2$, and 342.9 mM HEPES pH 7.2) was mixed with 1 μL of crystallization buffer (90 mM Bis-Tris pH 6.3, 9% PEG 5 *k*MME). A large single crystal was transferred to a cryobuffer (15% glycerol, 19% PEG 5*k*MME, 1 mM zoliflodacin, 5% DMSO, 81 mM Bis-Tris pH 6.3) before flash-cooling in liquid nitrogen for data collection.

4.2. Data Collection, Structural Determination, and Refinement

The data were collected (3600 x 0.1° degree images) on beamline I24 at Diamond Light Source. The data were processed and merged with dials [66–68], as shown in Table 2. A low-resolution cutoff of 25 Å was applied when manually reprocessing the data with dials to avoid problems with the backstop shadow. The high-resolution cutoff was determined by having a $CC_{1/2} > 0.30$ [69]. The structure was refined starting from the 2.5 Å complex with the same DNA and the related compound QPT-1 (PDB code: 5CDM) [34,38]. The data, which are not twinned and are in space group $P6_1$, were reindexed (H = k, K = h, L = -l) to be in the same hand and of the same origin as other liganded structures in the same space group (e.g., PDB codes: 2XCS, 4BUL, 5IWI, 5IWM, 5NPP, 6QTK, 6QX1, and 5CDM). The initial rigid body refinement of 5cdm-BA-x.pdb ($P6_1$ cell: a = b = 93.88 Å, c = 412.48 Å) reduced the R-factor (R-free) from 0.3900 (0.3899) to 0.2684 (0.2743). Further refinement with phenix.refine [70,71] and refmac [43,72] gave the final structure (Table 3), which had a reasonable geometry. Restraints for zoliflodacin were generated in Acedrg [73]. As we were interested in structures with ligands and inhibitors, we used the standard BA-x numbering scheme throughout [10] (the coordinates, 8bp2-BA-x.pdb, are available from a table of structures from the 'research' tab of Ben Bax's website at Cardiff (https://www.cardiff.ac.uk/people/view/1141625-bax-ben, accessed on 4 January 2023). This means zoliflodacin inhibitors in sites 1 and 1′ have CHAINID I (for the inhibitor) and residue numbers 1 and 201 (see Figure 3 in [10]). In this 2.8 Å zoliflodacin *S. aureus* DNA gyrase structure, the chains are named as B (GyrB) and A (GyrA) from the first fusion truncate subunit and D (GyrB) and C (GyrA) from the second subunit (the BA-x nomenclature stands for GyrB/GyrA extended numbering). The DNA strands have CHAINIDs E and F (see [10] for further details). The electron density maps for the inhibitors are shown in Figure 2. The water structure near the inhibitors was based on that in the 2.5 Å structure with QPT-1 (PDB code: 5CDM). The water and glycerol structures were based on the electron density maps and higher resolution structures (the 1.98 Å *S. aureus* complex PDB code 5IWI, which contains over 940 water molecules, was superposed).

Table 2. Data collection statistics.

	ZOL-2.8
PDB code	8BP2
Diffraction source	I24, DLS
Wavelength (Å)	0.9999
Resolution range (Å)	24.85–2.78 (2.83–2.78) *
Space group	$P6_1$
Unit cell	94.54, 94.54, 417.13, 90, 90, 120
Total reflections	974,457 (46,716) *
Unique reflections	52,704 (2554) *
Multiplicity	18.5 (18.3) *
Completeness (%)	100.0 (100.0) *
Mean I/sigma(I)	5.9 (0.4) *
Wilson B_{factor}	62.4
R_{merge}	0.286 (3.872) *
R_{meas}	0.294 (3.980) *
R_{pim}	0.067 (1.297) *
$CC_{1/2}$	0.997 (0.318) *

* Numbers in brackets are in the outer (2.83–2.78) resolution shell.

Table 3. Refinement statistics.

	ZOL-2.8
PDB code	8BP2
Resolution range (Å)	24.85–2.80 (2.87–2.80)
Completeness (%)	99.41 (94.60)
No. of reflections, working set	48,785 (3421)
No. of reflections, test set	2538 (185)
Final Rcryst	0.1957 (0.372)
Final Rfree	0.2375 (0.374)
Cruickshank DPI (Å) *	0.325
No. of non-H atoms (total)	11,713
Protein	10,574
DNA	801
Zoliflodacin	70
Other ligands (Mn, glycerol etc.)	42
Water molecules	226
RMS deviations	
Bonds (Å)	0.009
Angles (°)	1.569
Average B factors (Å2)	
Protein	96.024
DNA	85.103
Zoliflodacin	84.491
Other ligands (Mn, glycerol etc.)	88.911
Waters **	75.396
Ramachandran plot	
Favored regions	96%
Additionally allowed	4%
Outliers	0%

* The Cruickshank DPI (Å) was calculated using the Online_DPI server [74] (Kumar et al., 2015). ** Water molecules were placed where there were water molecules in higher resolution structures (e.g., 5CDM and 5IWI).

At each DNA cleavage site, a single catalytic Mn^{2+} ion is seen at the B-site [10]. The electron density on His C 391 was interpreted as being due to a Mn^{2+} ion coordinated by a Bis-Tris buffer molecule, which mediates a crystal contact with one end of the DNA. This interpretation of the electron density was confirmed by re-refining the original 2.1 Å structure of GSK299423 with the *S. aureus* gyrase^{CORE} structure [13]; originally this electron density had been misinterpreted as being due to the DNA. The new interpretation explains why both Bis-Tris and Mn^{2+} ions are needed in the crystallization buffer when growing $P6_1$ crystals of *S. aureus* gyrase^{CORE} with ligands.

4.3. Structural Analysis

The van der Waals contacts with the ligands (defined as 3.8 Å or less) were calculated with 'contact' from the CCP4 suite of programs [75]. The structures were superposed using coot [76] or with limited sets of defined Cαs using LsqKab from the CCP4 suite [75].

4.4. Minimum Inhibitory Concentration Assay

The MICs of zoliflodacin against two carbapenem-resistant outbreak strains of *A. baumannii* (BCC 807, BCC 810) [58] were determined in triplicate using the modified broth microdilution reference method ISO 20776-1:2019 [77], as recommended by the EUCAST (European Committee on Antimicrobial Susceptibility Testing) [78]. The concentration range tested was between 40 and 0.313 μg/mL in two-fold serial dilutions. *S. aureus* NCTC 12981 was used as a quality control strain, as the MICs for *S. aureus* have previously been reported.

Author Contributions: Conceptualization, B.D.B. and H.M.; protein purification, J.H. and N.P.B.; complex preparation and crystallization, H.M. and M.L.-L.; data collection and processing, H.M. and A.J.W.; structural determination and analysis, H.M. and B.D.B.; microbiology and MIC determinations, N.H. and E.M.; writing—original draft preparation, B.D.B. and H.M.; writing—review and editing, all authors. All authors have read and agreed to the published version of the manuscript.

Funding: This research received no external funding.

Institutional Review Board Statement: Not applicable.

Informed Consent Statement: Not applicable.

Data Availability Statement: The 2.8 Å zoliflodacin structure has been deposited with the protein databank (the PDB) with the code 8BP2. The PDB nomenclature for compounds and metal ions is inconsistent between different structures. The sensibly named PDB files, following the BA-x nomenclature [10], are available from a table of structures from the 'research' tab on Ben Bax's website (https://www.cardiff.ac.uk/people/view/1141625-bax-ben, accessed on 4 January 2023).

Acknowledgments: We thank Simon Ward and the Medicines Discovery Institute (MDI) in Cardiff for supporting this work. We thank Diamond Light Source (DLS) for the beamtime and general support. H.M. is a joint PhD student between the MDI and DLS.

Conflicts of Interest: The authors declare no obvious conflict of interest.

References

1. Jacobsson, S.; Golparian, D.; Oxelbark, J.; Alirol, E.; Franceschi, F.; Gustafsson, T.N.; Brown, D.; Louie, A.; Drusano, G.; Unemo, M. Pharmacodynamic Evaluation of Dosing, Bacterial Kill, and Resistance Suppression for Zoliflodacin Against Neisseria gonorrhoeae in a Dynamic Hollow Fiber Infection Model. *Front. Pharm.* **2021**, *12*, 682135. [CrossRef]
2. Unemo, M.; Ahlstrand, J.; Sanchez-Buso, L.; Day, M.; Aanensen, D.; Golparian, D.; Jacobsson, S.; Cole, M.J.; European Collaborative, G. High susceptibility to zoliflodacin and conserved target (GyrB) for zoliflodacin among 1209 consecutive clinical *Neisseria gonorrhoeae* isolates from 25 European countries, 2018. *J. Antimicrob. Chemother.* **2021**, *76*, 1221–1228. [CrossRef]
3. Bradford, P.A.; Miller, A.A.; O'Donnell, J.; Mueller, J.P. Zoliflodacin: An Oral Spiropyrimidinetrione Antibiotic for the Treatment of *Neisseria gonorrheae*, Including Multi-Drug-Resistant Isolates. *ACS Infect. Dis.* **2020**, *6*, 1332–1345. [CrossRef] [PubMed]
4. Jacobsson, S.; Kularatne, R.; Kittiyaowamarn, R.; Maseko, V.; Paopang, P.; Sangprasert, P.; Sirivongrangson, P.; Piddock, L.; Wi, T.; Alirol, E.; et al. High in vitro susceptibility to the first-in-class spiropyrimidinetrione zoliflodacin among consecutive clinical *Neisseria gonorrhoeae* isolates from Thailand (2018) and South Africa (2015–2017). *Antimicrob. Agents Chemother.* **2019**, *63*, e01479-19. [CrossRef] [PubMed]
5. Miller, A.A.; Bundy, G.L.; Mott, J.E.; Skepner, J.E.; Boyle, T.P.; Harris, D.W.; Hromockyj, A.E.; Marotti, K.R.; Zurenko, G.E.; Munzner, J.B.; et al. Discovery and characterization of QPT-1, the progenitor of a new class of bacterial topoisomerase inhibitors. *Antimicrob. Agents Chemother.* **2008**, *52*, 2806–2812. [CrossRef]
6. Bisacchi, G.S. Origins of the quinolone class of antibacterials: An expanded "discovery story" miniperspective. *J. Med. Chem.* **2015**, *58*, 4874–4882. [CrossRef] [PubMed]

7. Schoeffler, A.J.; Berger, J.M. DNA topoisomerases: Harnessing and constraining energy to govern chromosome topology. *Q. Rev. Biophys.* **2008**, *41*, 41–101. [CrossRef]
8. Pommier, Y.; Sun, Y.; Huang, S.N.; Nitiss, J.L. Roles of eukaryotic topoisomerases in transcription, replication and genomic stability. *Nat. Rev. Mol. Cell Biol.* **2016**, *17*, 703–721. [CrossRef]
9. Bates, A.D.; Maxwell, A. *DNA Topology*; Oxford University Press: Cary, NC, USA, 2005.
10. Bax, B.D.; Murshudov, G.; Maxwell, A.; Germe, T. DNA Topoisomerase Inhibitors: Trapping a DNA-Cleaving Machine in Motion. *J. Mol. Biol.* **2019**, *431*, 3427–3449. [CrossRef] [PubMed]
11. Bush, N.G.; Diez-Santos, I.; Abbott, L.R.; Maxwell, A. Quinolones: Mechanism, lethality and their contributions to antibiotic resistance. *Molecules* **2020**, *25*, 5662. [CrossRef]
12. Aldred, K.J.; Kerns, R.J.; Osheroff, N. Mechanism of quinolone action and resistance. *Biochemistry* **2014**, *53*, 1565–1574. [CrossRef]
13. Bax, B.D.; Chan, P.F.; Eggleston, D.S.; Fosberry, A.; Gentry, D.R.; Gorrec, F.; Giordano, I.; Hann, M.M.; Hennessy, A.; Hibbs, M.; et al. Type IIA topoisomerase inhibition by a new class of antibacterial agents. *Nature* **2010**, *466*, 935–940. [CrossRef] [PubMed]
14. Vanden Broeck, A.; Lotz, C.; Ortiz, J.; Lamour, V. Cryo-EM structure of the complete *E. coli* DNA gyrase nucleoprotein complex. *Nat. Commun.* **2019**, *10*, 4935. [CrossRef] [PubMed]
15. Gupta, R.S. Origin of diderm (Gram-negative) bacteria: Antibiotic selection pressure rather than endosymbiosis likely led to the evolution of bacterial cells with two membranes. *Antonie Leeuwenhoek* **2011**, *100*, 171–182. [CrossRef] [PubMed]
16. Butler, M.S.; Paterson, D.L. Antibiotics in the clinical pipeline in October 2019. *J. Antibiot.* **2020**, *73*, 329–364. [CrossRef]
17. Butler, M.S.; Blaskovich, M.A.; Cooper, M.A. Antibiotics in the clinical pipeline in 2013. *J. Antibiot.* **2013**, *66*, 571–591. [CrossRef]
18. Butler, M.S.; Cooper, M.A. Antibiotics in the clinical pipeline in 2011. *J. Antibiot.* **2011**, *64*, 413–425. [CrossRef]
19. Payne, D.J.; Gwynn, M.N.; Holmes, D.J.; Pompliano, D.L. Drugs for bad bugs: Confronting the challenges of antibacterial discovery. *Nat. Rev. Drug Discov.* **2007**, *6*, 29–40. [CrossRef]
20. Tommasi, R.; Brown, D.G.; Walkup, G.K.; Manchester, J.I.; Miller, A.A. ESKAPEing the labyrinth of antibacterial discovery. *Nat. Rev. Drug Discov.* **2015**, *14*, 529–542. [CrossRef]
21. Scangarella-Oman, N.E.; Hossain, M.; Hoover, J.L.; Perry, C.R.; Tiffany, C.; Barth, A.; Dumont, E.F. Dose Selection for Phase III Clinical Evaluation of Gepotidacin (GSK2140944) in the Treatment of Uncomplicated Urinary Tract Infections. *Antimicrob. Agents Chemother.* **2022**, *66*, e01492-21. [CrossRef]
22. Miles, T.J.; Hennessy, A.J.; Bax, B.; Brooks, G.; Brown, B.S.; Brown, P.; Cailleau, N.; Chen, D.; Dabbs, S.; Davies, D.T. Novel tricyclics (eg, GSK945237) as potent inhibitors of bacterial type IIA topoisomerases. *Bioorg. Med. Chem. Lett.* **2016**, *26*, 2464–2469. [CrossRef]
23. Miles, T.J.; Hennessy, A.J.; Bax, B.; Brooks, G.; Brown, B.S.; Brown, P.; Cailleau, N.; Chen, D.; Dabbs, S.; Davies, D.T. Novel hydroxyl tricyclics (eg, GSK966587) as potent inhibitors of bacterial type IIA topoisomerases. *Bioorg. Med. Chem. Lett.* **2013**, *23*, 5437–5441. [CrossRef] [PubMed]
24. Gibson, E.G.; Bax, B.; Chan, P.F.; Osheroff, N. Mechanistic and Structural Basis for the Actions of the Antibacterial Gepotidacin against *Staphylococcus aureus* Gyrase. *ACS Infect. Dis.* **2019**, *5*, 570–581. [CrossRef] [PubMed]
25. Kolaric, A.; Germe, T.; Hrast, M.; Stevenson, C.E.M.; Lawson, D.M.; Burton, N.P.; Voros, J.; Maxwell, A.; Minovski, N.; Anderluh, M. Potent DNA gyrase inhibitors bind asymmetrically to their target using symmetrical bifurcated halogen bonds. *Nat. Commun.* **2021**, *12*, 150. [CrossRef] [PubMed]
26. Singh, S.B.; Kaelin, D.E.; Wu, J.; Miesel, L.; Tan, C.M.; Black, T.; Nargund, R.; Meinke, P.T.; Olsen, D.B.; Lagrutta, A.; et al. Tricyclic 1,5-naphthyridinone oxabicyclooctane-linked novel bacterial topoisomerase inhibitors as broad-spectrum antibacterial agents-SAR of left-hand-side moiety (Part-2). *Bioorg. Med. Chem. Lett.* **2015**, *25*, 1831–1835. [CrossRef] [PubMed]
27. Singh, S.B.; Kaelin, D.E.; Wu, J.; Miesel, L.; Tan, C.M.; Meinke, P.T.; Olsen, D.; Lagrutta, A.; Bradley, P.; Lu, J.; et al. Oxabicyclooctane-linked novel bacterial topoisomerase inhibitors as broad spectrum antibacterial agents. *ACS Med. Chem. Lett.* **2014**, *5*, 609–614. [CrossRef] [PubMed]
28. Boucher, H.W.; Talbot, G.H.; Bradley, J.S.; Edwards, J.E.; Gilbert, D.; Rice, L.B.; Scheld, M.; Spellberg, B.; Bartlett, J. Bad bugs, no drugs: No ESKAPE! An update from the Infectious Diseases Society of America. *Clin. Infect. Dis.* **2009**, *48*, 1–12. [CrossRef] [PubMed]
29. Tashiro, S.; Mihara, T.; Sasaki, M.; Shimamura, C.; Shimamura, R.; Suzuki, S.; Yoshikawa, M.; Hasegawa, T.; Enoki, Y.; Taguchi, K. Oral fidaxomicin versus vancomycin for the treatment of Clostridioides difficile infection: A systematic review and meta-analysis of randomized controlled trials. *J. Infect. Chemother.* **2022**, *28*, 1536–1545. [CrossRef]
30. Butler, M.S.; Blaskovich, M.A.; Cooper, M.A. Antibiotics in the clinical pipeline at the end of 2015. *J. Antibiot.* **2017**, *70*, 3–24. [CrossRef]
31. Lockshon, D.; Morris, D.R. Sites of reaction of Escherichia coli DNA gyrase on pBR322 in vivo as revealed by oxolinic acid-induced plasmid linearization. *J. Mol. Biol.* **1985**, *181*, 63–74. [CrossRef]
32. Leo, E.; Gould, K.A.; Pan, X.-S.; Capranico, G.; Sanderson, M.R.; Palumbo, M.; Fisher, L.M. Novel symmetric and asymmetric DNA scission determinants for *Streptococcus pneumoniae* topoisomerase IV and gyrase are clustered at the DNA breakage site. *J. Biol. Chem.* **2005**, *280*, 14252–14263. [CrossRef]
33. Wohlkonig, A.; Chan, P.F.; Fosberry, A.P.; Homes, P.; Huang, J.; Kranz, M.; Leydon, V.R.; Miles, T.J.; Pearson, N.D.; Perera, R.L.; et al. Structural basis of quinolone inhibition of type IIA topoisomerases and target-mediated resistance. *Nat. Struct. Mol. Biol.* **2010**, *17*, 1152–1153. [CrossRef]

34. Chan, P.F.; Srikannathasan, V.; Huang, J.; Cui, H.; Fosberry, A.P.; Gu, M.; Hann, M.M.; Hibbs, M.; Homes, P.; Ingraham, K.; et al. Structural basis of DNA gyrase inhibition by antibacterial QPT-1, anticancer drug etoposide and moxifloxacin. *Nat. Commun.* **2015**, *6*, 10048. [CrossRef]
35. Blower, T.R.; Williamson, B.H.; Kerns, R.J.; Berger, J.M. Crystal structure and stability of gyrase-fluoroquinolone cleaved complexes from *Mycobacterium tuberculosis*. *Proc. Natl. Acad. Sci. USA* **2016**, *113*, 1706–1713. [CrossRef] [PubMed]
36. Laponogov, I.; Sohi, M.K.; Veselkov, D.A.; Pan, X.S.; Sawhney, R.; Thompson, A.W.; McAuley, K.E.; Fisher, L.M.; Sanderson, M.R. Structural insight into the quinolone-DNA cleavage complex of type IIA topoisomerases. *Nat. Struct. Mol. Biol.* **2009**, *16*, 667–669. [CrossRef] [PubMed]
37. Laponogov, I.; Pan, X.S.; Veselkov, D.A.; McAuley, K.E.; Fisher, L.M.; Sanderson, M.R. Structural basis of gate-DNA breakage and resealing by type II topoisomerases. *PLoS ONE* **2010**, *5*, e11338. [CrossRef]
38. Srikannathasan, V.; Wohlkonig, A.; Shillings, A.; Singh, O.; Chan, P.F.; Huang, J.; Gwynn, M.N.; Fosberry, A.P.; Homes, P.; Hibbs, M.; et al. Crystallization and preliminary X-ray crystallographic analysis of covalent DNA cleavage complexes of *Staphyloccocus Aureus* DNA Gyrase with QPT-1, Moxilfloxacin and Etoposide. *Acta Crystallogr. Sect. F Struct. Biol. Cryst. Commun.* **2015**, *71*, 1242–1246. [CrossRef]
39. Tacconelli, E.; Carrara, E.; Savoldi, A.; Harbarth, S.; Mendelson, M.; Monnet, D.L.; Pulcini, C.; Kahlmeter, G.; Kluytmans, J.; Carmeli, Y. Discovery, research, and development of new antibiotics: The WHO priority list of antibiotic-resistant bacteria and tuberculosis. *Lancet Infect. Dis.* **2018**, *18*, 318–327. [CrossRef] [PubMed]
40. Basarab, G.S.; Ghorpade, S.; Gibhard, L.; Mueller, R.; Njoroge, M.; Peton, N.; Govender, P.; Massoudi, L.M.; Robertson, G.T.; Lenaerts, A.J. Spiropyrimidinetriones: A class of DNA gyrase inhibitors with activity against *Mycobacterium tuberculosis* and without cross-resistance to fluoroquinolones. *Antimicrob. Agents Chemother.* **2022**, *66*, e02192-21. [CrossRef]
41. Govender, P.; Müller, R.; Singh, K.; Reddy, V.; Eyermann, C.J.; Fienberg, S.; Ghorpade, S.R.; Koekemoer, L.; Myrick, A.; Schnappinger, D. Spiropyrimidinetrione DNA Gyrase Inhibitors with Potent and Selective Antituberculosis Activity. *J. Med. Chem.* **2022**, *65*, 6903–6925. [CrossRef] [PubMed]
42. Al-Kadmy, I.M.; Ibrahim, S.A.; Al-Saryi, N.; Aziz, S.N.; Besinis, A.; Hetta, H.F. Prevalence of genes involved in colistin resistance in *Acinetobacter baumannii*: First report from Iraq. *Microb. Drug Resist.* **2020**, *26*, 616–622. [CrossRef]
43. Murshudov, G.N.; Skubak, P.; Lebedev, A.A.; Pannu, N.S.; Steiner, R.A.; Nicholls, R.A.; Winn, M.D.; Long, F.; Vagin, A.A. REFMAC5 for the refinement of macromolecular crystal structures. *Acta Cryst. D Biol. Cryst.* **2011**, *67*, 355–367. [CrossRef]
44. Pettersen, E.F.; Goddard, T.D.; Huang, C.C.; Meng, E.C.; Couch, G.S.; Croll, T.I.; Morris, J.H.; Ferrin, T.E. UCSF ChimeraX: Structure visualization for researchers, educators, and developers. *Protein Sci.* **2021**, *30*, 70–82. [CrossRef]
45. Goddard, T.D.; Huang, C.C.; Meng, E.C.; Pettersen, E.F.; Couch, G.S.; Morris, J.H.; Ferrin, T.E. UCSF ChimeraX: Meeting modern challenges in visualization and analysis. *Protein Sci.* **2018**, *27*, 14–25. [CrossRef]
46. Aldred, K.J.; McPherson, S.A.; Turnbough, C.L., Jr.; Kerns, R.J.; Osheroff, N. Topoisomerase IV-quinolone interactions are mediated through a water-metal ion bridge: Mechanistic basis of quinolone resistance. *Nucleic Acids Res.* **2013**, *41*, 4628–4639. [CrossRef] [PubMed]
47. Aldred, K.J.; McPherson, S.A.; Wang, P.; Kerns, R.J.; Graves, D.E.; Turnbough, C.L., Jr.; Osheroff, N. Drug interactions with *Bacillus anthracis* topoisomerase IV: Biochemical basis for quinolone action and resistance. *Biochemistry* **2012**, *51*, 370–381. [CrossRef] [PubMed]
48. Alm, R.A.; Lahiri, S.D.; Kutschke, A.; Otterson, L.G.; McLaughlin, R.E.; Whiteaker, J.D.; Lewis, L.A.; Su, X.; Huband, M.D.; Gardner, H.; et al. Characterization of the Novel DNA Gyrase Inhibitor AZD0914: Low Resistance Potential and Lack of Cross-Resistance in *Neisseria gonorrhoeae*. *Antimicrob. Agents Chemother.* **2015**, *59*, 1478–1486. [CrossRef] [PubMed]
49. Bax, B.; Chung, C.W.; Edge, C. Getting the chemistry right: Protonation, tautomers and the importance of H atoms in biological chemistry. *Acta Cryst. D Struct. Biol.* **2017**, *73*, 131–140. [CrossRef]
50. Fogg, J.M.; Randall, G.L.; Pettitt, B.M.; Sumners, W.L.D.; Harris, S.A.; Zechiedrich, L. Bullied no more: When and how DNA shoves proteins around. *Q. Rev. Biophys.* **2012**, *45*, 257–299. [CrossRef]
51. Gibson, E.G.; Oviatt, A.A.; Cacho, M.; Neuman, K.C.; Chan, P.F.; Osheroff, N. Bimodal actions of a naphthyridone/aminopiperidine-based antibacterial that targets gyrase and topoisomerase IV. *Biochemistry* **2019**, *58*, 4447–4455. [CrossRef]
52. Sutormin, D.; Rubanova, N.; Logacheva, M.; Ghilarov, D.; Severinov, K. Single-nucleotide-resolution mapping of DNA gyrase cleavage sites across the *Escherichia coli* genome. *Nucleic Acids Res.* **2019**, *47*, 1601. [CrossRef] [PubMed]
53. Adamson, P.C.; Lin, E.Y.; Ha, S.M.; Klausner, J.D. Using a public database of *Neisseria gonorrhoeae* genomes to detect mutations associated with zoliflodacin resistance. *J. Antimicrob. Chemother.* **2021**, *76*, 2847–2849. [CrossRef] [PubMed]
54. Foerster, S.; Drusano, G.; Golparian, D.; Neely, M.; Piddock, L.J.V.; Alirol, E.; Unemo, M. In vitro antimicrobial combination testing of and evolution of resistance to the first-in-class spiropyrimidinetrione zoliflodacin combined with six therapeutically relevant antimicrobials for *Neisseria gonorrhoeae*. *J. Antimicrob. Chemother.* **2019**, *74*, 3521–3529. [CrossRef] [PubMed]
55. Foerster, S.; Golparian, D.; Jacobsson, S.; Hathaway, L.J.; Low, N.; Shafer, W.M.; Althaus, C.L.; Unemo, M. Genetic resistance determinants, in vitro time-kill curve analysis and pharmacodynamic functions for the novel topoisomerase II inhibitor ETX0914 (AZD0914) in *Neisseria gonorrhoeae*. *Front. Microbiol.* **2015**, *6*, 1377. [CrossRef]
56. Scangarella-Oman, N.E.; Ingraham, K.A.; Tiffany, C.A.; Tomsho, L.; Van Horn, S.F.; Mayhew, D.N.; Perry, C.R.; Ashton, T.C.; Dumont, E.F.; Huang, J.; et al. In Vitro Activity and Microbiological Efficacy of Gepotidacin from a Phase 2, Randomized, Multicenter, Dose-Ranging Study in Patients with Acute Bacterial Skin and Skin Structure Infections. *Antimicrob. Agents Chemother.* **2020**, *64*, e01302-19. [CrossRef]

57. Hooper, D.C.; Rubinstein, E. (Eds.) *Quinolone Antimicrobial Agents*; ASM Press: Washington, DC, USA, 2003.
58. Coelho, J.M.; Turton, J.F.; Kaufmann, M.E.; Glover, J.; Woodford, N.; Warner, M.; Palepou, M.-F.; Pike, R.; Pitt, T.L.; Patel, B.C. Occurrence of carbapenem-resistant Acinetobacter baumannii clones at multiple hospitals in London and Southeast England. *J. Clin. Microbiol.* **2006**, *44*, 3623–3627. [CrossRef] [PubMed]
59. Damier-Piolle, L.; Magnet, S.; Brémont, S.; Lambert, T.; Courvalin, P. AdeIJK, a resistance-nodulation-cell division pump effluxing multiple antibiotics in *Acinetobacter baumannii*. *Antimicrob. Agents Chemother.* **2008**, *52*, 557–562. [CrossRef] [PubMed]
60. Colclough, A.L.; Alav, I.; Whittle, E.E.; Pugh, H.L.; Darby, E.M.; Legood, S.W.; McNeil, H.E.; Blair, J.M. RND efflux pumps in Gram-negative bacteria; regulation, structure and role in antibiotic resistance. *Future Microbiol.* **2020**, *15*, 143–157. [CrossRef]
61. Lowrence, R.C.; Subramaniapillai, S.G.; Ulaganathan, V.; Nagarajan, S. Tackling drug resistance with efflux pump inhibitors: From bacteria to cancerous cells. *Crit. Rev. Microbiol.* **2019**, *45*, 334–353. [CrossRef]
62. Imai, Y.; Hauk, G.; Quigley, J.; Liang, L.; Son, S.; Ghiglieri, M.; Gates, M.F.; Morrissette, M.; Shahsavari, N.; Niles, S. Evybactin is a DNA gyrase inhibitor that selectively kills *Mycobacterium tuberculosis*. *Nat. Chem. Biol.* **2022**, *18*, 1236–1244. [CrossRef]
63. Chan, P.F.; Germe, T.; Bax, B.D.; Huang, J.; Thalji, R.K.; Bacqué, E.; Checchia, A.; Chen, D.; Cui, H.; Ding, X. Thiophene antibacterials that allosterically stabilize DNA-cleavage complexes with DNA gyrase. *Proc. Natl. Acad. Sci. USA* **2017**, *114*, E4492–E4500. [CrossRef]
64. Thalji, R.K.; Raha, K.; Andreotti, D.; Checchia, A.; Cui, H.; Meneghelli, G.; Profeta, R.; Tonelli, F.; Tommasi, S.; Bakshi, T.; et al. Structure-guided design of antibacterials that allosterically inhibit DNA gyrase. *Bioorg. Med. Chem. Lett.* **2019**, *29*, 1407–1412. [CrossRef]
65. Gubaev, A.; Weidlich, D.; Klostermeier, D. DNA gyrase with a single catalytic tyrosine can catalyze DNA supercoiling by a nicking-closing mechanism. *Nucleic Acids Res.* **2016**, *44*, 10354–10366. [CrossRef]
66. Winter, G.; Beilsten-Edmands, J.; Devenish, N.; Gerstel, M.; Gildea, R.J.; McDonagh, D.; Pascal, E.; Waterman, D.G.; Williams, B.H.; Evans, G. DIALS as a toolkit. *Protein Sci.* **2022**, *31*, 232–250. [CrossRef] [PubMed]
67. Beilsten-Edmands, J.; Winter, G.; Gildea, R.; Parkhurst, J.; Waterman, D.; Evans, G. Scaling diffraction data in the DIALS software package: Algorithms and new approaches for multi-crystal scaling. *Acta Crystallogr. Sect. D Struct. Biol.* **2020**, *76*, 385–399. [CrossRef]
68. Winter, G.; Waterman, D.G.; Parkhurst, J.M.; Brewster, A.S.; Gildea, R.J.; Gerstel, M.; Fuentes-Montero, L.; Vollmar, M.; Michels-Clark, T.; Young, I.D.; et al. DIALS: Implementation and evaluation of a new integration package. *Acta Cryst. D Struct. Biol.* **2018**, *74*, 85–97. [CrossRef] [PubMed]
69. Evans, P.R.; Murshudov, G.N. How good are my data and what is the resolution? *Acta Crystallogr. D Biol. Cryst.* **2013**, *69*, 1204–1214. [CrossRef] [PubMed]
70. Afonine, P.V.; Poon, B.K.; Read, R.J.; Sobolev, O.V.; Terwilliger, T.C.; Urzhumtsev, A.; Adams, P.D. Real-space refinement in PHENIX for cryo-EM and crystallography. *Acta Cryst. D Struct. Biol.* **2018**, *74*, 531–544. [CrossRef]
71. Afonine, P.V.; Grosse-Kunstleve, R.W.; Echols, N.; Headd, J.J.; Moriarty, N.W.; Mustyakimov, M.; Terwilliger, T.C.; Urzhumtsev, A.; Zwart, P.H.; Adams, P.D. Towards automated crystallographic structure refinement with phenix.refine. *Acta Cryst. D Biol. Cryst.* **2012**, *68*, 352–367. [CrossRef]
72. Kovalevskiy, O.; Nicholls, R.A.; Long, F.; Carlon, A.; Murshudov, G.N. Overview of refinement procedures within REFMAC5: Utilizing data from different sources. *Acta Cryst. D Struct. Biol.* **2018**, *74*, 215–227. [CrossRef]
73. Long, F.; Nicholls, R.A.; Emsley, P.; Graaeulis, S.; Merkys, A.; Vaitkus, A.; Murshudov, G.N. AceDRG: A stereochemical description generator for ligands. *Acta Cryst. D Struct. Biol.* **2017**, *73*, 112–122. [CrossRef] [PubMed]
74. Kumar, K.S.D.; Gurusaran, M.; Satheesh, S.N.; Radha, P.; Pavithra, S.; Thulaa Tharshan, K.P.S.; Helliwell, J.R.; Sekar, K. Online_DPI: A web server to calculate the diffraction precision index for a protein structure. *J. Appl. Crystallogr.* **2015**, *48*, 939–942. [CrossRef]
75. Winn, M.D.; Ballard, C.C.; Cowtan, K.D.; Dodson, E.J.; Emsley, P.; Evans, P.R.; Keegan, R.M.; Krissinel, E.B.; Leslie, A.G.; McCoy, A.; et al. Overview of the CCP4 suite and current developments. *Acta Crystallogr. D. Biol. Cryst.* **2011**, *67*, 235–242. [CrossRef] [PubMed]
76. Emsley, P.; Cowtan, K. Coot: Model-building tools for molecular graphics. *Acta Cryst. D Biol. Cryst.* **2004**, *60*, 2126–2132. [CrossRef] [PubMed]
77. International Organization for Standardization. *Susceptibility Testing of Infectious Agents and Evaluation of Performance of Antimicrobial Susceptibility Test Devices—Part 1: Broth Micro-Dilution Reference Method for Testing the In Vitro Activity of Antimicrobial Agents against Rapidly Growing Aerobic Bacteria Involved in Infectious Diseases*; International Organization for Standardization: Geneva, Switzerland, 2019.
78. EUCAST. MIC Determination of Non-fastidious and Fastidious Organisms. 2022. Available online: https://www.eucast.org/ast_of_bacteria/mic_determination (accessed on 9 November 2022).

MDPI
St. Alban-Anlage 66
4052 Basel
Switzerland
www.mdpi.com

International Journal of Molecular Sciences Editorial Office
E-mail: ijms@mdpi.com
www.mdpi.com/journal/ijms

www.ingramcontent.com/pod-product-compliance
Lightning Source LLC
Chambersburg PA
CBHW061600210926
51850CB00009B/2166